DATE DUE			

ESSENTIALS OF Molecular Biology

THE JONES AND BARTLETT SERIES IN BIOLOGY

Genetics
John R. S. Fincham
University of Edinburgh

Genetics of Populations
Philip W. Hedrick
University of Kansas

Genetic Principles: Human and Social Consequences (nonmajors text)
Gordon Edlin
University of California, Davis

Molecular Biology: A Comprehensive Introduction to Prokaryotes and Eukaryotes
David Freifelder
University of California, San Diego

Essentials of Molecular Biology
David Freifelder
University of California, San Diego

Molecular Evolution: An Annotated Reader
Eric Terzaghi, Adam S. Wilkins, and David Penny
All of Massey University, New Zealand

Viral Assembly
Sherwood Casjens
University of Utah College of Medicine

The Molecular Biology of Bacterial Growth (a symposium volume)
M. Schaechter, Tufts University Medical School; F. Neidhardt, University of Michigan; J. Ingraham, University of California, Davis; N.O. Kjeldgaard, University of Aarhus, Denmark, editors

Population Biology
Philip W. Hedrick
University of Kansas

General Genetics
Leon A. Snyder, University of Minnesota, St. Paul; David Freifelder, University of California, San Diego; Daniel L. Hartl, Washington University School of Medicine

ESSENTIALS OF
Molecular Biology

DAVID FREIFELDER
University of California, San Diego

Jones and Bartlett Publishers, Inc.
BOSTON PORTOLA VALLEY

Editorial offices: 30 Granada Court, Portola Valley, California 94025
Sales and customer service offices: 20 Park Plaza Boston, Mass. 02116

Library of Congress Cataloging in Publication Data

Freifelder, David Michael,
Essentials of molecular biology.
Rev., condensed ed. of: Molecular biology. c1983.
Includes index.
1. Molecular biology. I. Freifelder, David Michael Molecular biology. II. Title.
QH506.F37 1985 574.8'8 84-29744
ISBN 0-86720-051-0

ISBN 0-86720-051-0

Production: Unicorn Production Services Inc.
Composition: Polyglot, Pte, Ltd
Printing & binding: Halliday Lithograph

Printed in the United States of America
Printing (last digit): 10 9 8 7 6 5 4 3 2

Preface

In 1983 *Molecular Biology, A Comprehensive Introduction to Prokaryotes and Eukaryotes*, was published—a 979-page book designed for a complete course in molecular biology. I was pleased to receive many favorable comments about the book in the following year. However, many instructors said that the book was too large to require for use in some courses and suggested that I prepare a short version.

In mid-1984 I began the job, cutting and reassembling existing material, but I quickly found this method to be impossible: in the parts I kept, there were too many references to deleted material, and the flow of thought was often interrupted. Thus, it was clear that many paragraphs had to be rewritten, and connecting sentences were needed. While attending to this, I took the opportunity to update existing material, where needed, and added new information in the interest of making a better "short book." The most advanced material has been omitted, along with descriptions of techniques and some experimental support for the facts. This information will have to be provided by the instructor, or the student can be directed to the more detailed text of the larger book. Thus the reader will occasionally find in this book phrases such as "For more information, see *MB*, pages 000-000." *MB* refers to the first edition of the original book, *Molecular Biology*.

Molecular Biology had no exercises and problems for the student; instead a separate problems book, *Problems for Molecular Biology*, was published. *Essentials of Molecular Biology* contains problems and answers, taken from this book, so a second purchase is not necessary.

I hope the short book will prove to be useful to others as well as to those who requested it.

November, 1984 David Freifelder
San Diego, California

Contents

2 Macromolecules 16

3 Nucleic Acids 27

4 The Physical Structure of Protein Molecules 47

5 Macromolecular Interactions and the Structure of Complex Aggregates 64

6 The Genetic Material 79

10 Translation 151

11 Mutagenesis, Mutations, and Mutants 172

12 Plasmids and Transposable Elements 189

13 Recombinant DNA and Genetic Engineering 210

14 Regulation of Gene Activity in Prokaryotes 231

15 Bacteriophages 250

16 Regulation of Gene Activity in Eukaryotes 278

1 Systems and Methods of Molecular Biology

The term molecular biology was first used in 1945 by William Astbury, who was referring to the study of the chemical and physical structure of biological macromolecules. By that time, biochemists had discovered many fundamental intracellular chemical reactions, and the importance of specific reactions and of protein structure in defining the numerous properties of cells was appreciated. However, the development of molecular biology had to await the realization that the most productive advances would be made by studying "simple" systems such as bacteria and bacteriophages (bacterial viruses), which yield information about basic biological processes more readily than animal cells. Although bacteria and bacteriophages are themselves quite complicated biologically, they enabled scientists to identify DNA as the molecule that contains most, if not all, of the genetic information of a cell. Following this discovery, the new field of molecular genetics moved ahead rapidly in the late 1950's and early 1960's and provided new concepts at a rate that can be matched only by the development of quantum mechanics in the 1920's. The initial success and the accumulation of an enormous body of information enabled researchers to apply the techniques and powerful logical methods of molecular genetics to the subjects of muscle and nerve function, membrane structure, the mode of action of antibiotics, cellular differentiation and development, immunology, and so forth. Faith in the basic uniformity of life processes was an important factor in this rapid growth. That is, it was believed that the fundamental biological principles that govern the activity of simple organisms, such as bacteria and viruses, must apply to more complex cells; only the details should vary. This faith has been amply justified by experimental results.

In this book prokaryotes and eukaryotes will be discussed separately and compared and contrasted. Usually prokaryotes will be discussed first because they are simpler. In keeping with this, we begin by describing the properties of bacteria that are important in molecular biological research.

BACTERIA

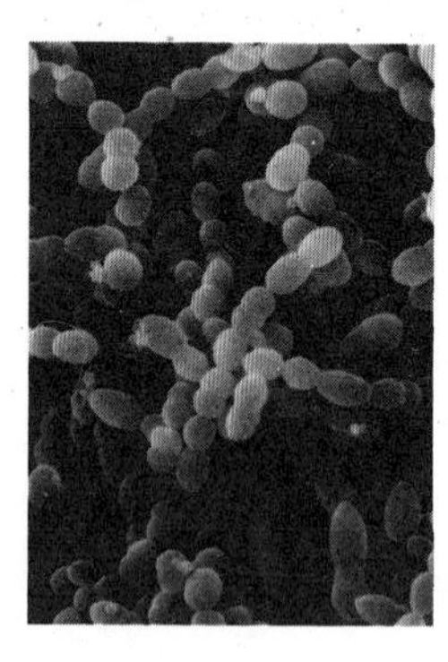

Bacteria are free-living unicellular organisms. They have a single chromosome, which is not enclosed in a nucleus (they are prokaryotes), and compared to eukaryotes they are simple in their physical organization. For all practical purposes, a bacterium can be thought of as a solution of several thousand chemicals and a few organized particles enclosed in a rigid cell wall.

Bacteria have many features that make them suitable objects for the study of fundamental biological processes. For example, they are grown easily and rapidly and, compared to cells in multicellular organisms, they are relatively simple in their needs. The bacterium that has served the field of molecular biology best is *Escherichia coli* (usually referred to as *E. coli*), which divides every 20 minutes at 37°C under optimal conditions. Thus, a single cell becomes 10^9 bacteria in about 20 hours. Bacteria can be grown in a **liquid growth medium** or on a solid surface. A population growing in a liquid medium is called a bacterial **culture.** If the liquid is a complex extract of biological material, it is called a **broth.** If the growth medium is a simple mixture containing no organic compounds other than a carbon source such as a sugar, it is called a **minimal medium.** A typical minimal medium contains each of the ions Na^+, K^+, Mg^{2+}, Ca^{2+}, NH_4^+, Cl^-, HPO_4^{2-}, and SO_4^{2-}, and a source of carbon such as glucose, glycerol, or lactate. If a bacterium can grow in a minimal medium—that is, if it can synthesize *all* necessary organic substances such as amino acids, vitamins, and lipids—the bacterium is said to be a **prototroph.** If any organic substances other than a carbon source must be added for growth to occur, the bacterium is termed an **auxotroph.** For example, if the amino acid leucine is required in the growth medium, the bacterium is a leucine auxotroph; the genetic symbol for such a bacterium is Leu^-. A prototroph would be indicated Leu^+. Bacteria are frequently grown on solid surfaces. The earliest surface used for growing bacteria was a slice of raw potato. This was later replaced by media solidified by gelatin. Because many bacteria excrete enzymes that digest gelatin, an inert gelling agent was sought. **Agar,** which is a gelling agent obtained from a variety of seaweed and used extensively as a thickening agent in Japanese cuisine, is resistant to bacterial enzymes and has been universally used. A solid growth medium is called a nutrient agar, if the liquid medium is a broth, or a minimal agar, if a minimal medium is gelled. Solid media are typically placed in a **petri dish.** In lab jargon a petri dish containing a solid medium is called a **plate** and the act of depositing bacteria on the agar surface is called **plating.**

A bacterium growing on an agar surface divides. Since most bacteria are not very motile on a solid surface, the progeny bacteria remain very near the location of the original bacterium. The number of progeny increases so much that a visible cluster of bacteria appears. This cluster is called a **bacterial colony** (Figure 1-1). Colony formation allows one to determine the number of bacteria in a culture. For instance, if 100 cells are plated, 100 colonies will appear the next day.

Plating is a method for determining if a bacterium is an auxotroph. This is done in the following way. Minimal agar and nutrient agar plates are prepared. Several hundred bacteria are plated on each plate and the plates are incubated overnight in an oven. Several hundred colonies are subsequently found on the nutrient agar because it contains so many substances that it can satisfy the requirements of nearly any bacterium. If colonies are also found on the minimal agar, the bacterium is a prototroph; if no colonies are found, it is an auxotroph and some required substance is not present in the minimal agar. Minimal plates are then prepared with various supplements. If the bacterium is a leucine auxotroph, the addition of leucine alone will

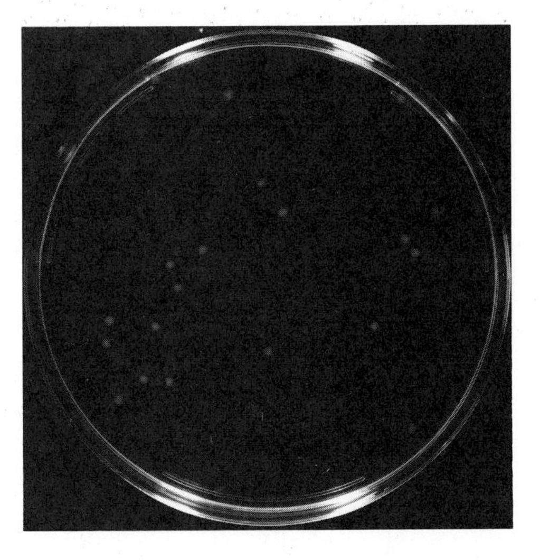

Figure 1-1
A petri dish containing colonies of *E. coli* grown on agar. (Courtesy of Robert Haynes.)

enable a colony to form. If both leucine and histidine must be added, the bacterium is auxotrophic for both of these substances.

Metabolic Regulation in Bacteria

Bacteria are well-regulated and highly efficient organisms. For example, they rarely synthesize substances that are not needed. Thus the enzymatic system for synthesizing the amino acid tryptophan is not formed if tryptophan is present in the growth medium, but when the tryptophan in the medium is used up, the enzymatic system will be rapidly formed. The systems responsible for utilization of various energy sources are also efficiently regulated. A well-studied example is the metabolism of the sugar lactose as an alternate carbon source to glucose. Control of both tryptophan synthesis and lactose degradation are two examples of **metabolic regulation**. This very general phenomenon will be explored extensively throughout the book. Both simple and complex regulatory systems will be seen, all of which act to determine how much of a particular compound is utilized and how much of each intracellular compound is synthesized at different times and in different circumstances. This will demonstrate the length to which the so-called simple cells have gone to utilize limited resources efficiently and to optimize their metabolic pathways for efficient growth.

BACTERIOPHAGE

Bacteria are subject to attack by smaller organisms called **bacteriophage** or simply **phage.** These are small particles, part of the general class of particles called viruses, and they are capable of growing only inside bacteria. Phage* have been the object of choice for a great many types of experiments because they are much simpler than bacteria in their structures (usually having between two and ten components) as well as their life cycles and yet possess the most essential, if minimal, attributes of life.

Most phages contain only a few different types of molecules, usually several hundred protein molecules of one to ten types (depending on the complexity of the phage) and one nucleic acid molecule. The protein molecules are organized in one of three ways. In the most

*The plural word phages refers to different species; the word phage is both singular and plural and in the plural sense refers to particles of the same type. Thus, T4 and T7 are both phages, but a test tube might contain either 1 T7 phage or 100 T7 phage.

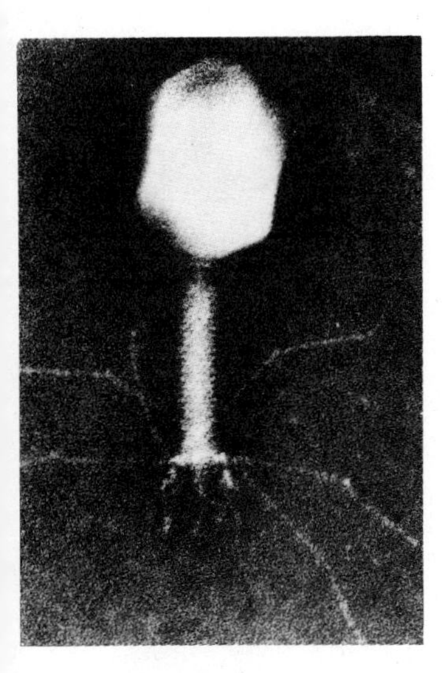

Figure 1-2
An *E. coli* T4 phage. The DNA is contained in the head. Tail fibers come from the pronged plate at the tip of the tail. (Courtesy of Robley Williams.)

common mode the protein molecules form a protein shell called the **coat** or **phage head,** to which a **tail** is generally attached; the nucleic acid molecule is contained in the head (Figure 1-2). Another form of a phage is a tailless head. The least common form is a filament in which the protein molecules form a tubular structure in which the nucleic acid is embedded. Phages are known that contain double-stranded DNA (the most common variety), single-stranded DNA, single-stranded RNA, and double-stranded RNA (least common).

Phage are parasites and cannot multiply except in a host bacterium. Thus, a phage must be able to enter a bacterium, multiply, and then escape. There are many ways by which this can be accomplished. However, a basic life cycle is outlined below and depicted in Figure 1-3.

The life cycle of a phage begins when a phage particle adsorbs to the surface of a susceptible bacterium. The phage nucleic acid then leaves the phage particle through the phage tail (if the phage has a tail) and enters the bacterium through the bacterial cell wall. In a complicated but understandable way the phage converts the bacterium to a phage-synthesizing factory. Within about an hour, the time varying with the phage species, the infected bacterium bursts or **lyses** and several hundred progeny phage are released. The suspension of newly synthesized phage is called a **phage lysate.**

Phage multiply faster than bacteria. A typical bacterium doubles in about half an hour, while a single phage particle gives rise to more than 100 progeny in the same time period. Each of these phage can then infect more bacteria, and those released in this second cycle of infection can infect even more. Thus, in two hours there are four cycles of infection for both a bacterium and a phage yet a single bacterium has

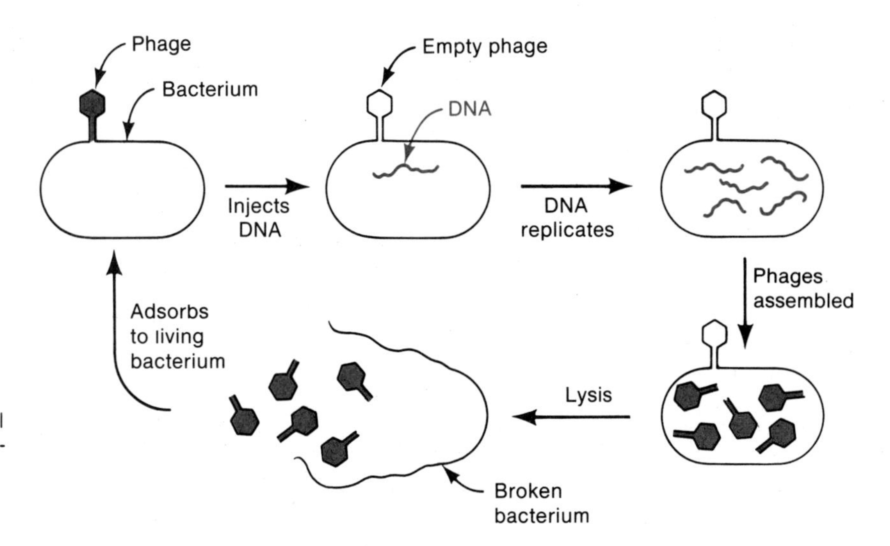

Figure 1-3
A schematic diagram of the life cycle of a typical bacteriophage. The number of phage released usually ranges from 20 to 500.

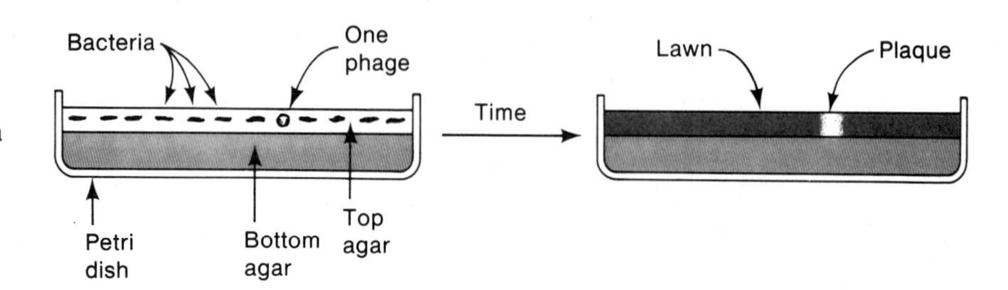

Figure 1-4
Schematic drawing of plaque formation. Bacteria grow and form a translucent lawn. There are no bacteria in the vicinity of the plaque, which remains transparent.

become $2^4 = 16$ bacteria and a single phage becomes $100^4 = 10^8$ phage particles.

By using procedures that make the bacterial cell membrane and cell wall permeable, infection can be initiated with free DNA rather than by phage particles. This technique, which is called **transfection,** is exceedingly useful, as it allows an experimenter to alter the DNA molecule either chemically or physically and then study the effect of the change by infecting a cell with the altered DNA. Transfection is an essential procedure in genetic engineering and has been widely used in studying animal and plant viruses.

The life cycle of a phage is highly regulated, but in a slightl different way from the metabolic regulation of a bacterium. Phages are totally dependent on the metabolism of their host bacteria, so that usually the regulatory systems of the hosts control the basic metabolic processes such as energy generation and synthesis of the precursors of DNA, RNA, and proteins. The job of an infecting phage is to reproduce itself by synthesizing its own nucleic acid and structural proteins, and finally to cause the bacterial cell wall to break so that progeny phage can escape. This requires that the various steps in phage production be regulated in time. Study of this regulation has been an important part of molecular biological research and has yielded a great deal of information about basic processes in all cells. Some examples will be seen in Chapter 14.

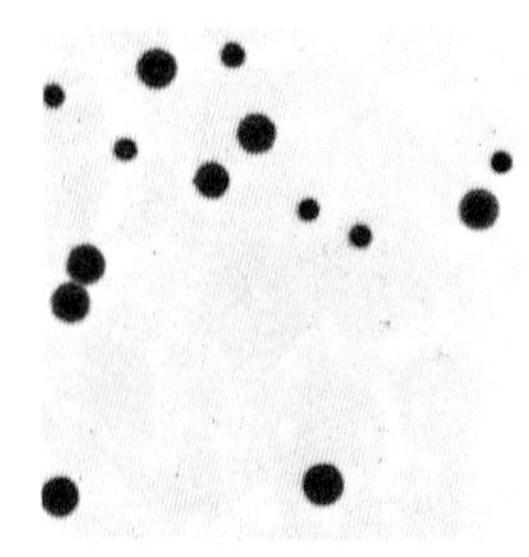

Figure 1-5
Plaques of *E. coli* phage T4. Two types of plaques are present. The smaller plaques are made by wild-type phage; the larger plaques are those of an *rII* mutant. (Courtesy of A. H. Doermann.)

Phage are counted by a technique called the **plaque assay** (Figure 1-4). About 10^8 bacteria plus a phage sample are added to warm liquid agar, which is then poured onto solid agar where it hardens. The bacteria multiply, forming a turbid layer in the agar called a lawn. While the bacteria are multiplying, each phage adsorbs to one cell and initiates an infection, resulting in the release of about 100 phage per cell, which remain localized in the agar. These progeny phage adsorb to nearby bacteria and multiply again; several cycles of infection occur, giving rise to a clear transparent hole in the turbid layer. This clear area is called a **plaque** (Figure 1-5). Since each phage forms one plaque, the individual phage can be counted. Different phage mutants often produce characteristic identifiable plaques, making the technique useful for genetic analysis.

YEASTS

Yeasts are unicellular organisms that have been used for millennia for producing wine and beer (Figure 1-6 (a,b)). A great deal of early biochemical research was carried out with yeasts rather than bacteria, work stimulated mainly by interest in understanding and improving beer.

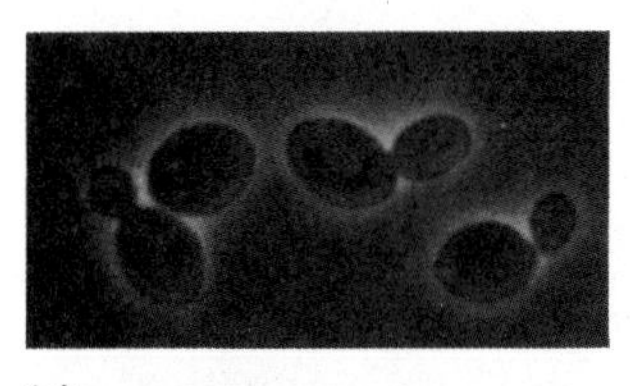

(a)

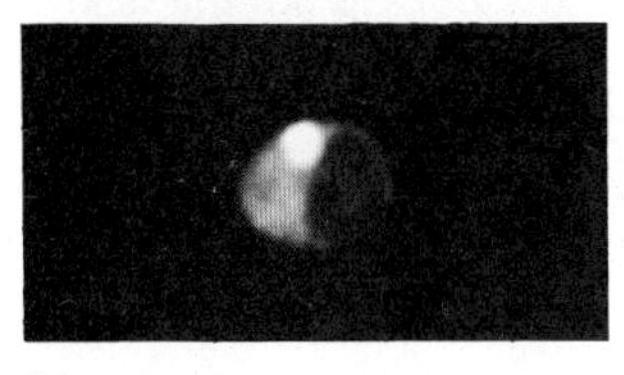

(b)

Figure 1-6
(a) A light of the yeast *Saccharomyces cerevisiae*. Many cells are budding. (Courtesy of Breck Byers.) (b) A fluorescence micrograph of a single cell stained with a fluorescent dye that binds to nucleic acids. The bright spot is the nucleus. The dark region is the vacuole, a liquid-filled sac that is free of nucleic acids.

Yeast cells are propagated in the laboratory and counted in much the same way as bacteria. They grow in liquid suspensions in either chemically defined media or in complex broths. They also grow on a solid surface to form colonies. The multiplication mechanism of all but the fission yeasts differs from the simple splitting of a mature bacterium in that yeast cells do not divide but multiply by budding (Figure 1-6(a)). That is, each mother cell produces a daughter cell (a bud) by outgrowth from the cell wall of the mother. The daughter cell grows and matures and can then also produce progeny by budding. The mother cell can bud many times.

In the past, most of the research effort in molecular biology has been with bacteria and viruses, because of their simplicity compared to eukaryotes. In recent years technological advances have made possible efficient and informative study of eukaryotes. The yeast *Saccharomyces cerevisiae* has been an important object of study: it has the genetic organization of eukaryotes and uses many regulatory strategies peculiar to higher organisms, yet the ease of handling and speed of growth are those encountered with typical microorganisms. Of particular interest also is the fact that *Saccharomyces* has both haploid and diploid phases. Haploid cells are of two types, which can mate sexually to form a stable diploid cell line. In particular nutritional conditions diploids undergo meiosis, generating haploid sexual spores. This mating system allows detailed genetic experiments to be performed and makes yeast a useful system for the study of both genetic recombination and the mechanism of meiosis.

Other unicellular eukaryotes, namely the alga *Chlamydomonas* and the protozoan *Tetrahymena*, also possess many of the attributes of yeast and are being used more frequently in eukaryotic molecular biology.

ANIMAL CELLS

As more information about fundamental mechanisms in bacteria has been gained, increasing efforts have been made to study animal cells. Research with animal cells is exciting because we are beginning to understand such complex processes as hormonal regulation and the development of an egg into an adult organism, and to gain some insight into the differences between normal cells and cancer cells. However, research with animal cells proceeds much more slowly than work with bacteria. There are two reasons for this. First, animal cells divide every 24 to 48 hours, whereas many bacteria divide every 25 to 50 minutes, so experiments with animal cells often take much longer than experiments with bacteria. Second, bacteria growing in culture are not significantly different from bacteria in nature but experiments with animal cells require that the cells be removed from the animal and often separated from one another. Cells treated in this way have lost the normal route for receiving nutrients and are definitely in an unnatural state. Many growth media that enable cells to grow in culture have been developed. They have been designed to keep the cells alive as long as possible but, they do not always maintain a normal state for the cell. The most obvious difference is that most cells taken from an organism die within a few weeks. Furthermore, during that period of time the cells grow and divide, which differs from the normal quiescent state of cells in living organisms. A few cells survive and grow indefinitely; these are said to have been *immortalized* and have generated an **established cell line.** In time, these cells develop abnormalities in chromosome number (often extra chromosomes) and behave somewhat like tumor cells. A complete discussion of the problems of working with animal cells is beyond the scope of this book; suffice it to say that a variety of techniques are available that make possible experiments with animal cells. A discussion of this phenomenon can be found in *MB*, pages 25–28, and in almost any text on cell biology. Figure 1-7 shows cells growing in culture.

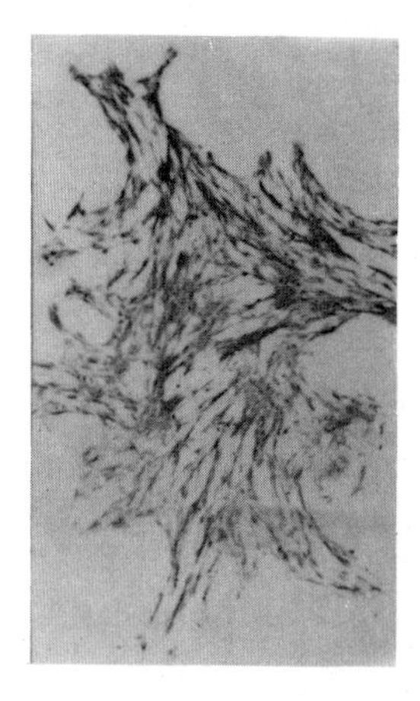

Figure 1-7
A microcolony of Chinese hamster fibroblasts which have been growing for a few days on a glass surface. (Courtesy of Theodore Puck.)

GENETICS AND MOLECULAR BIOLOGY

Genetics is an abstract and logical system by which the genetic factors determining certain properties of living cells can be located with respect to one another and by which changes in these properties can be expressed in quantitative terms. Genetics alone cannot prove anything about molecular mechanisms because the conclusions it leads to are independent of the molecular basis of biological phenomena. However, genetic analyses have been the major source of intuition about molecular mechanisms and have forced scientists to consider phenomena that might otherwise be ignored. Genetics-based arguments can be so suggestive of a particular mechanism, and the suggested mechanism has so often turned out to be the correct one, that molecular biologists often view alternative and conflicting ideas as not even worthy of consideration. The following sections describe a few of the more common uses of genetics in molecular biology.

Uses of Mutants: Some Examples

Some of the most significant advances in molecular biology have come about through the use of mutants. In the following the kinds of approaches that have been taken are described.

1. *A mutant defines a function.* For example, the intake of Fe^{3+} ions by bacteria might be by passive diffusion through the cell membrane, or some system might be responsible for the process. Wild-type *E. coli* can take in the Fe^{3+} ion from a 10^{-5} *M* solution but mutants have been found that cannot do so unless the ion concentration is very high. This finding indicates that a genetically determined system for Fe^{3+} intake exists, though the observation does not tell what this system is. Temperature-sensitive mutants (mutants that have a defect only above a particular temperature) are especially useful in defining functions. For example, temperature-sensitive mutants of *E. coli* have been isolated that fail to synthesize DNA. These mutants fall into at least ten distinct classes, suggesting that there may be at least ten different proteins required for DNA synthesis. One of these mutants lacks a particular protein that is normally located in the cell membrane; the interpretation of such an experimental result is not unambiguous but does suggest that there is some connection between DNA synthesis and the cell membrane.

2. *Mutants can introduce biochemical blocks that aid in the elucidation of metabolic pathways.* The metabolism of the sugar galactose, for example, requires the activity of three distinct genes called *galK*, *galT*, and *galE*. If radioactive galactose ([^{14}C]Gal) is added to a culture of gal^+ cells, many different radioactive compounds can be found as the galactose is metabolized. At very early times after addition of [^{14}C]Gal, three related compounds are detectable: [^{14}C]galactose 1-phosphate (Gal-1-P), [^{14}C]uridine diphosphogalactose (UDP-Gal), and [^{14}C]uridine diphosphoglucose (UDP-Glu). Different mutant genes will block different steps of the metabolic pathway. If the cell is a $galK^-$ mutant, the [^{14}C]Gal label is found only in galactose. Thus the *galK* gene is known to be responsible for the first metabolic step. If the mutant $galT^-$ is used, Gal-1-P accumulates. Thus, the first step in the reaction sequence is found to be the conversion of galactose to Gal-1-P by the *galK* gene product (namely, the enzyme galactokinase). If a $galE^-$ mutant is used, some Gal-1-P is found but the principal radiochemical is

UDP-Gal. Thus the biochemical pathway must be

$$\text{Gal} \xrightarrow[\substack{galK \\ \text{product}}]{} \text{Gal-1-P} \xrightarrow[\substack{galT \\ \text{product}}]{} \text{UDP-Gal} \xrightarrow[\substack{galE \\ \text{product}}]{} \text{X}$$

The identity of X cannot be determined from these genetic experiments.

3. *Mutants enable one to learn about metabolic regulation.* Many mutants have been isolated that alter the amount of a particular protein that is synthesized or the way the amount synthesized responds to external signals. These mutants define regulatory systems. For example, the enzymes corresponding to the *galK*, *galT*, and *galE* genes are normally not present in bacteria but appear only after galactose is added to the growth medium. However, mutants have been isolated in which these enzymes are always present, whether or not galactose is also present. This indicates that some gene is responsible for turning the system of enzyme production on and off and this regulatory gene must be responsive to the presence and absence of galactose.

4. *Mutants enable a biochemical entity to be matched with a biological function or an intracellular protein.* For many years an *E. coli* enzyme called DNA polymerase I was studied in great detail. Purified polymerase I is capable of synthesizing DNA *in vitro*, so it was believed that this enzyme was also responsible for *in vivo* bacterial DNA synthesis. However, an *E. coli* mutant ($polA^-$) was isolated in which the activity of polymerase I was reduced 50-fold yet the mutant bacterium grew and synthesized DNA normally. This observation suggested strongly that polymerase I could not be the only enzyme that synthesizes intracellular DNA. Indeed, biochemical analysis of cell extracts of the $polA^-$ mutant showed the existence of two other enzymes, polymerase II and polymerase III, which could, when purified, also synthesize DNA. In further study, a temperature-sensitive mutation in a gene called *dnaE* was found to be unable to synthesize DNA at 42°C, although synthesis was normal at 30°C. The three enzymes, polymerases I, II, and III, were isolated from cultures of the $dnaE^-$ mutant and each enzyme was assayed. Although polymerases I and II were active at both 30°C and 42°C, polymerase III was active at 30°C but not at 42°C, so that polymerase III was determined to be the product of the *dnaE* gene and the enzyme responsible for intracellular DNA synthesis.

5. *Mutants locate the site of action of external agents.* The antibiotic rifampicin prevents synthesis of RNA. When first discovered, it was not known whether rifampicin might act by preventing synthesis

of precursor molecules (by binding to DNA and thereby preventing the DNA from being transcribed into RNA) or by binding to RNA polymerase, the enzyme responsible for synthesizing RNA. Mutants were isolated that were resistant to rifampicin. These mutants were of two types—those in which the bacterial cell wall was altered so that rifampicin could not enter the cell (an uninformative type of mutant) and those in which the RNA polymerase was slightly altered. The finding of the latter mutants proved that the antibiotic acts by binding to RNA polymerase.

6. *Mutants can indicate relations between apparently unrelated systems.* Bacteriophage λ, which normally adsorbs to and grows in *E. coli,* fails to adsorb to a bacterial mutant unable to metabolize the sugar maltose. Such failure is not associated with mutants incapable of metabolizing other sugars or with any other phages, and this knowledge implicated some product or agent of maltose metabolism in the adsorption of λ.

GENETIC ANALYSIS OF MUTANTS

A great deal of information about molecular mechanisms can be obtained by combining separately isolated mutations in various combinations. This is done by **genetic recombination.** Another important use of genetic recombination is the construction of an array that indicates the positions of genes on a chromosome with respect to one another. When this is done by genetic techniques, the array is called a **genetic map. Physical maps** have also been constructed by using various physical techniques, and some elegant experiments have shown that the gene positions in the two maps are identical. Another genetic technique is **complementation.** By this procedure it is possible to determine the number of genes responsible for a particular phenotype and to distinguish genes from regulatory sites. In the following sections both genetic recombination and complementation will be reviewed.

Genetic Recombination and Genetic Mapping

Genetic recombination includes a variety of phenomena in which genetic loci are rearranged. Here we will be concerned only with the process of combining two genes or two mutations, initially on two chromosomes, onto a single chromosome. The molecular basis is quite complex and is discussed in *MB*, Chapter 18; for our purposes it is sufficient to refer to the "scissors-and-tape" mechanism. In this, one

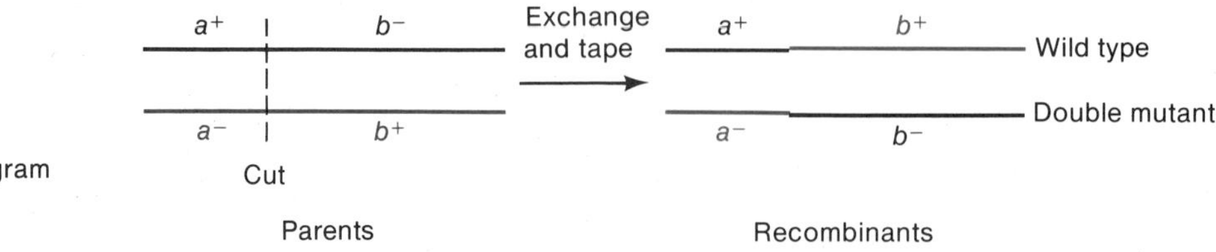

Figure 1-8
A schematic diagram showing genetic exchange.

assumes that two chromosomes align with one another, that a cut is made in both chromosomes at random but matching points, and that the four fragments are then joined together to form two new combinations of genes. This is a naive and incorrect model; it enables one to account for only the simpler features of gene exchange, but these features are in fact the only ones that are of concern at this time. The process is depicted in Figure 1-8. Two parental chromosomes having the genotypes a^+b^- and a^-b^+ pair, are cut, and are then joined to form two recombinant chromosomes whose genotypes are a^+b^+ (wild type) and a^-b^- (double mutant).

In genetic mapping, distance along a chromosome between two recombining genetic loci (or mutations) determines the recombination frequency. As long as the two loci are not too near one another, the recombination frequency is proportional to distance, because chromosomal cuts are made at random. Thus, in the following crosses between chromosomes, the genotypes of which are $a^+b^-c^-$ and $a^-b^+c^+$, and the genes of which are in alphabetical order and equally spaced,

$$\frac{a^+ \quad b^- \quad c^-}{\times}\\ \overline{a^- \quad b^+ \quad c^+}$$

there will be twice as many a^+c^+ recombinants as a^+b^+ recombinants, because loci a and c are twice as far apart as loci a and b.

Because the recombination frequency is proportional to distance, recombination frequency can be used to determine the arrangement of genes on the chromosome. This can be seen in a simple example (Figure 1-9). Consider three genes a, b, and c, whose arrangement is unknown.

Figure 1-9
Two arrangements of three genes for which $a \times b = 1$ percent and $b \times c = 2$ percent. See text for discussion.

Using the notation $p \times q = m$ percent to denote a recombination frequency of m percent between genes p and q, we assume it has been shown that $a \times b = 1$ percent and that $b \times c = 2$ percent. The two arrangements shown in the figure are consistent with these values and can be distinguished by determining the recombination frequency between a and c. Let us assume that this is 1 percent. If that is the case, only arrangement II is possible. The order $c\ a\ b$ for these genes and the relative separation constitute a genetic map.

Any number of genes can be mapped in this way. For instance, consider a fourth gene, d, in the preceding example. If $d \times b = 0.5$ percent, d must be located 0.5 unit either to the left or to the right of b. If $a \times d = 1.5$ percent, d is clearly to the right of b and the gene order is $c\ a\ b\ d$. If $a \times d = 0.5$ percent, the gene order would be $c\ a\ d\ b$.

Complementation

As mentioned earlier, a particular phenotype is frequently the result of the activity of many genes. In the study of any genetic system it is always important to know the number of genes and regulatory elements that constitute the system. The genetic test used to evaluate this number is called **complementation.**

Complementation is best explained by example. The test requires that two copies of the genetic unit to be tested be present in the same cell. In bacteria this can be done by constructing a **partial diploid**—that is, a cell containing one complete set of genes and duplicates of some of these genes. How cells of this type are constructed is described in a later section. A partial diploid is described by writing the genotype of each set of genes on either side of a diagonal line. As an example of this, $b^+c^+d^+/a^+b^-c^-d^-e^+ \cdots z^+$ indicates that a chromosomal segment containing genes b, c, and d is present in a cell whose single chromosome contains all of the genes $a, b, c, \ldots, z$. Usually, only the duplicated genes are indicated, so that this partial diploid would be designated $b^+c^+d^+/b^-c^-d^-$. We now consider a hypothetical bacterium that synthesizes a green pigment from the combined action of genes a, b, and c. The genes encode the enzymes A, B, C, which we assume to act sequentially to form the pigment. If there is a mutation in any of these genes, no pigment is made. Pigment is made by the partial diploid $a^-b^+c^+/a^+b^-c^+$, however, because the cell contains a set of genes that produce functional proteins A, B, and C; B will be made from the $a^-b^+c^+$ chromosome and A from the $a^+b^-c^+$ chromosome. In a partial diploid $a_1^-b^+c^+/a_2^-b^+c^+$, in which a_1^- and a_2^- are two different mutations in gene a, no pigment can be made because the bacterium will not contain a functional A protein. The two mutations a^- and b^- of the diploid are said to complement one another because the phenotype of the partial diploid containing them is A^+B^+; the mutations a_1^- and a_2^- do not complement one another because the phenotype of the partial diploid containing these mutations is A^-.

Suppose that now a mutation x^- has been isolated but the gene in which the mutation has occurred has not been ascertained. By constructing a set of partial diploids, this gene can be identified by a complementation test. As a start, we might test the genes *a*, *b*, and *c* with the partial diploids a^-/x^- (I), b^-/x^- (II), and c^-/x^- (III). If diploids I and II make pigment, the mutation cannot be in genes *a* or *b*; if no pigment is made by diploid III, the mutation must be in gene *c*. If pigment were made in all three diploids, then the important conclusion that mutation x^- is in none of the genes *a*, *b*, or *c* could be drawn; furthermore, since we have assumed that *a*, *b*, and *c* are each pigment genes, the fact that x is not in any of these genes and yet prevent pigment formation would be evidence that pigment formation requires at least four genes ("at least" because more genes might be discovered).

A common approach to the initial characterization of a genetic system is to isolate about fifty mutants and perform complementation tests between them, as will be shown in the example of the next section. The analysis, which may be tedious, is nonetheless a straightforward one. The basic rule is the following:

Rule I. If x_1^- and x_2^- complement, they are in different genes. The converse statement—that is, if two mutations do not complement, they are in the same gene—does not always follow.

There are three explanations for the lack of complementation of two mutations; these are stated in the following rule:

Rule II. If mutations x_1^- and x_2^- fail to complement, then one of the following is true:

(a) They are in the same gene, or
(b) At least one of the mutations is in a regulatory site for the other gene, or
(c) At least one of the mutations yields an inhibitory gene product.

A detailed example of the application of these rules can be found in *MB*, pages 43–45.

So far we have discussed complementation only in bacteria. The concept is also applicable to phages and is a common way to assign phage mutations to particular genes. The equivalent of the partial diploid test is the infection of a bacterium with two mutant phages, neither of which can grow alone; one then notes whether phage are produced by the doubly infected cells. This test is usually done with conditional mutants in the following way. Consider two temperature-sensitive mutations a^-(Ts) and b^-(Ts) in genes *a* and *b* of two respective phage; neither gene product is active at 42°C, so the mutants cannot grow at 42°C and phage production depends upon the complementation of separate Ts mutations. The phage are allowed to adsorb at 42°C to a host bacterium and then the infected cell is incubated at that temperature. If only a^-(Ts) or only b^-(Ts) phage have adsorbed, no phage will be produced. If both have adsorbed, however, and the Ts^- mutations are in different genes, the infected cell will contain a good copy of product A and a good copy of product B; complementation can then occur and progeny phage will be produced. Note that in addition to the parental types of a^-(Ts) and b^-(Ts) phage that are present in the resulting lysate, there are also some Ts^+ and a^-(Ts)b^-(Ts) recombinants.

2 Macromolecules

A typical cell contains 10^4–10^5 different kinds of molecules. Roughly half of these are small molecules—namely, inorganic ions and organic compounds whose molecular weights usually do not exceed several hundred. The others are polymers that are so massive (molecular weights from 10^4 to 10^{12}) that they are called **macromolecules.** These molecules are of three major classes: proteins, nucleic acids, and polysaccharides are polymers of amino acids, nucleotides, and sugars, respectively.

Knowledge of the properties of macromolecules is essential for understanding living processes. This and the next two chapters will emphasize nucleic acids and proteins. Information about other types of macromolecules will be given as needed in other chapters.

CHEMICAL STRUCTURES OF THE MAJOR CLASSES OF MACROMOLECULES

In this unit we examine the chemical structure of proteins, nucleic acids, and polysaccharides—in particular, the monomers and the chemical linkages by which they are joined. Physical properties of these macromolecules and the interactions that exist between various parts of a single macromolecule are described in later units.

Proteins

A protein is a polymer consisting of several amino acids (a **polypeptide**). Each amino acid can be thought of as a single carbon atom (the α carbon) to which there is attached one carboxyl group, one amino group, and a side chain denoted R (Figure 2-1). The side chains are generally carbon chains or rings to which various functional groups may be attached. The simplest side chains are those of glycine (a hydrogen atom) and alanine (a methyl group). The complete chemical structures of each amino acid are given in Figure 2-2.

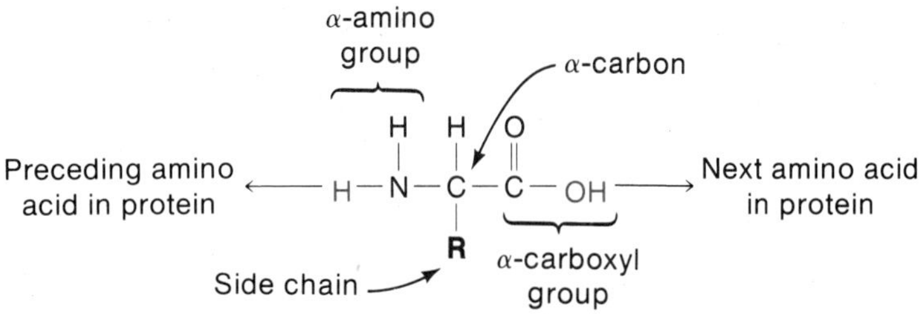

α-amino group
α-carbon
Preceding amino acid in protein
Next amino acid in protein
Side chain
α-carboxyl group

Figure 2-1
Basic structure of an α-amino acid. The NH_2 and COOH groups are used to connect amino acids to one another. The red OH of one amino acid and the red H of the next amino acid are removed when two amino acids are linked together (see Figure 2-3).

To form a protein the amino group of one amino acid reacts with the carboxyl group of another; the resulting chemical bond is called a **peptide bond** (Figure 2-3). Amino acids are joined together in succession to form a linear **polypeptide chain** (Figure 2-4). When the number of peptide bonds exceeds about 15 (the number is arbitrary), the polypeptide is called a **protein.**

The two ends of every protein molecule are distinct. One end has a free $—NH_2$ group and is called the **amino terminus**; the other end has a free —COOH group and is the **carboxyl terminus.** The ends are also called the N (or NH_2) and C termini, respectively.

The amino-acid side chains usually do not engage in covalent bond formation; an exception is the —SH group of cysteine, which often forms a disulfide (S—S) bond by reaction with a second cysteine in the same or different polypeptide chain.

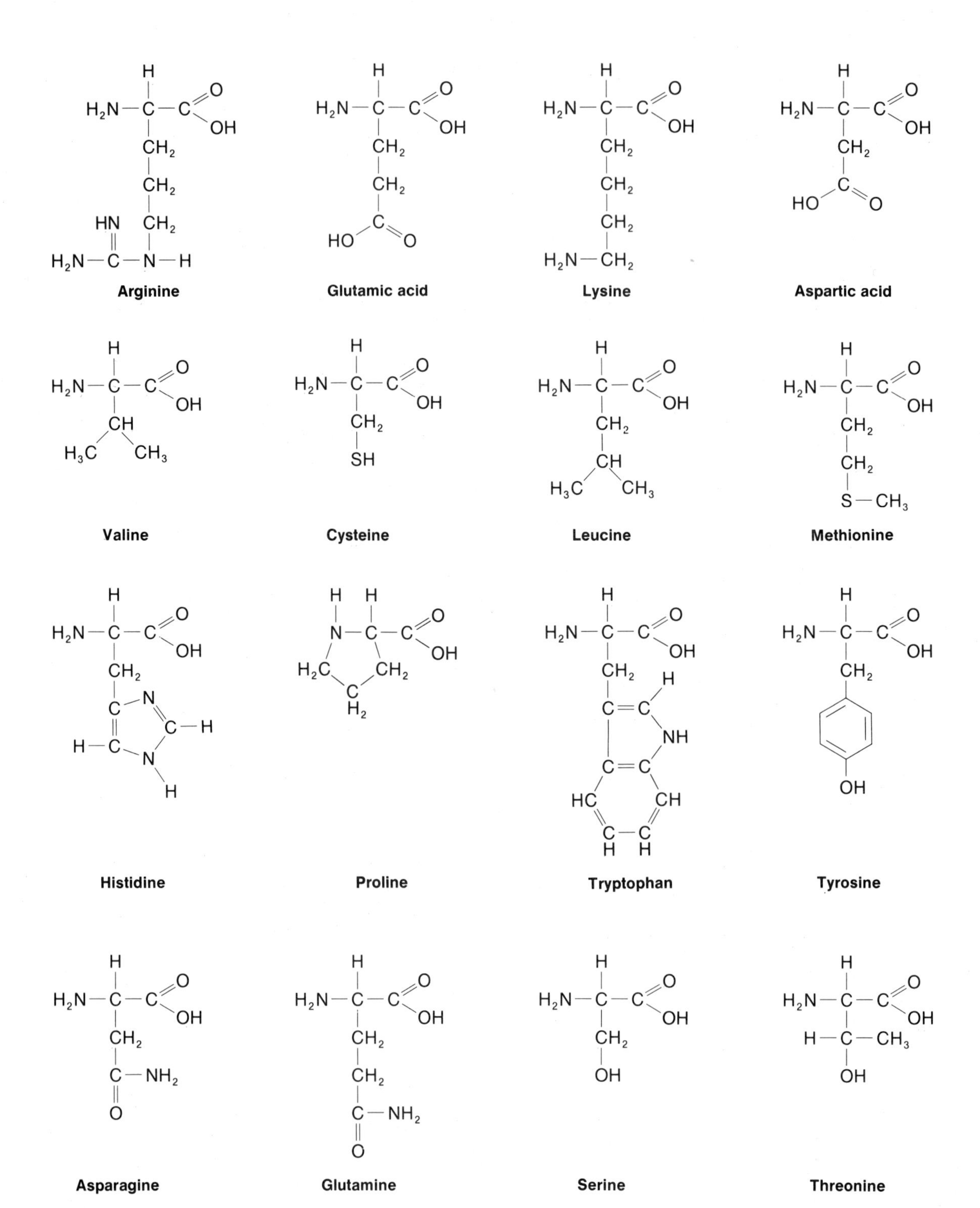

Figure 2-2
Chemical structures of the amino acids.

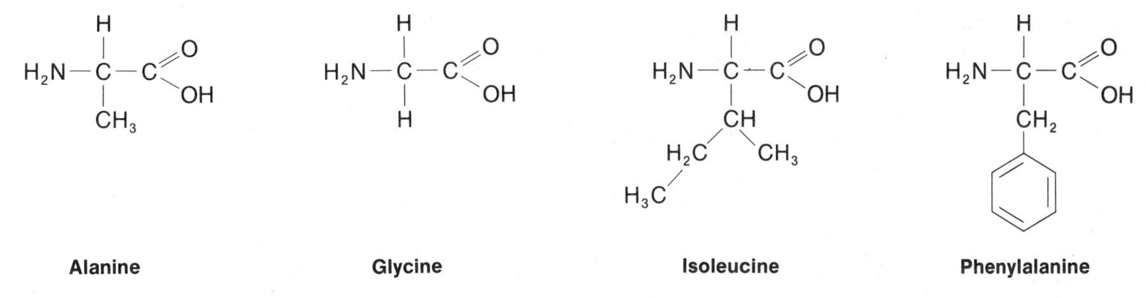

Figure 2-2 (continued)

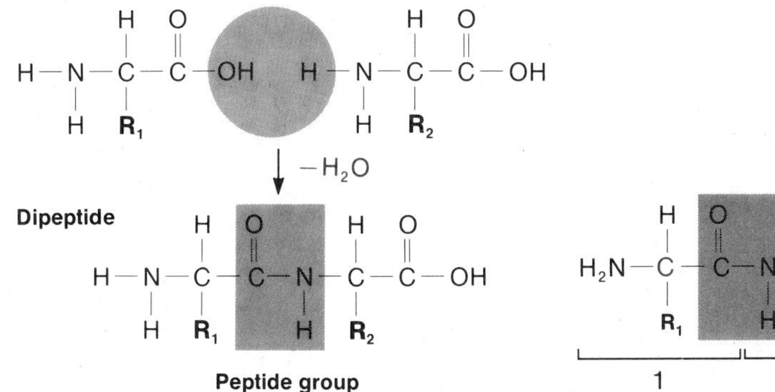

Figure 2-3
Formation of a dipeptide from two amino acids by elimination of water (shaded circle) to form a peptide group (shaded rectangle).

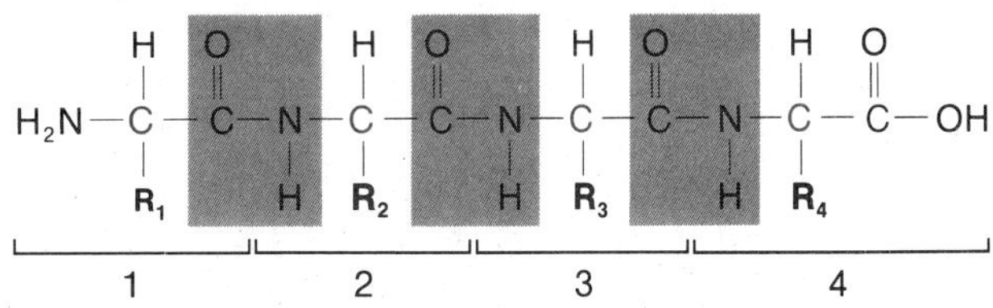

Figure 2-4
A tetrapeptide showing the alternation of α-carbon atoms (unshaded) and peptide groups (shaded). The four amino acids are numbered below.

Nucleic Acids

A nucleic acid is a polynucleotide—that is, a polymer consisting of nucleotides. Each nucleotide has the three following components (Figure 2-5):

1. A cyclic five-carbon sugar. This is ribose, in the case of ribonucleic acid (RNA), and deoxyribose, in deoxyribonucleic acid (DNA). The difference in the structure of ribose and 2′-deoxyribose is shown in Figure 2-5.
2. A purine or pyrimidine base attached to the 1′-carbon atom of the sugar by an *N*-glycosylic bond. The bases, which are shown in Figure 2-6, are the purines, adenine (A) and guanine (G), and the pyrimidines, cytosine (C), thymine (T), and uracil (U). DNA and RNA contain A, G, and C; however, T is found only in DNA and U is found only in RNA.

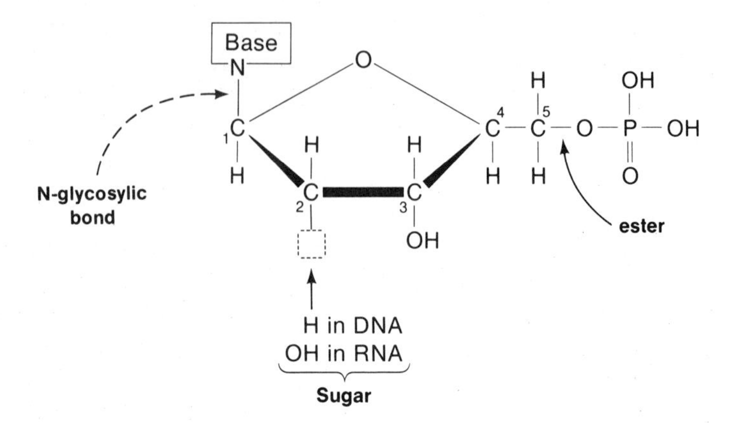

Figure 2-5
Structure of a mononucleotide. The carbon atoms in the sugar ring are numbered.

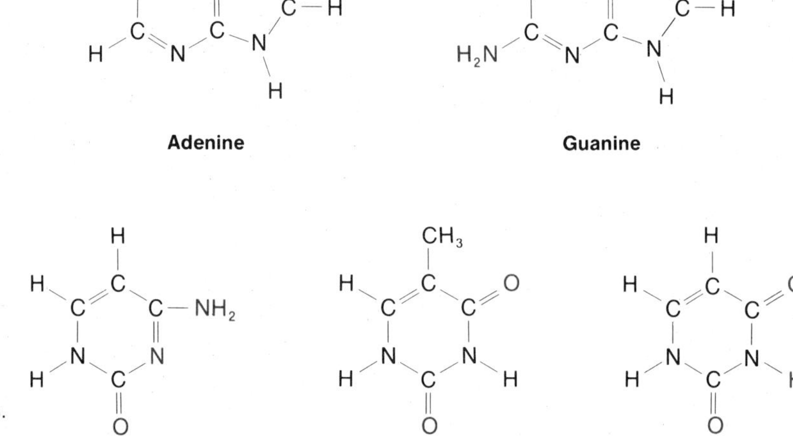

Figure 2-6
The bases found in nucleic acids. The weakly charged groups are shown in red.

3. A phosphate attached to the 5′ carbon of the sugar by a phosphoester linkage. This phosphate is responsible for the strong negative charge of both nucleotides and nucleic acids.

A base linked to a sugar is called a **nucleoside**; thus *a nucleotide is a nucleoside phosphate.*

The nucleotides in nucleic acids are covalently linked by a second phosphoester bond that joins the 5′-phosphate of one nucleotide and the 3′-OH group of the adjacent nucleotide (Figure 2-7). Thus the phosphate is esterified to both the 3′- and 5′-carbon atoms; this unit is often called a **phosphodiester group.**

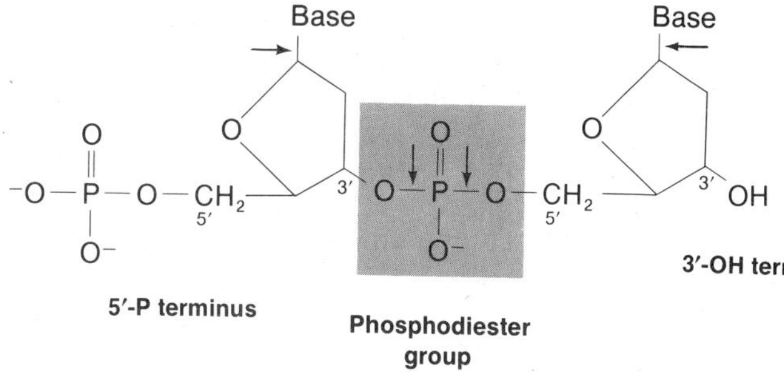

Figure 2-7
The structure of a dinucleotide. The vertical arrows show the bonds in the phosphodiester group about which there is free rotation. The horizontal arrows indicate the *N*-glycosylic bond about which the base can freely rotate. A polynucleotide would consist of many nucleotides linked together by phosphodiester bonds.

Polysaccharides

Polysaccharides are polymers of sugars (most often glucose) or sugar derivatives. These are very complex molecules because sometimes covalent bonds occur between many pairs of carbon atoms. This has the effect that one sugar unit can be joined to more than two other sugars, which results in the formation of highly branched macromolecules. These branched structures are sometimes so enormous that they are almost macroscopic. For example, the cell walls of many bacteria and plant cells are single gigantic polysaccharide molecules.

NONCOVALENT INTERACTIONS THAT DETERMINE THE THREE-DIMENSIONAL STRUCTURES OF PROTEINS AND NUCLEIC ACIDS

The biological properties of macromolecules are mainly determined by noncovalent interactions that result in each molecule acquiring a unique three-dimensional structure. In this section the noncovalent interactions that are important and the chemical groups responsible for them are described.

The Random Coil

Linear polypeptide and polynucleotide chains contain several bonds about which there is free rotation. In the absence of any intrastrand interactions* each monomer would be free to rotate with respect to its adjacent monomers; this is limited only by the fact that atoms cannot occupy the same space. The three-dimensional configuration of such a chain is called a **random coil**; it is a somewhat compact and globular

*An intrastrand interaction is an interaction between two regions of the same strand. An interstrand interaction is between different strands. It is important to remember the distinction between these similar words.

structure that *changes shape continually* owing to constant bombardment by solvent molecules.

Few, if any, nucleic acid or protein molecules exist in nature as random coils because there are many interactions between elements of the chains. These interactions are hydrogen bonds, hydrophobic interactions, ionic bonds, and van der Waals interactions.

Hydrogen-bonding

The most common hydrogen bonds found in biological systems are shown in Figure 2-8. In nucleic acids, hydrogen-bonding causes intrastrand pairing between nucleotide bases, which, if it were random, would cause a single polynucleotide strand to be more compact than a random coil. In DNA, interstrand hydrogen bonds are responsible for the double-stranded helical structure. In proteins, intrastrand hydrogen-bonding occurs between a hydrogen atom on a nitrogen adjacent to one peptide bond and an oxygen atom adjacent to a different peptide bond. This interaction gives rise to several particular polypeptide chain configurations, which will be discussed shortly.

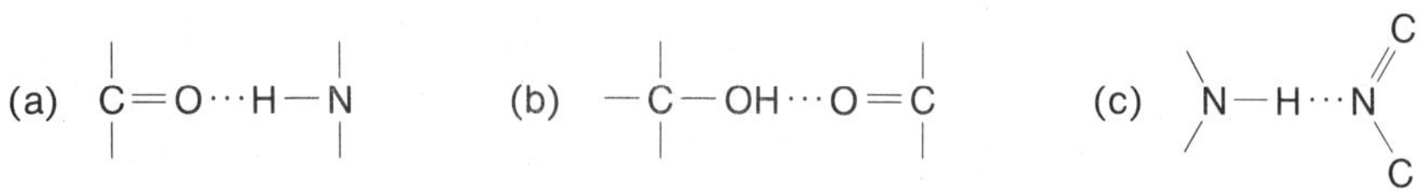

(a) C=O···H—N   (b) —C—OH···O=C   (c) N—H···N (with C attached to the second N)

Figure 2-8
Structures of three types of hydrogen bonds (indicated by three dots). (a) A type found in proteins and nucleic acids. (b) A weak bond found in proteins. (c) A type found in DNA and RNA.

The Hydrophobic Interaction

A hydrophobic interaction is an interaction between two molecules (or portions of molecules) that are somewhat insoluble in water. The phenomenon is simply that *two molecules (which may be different) that are poorly soluble in water tend to associate.* The explanation is a thermodynamic one and can be found in *MB*, page 84.

Many components in nucleic acids and proteins have the hydrophobic property just described. For example, the bases of nucleic acids are planar organic rings carrying localized weak charges (Figure 2-6). The localized charges are sufficient to maintain solubility, but the large, poorly soluble organic ring portion causes the ordering of water just described, so that the bases tend to cluster. The most efficient kind of clustering is one in which the faces of the rings are in contact, an array known as **base-stacking** (Figure 2-9). Note that since the bases are adjacent in the chain, stacking gives some rigidity to single polynucleotide strands and tends to extend them rather than allow a random coil to form. In the following chapter we will see how stacking is of major importance in determining nucleic acid structure. Many amino-acid side chains are very poorly soluble, and this causes (1) the benzene ring of phenylalanine to stack and (2) the hydrocarbon chains of alanine, leucine, isoleucine, and valine to form tight unstacked clusters. Since these amino acids are not necessarily adjacent, *hydrophobic interactions tend to bring distant hydrophobic parts of a polypeptide chain together.*

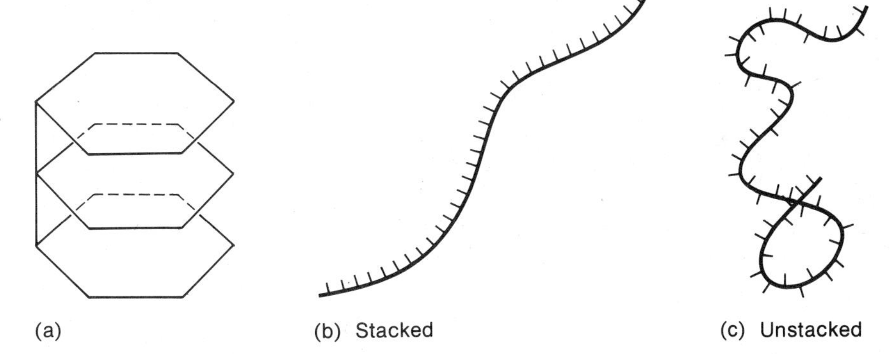

Figure 2-9
The meaning of stacking. (a) Three stacked bases. (b) Structures of stacked and unstacked polynucleotides. The stacked polynucleotide is more extended because the stacking tends to decrease the flexibility of the molecule.

Ionic Bonds

An ionic bond is the result of attraction between unlike charges. Several amino-acid side chains have ionized, negatively charged carboxyl groups (aspartic and glutamic acids) and positively charged amino groups (lysine, histidine, and arginine) that can form such bonds. This tends to bring together distant parts of the chain. Ionic interactions can also be repulsive—namely, between two like charges; thus, it would be unlikely for a polypeptide chain to fold in such a way that two aspartic acids are very near one another. Ionic bonds are the strongest of the noncovalent interactions. However, they are destroyed by extremes of pH, which can change the charge of the groups, and by high concentrations of salt, the ions of which shield the charged groups from one another.

Van der Waals Attraction

Van der Waals forces exist between all molecules and are a result of both permanent dipoles and the circulation of electrons. The dependence of the attractive force between two atoms is proportional to $1/r^6$, in which r is the distance between their nuclei. Thus, the attraction is a very weak force and is significant only if two atoms are very near one another (1 to 2 Å apart).

Van der Waals forces are very weak and are easily overcome by thermal motion; in general, the force between two atoms will not keep them in proximity. However, if interactions of *several* pairs of atoms are combined, the cumulative attractive force can be great enough to withstand being disrupted by thermal motion. Thus, two molecules can attract one another if several of their component atoms can mutually interact. However, because of the $1/r^6$-dependence, the intermolecular fit must be nearly perfect. What this means is that *two molecules can bind to one another if their shapes are complementary*. This is true also of two separate regions of a polymer—that is, the regions can hold together if their shapes match. Sometimes the van der Waals attraction between two regions is not large enough to effect this; however, it can significantly strengthen other weak interactions such as the hydrophobic interaction, if the fit is good.

Summary of the Effects of Noncovalent Interactions

The effect of noncovalent interactions is to constrain a linear chain to fold in such a way that different regions of the chain, which may be quite distant in a chain (if the chain were linear), are brought together. An example of a molecule showing the four kinds of noncovalent folding that have been described is shown in Figure 2-10.

In Chapters 3, 4 and 5 we will discuss the specific structures brought about by these interactions and the physical properties of nucleic acids and proteins.

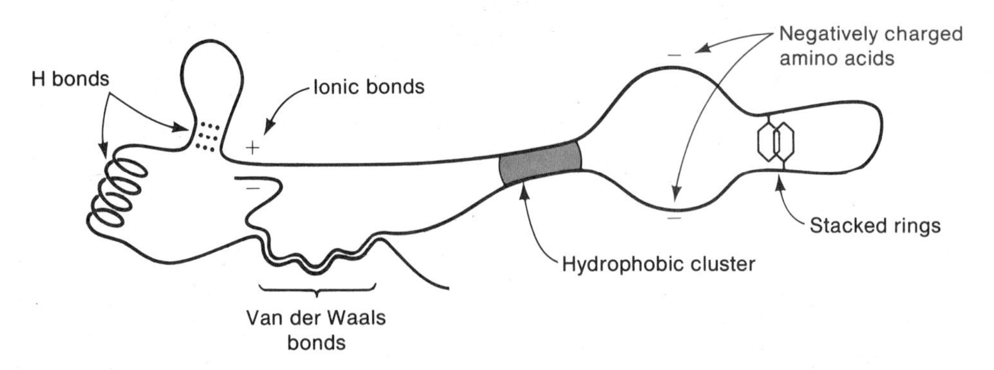

Figure 2-10
A hypothetical polypeptide chain showing attractive (black) and repulsive (red) interactions.

GEL ELECTROPHORESIS

Several methods of studying macromolecules are used repeatedly in molecular biology. Some of these—e.g., spectrophotometry and chromatography—are usually covered in elementary chemistry courses and will not be discussed here. Ultracentrifugation and electron microscopy are described in *MB*, pages 88–91, 93–96. One of the most prevalent techniques in use today is gel electrophoresis, which is described in this section.

Most biological macromolecules are electrically charged and thus move in an electric field. For example, if the terminals of a battery are connected to the opposite ends of a horizontal tube containing a solution of positively charged protein molecules, the molecules will move from the positive end of the tube to the negative end. The direction of motion obviously depends on the sign of the charge but the rate of movement depends on the magnitude of the charge and, as in sedimentation, on the shape of the molecule (that is, its frictional resistance). The mass of the molecule plays no direct role in the rate of migration (in contrast with sedimentation) and influences the rate only indirectly when the surface area of the molecule, which affects its frictional coefficient, is a function of its mass.

The most common type of electrophoresis used in molecular biology is zonal electrophoresis through a gel, or **gel electrophoresis.** This procedure can be performed so that the rate of movement depends only on the molecular weight of the molecule, as will be seen below.

An experimental arrangement for gel electrophoresis of DNA is shown in Figure 2-11. A thin slab of an agarose or polyacrylamide gel is prepared containing small slots ("wells") into which samples are placed. An electric field is applied and the negatively charged DNA molecules penetrate and move through the agarose. A gel is a complex network of molecules, and migrating macromolecules must squeeze through narrow, tortuous passages. The result is that smaller molecules pass through more easily and thus the migration rate increases as the molecular weight, M, decreases. For unknown reasons the distance moved, D, depends logarithmically on M, obeying the equation

$$D = a - b \log M,$$

in which a and b are empirically determined constants that depend on the buffer, the gel concentration, and the temperature. Figure 2-12 shows the result of electrophoresis of a collection of DNA molecules.

Gel electrophoresis of proteins is in principle carried out the same way. However, proteins can be either positively or negatively charged, and a sample containing several different proteins must be placed in a

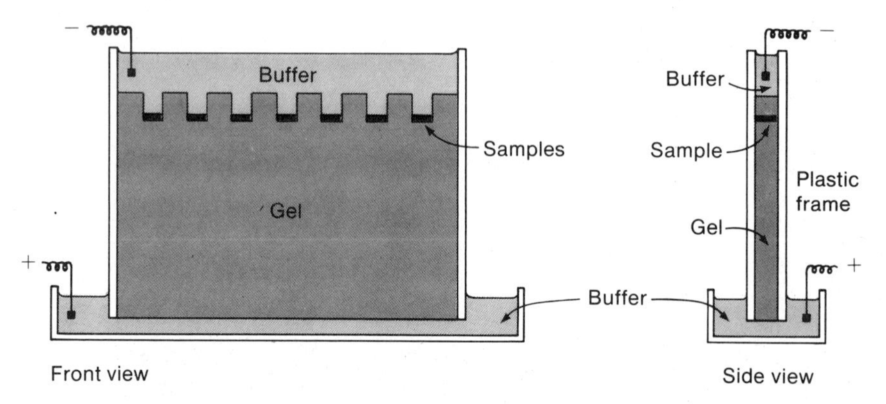

Figure 2-11
Apparatus for slab-gel electrophoresis capable of running seven samples simultaneously. The liquid gel is allowed to harden in place. An appropriately shaped mold is placed on top of the gel during the hardening in order to make wells for the samples (red). After electrophoresis, the slab is stained by removing the plastic frame and immersing the gel in a solution of the stain. Excess stain is removed by washing. The components of the sample appear as bands, which may be either visibly colored or fluorescent when irradiated with ultraviolet light. The region of the gel in which the components of one sample can move is called a *lane.* Thus, the gel shown has seven lanes. (After D. Freifelder. 1982. *Physical Biochemistry.* 2nd Ed. W. H. Freeman and Co.)

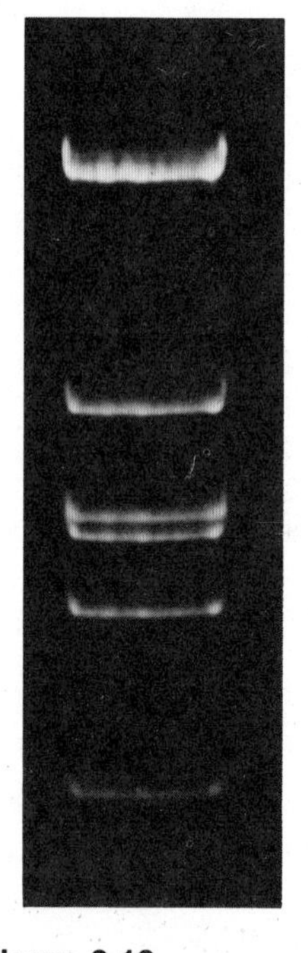

Figure 2-12
A gel electrophoregram of six fragments of *E. coli* phage λ DNA. The direction of movement is from top to bottom. The DNA is made visible by the fluorescence of bound ethidium bromide. (Courtesy of Arthur Landy and Wilma Ross.)

centrally located well in order that migration can occur in both directions. The charge per unit mass, which is very small because most amino acids are uncharged, varies from one protein to the next (in contrast with DNA or RNA, which have one negative charge per nucleotide) and proteins come in a variety of shapes, so there is no simple way to predict the migration rate. In one important technique the anionic detergent sodium dodecyl sulfate (SDS)

$$H_3C-(CH_2)_{10}-CH_2OSO_3^-Na^+$$

and the disulfide bond-breaking agent mercaptoethanol

```
     H   H
     |   |
HO — C — C — SH
     |   |
     H   H
```

are added to the protein solution. One molecule of SDS binds per amino acid, giving each protein molecule the same charge per amino acid residue. Furthermore, *when SDS is bound and there are no disulfide bonds, all proteins have nearly the same shape—namely, the near-random coil.* The net effect is that, as in the case of DNA, the migration rate increases as *M* decreases and the dependence is logarithmic and described by the equation given above.

3 Nucleic Acids

Deoxyribonucleic acid (DNA) is the single most important molecule in living cells and contains all of the information that specifies the cell. In this chapter its remarkable structure and many of its physical and chemical properties are described. Some properties of ribonucleic acid (RNA) will also be presented.

PHYSICAL AND CHEMICAL STRUCTURE OF DNA

In the previous chapter the structure of a nucleotide and how nucleotides are joined to form a polynucleotide, or a nucleic acid, were described. In this section we show how nucleotides interact with one another to produce a double helix from two polynucleotides having complementary base sequences, and we present the Watson-Crick model for the DNA double helix.

In early physical studies of DNA many experiments indicated that the molecule is an extended chain having a highly ordered structure. The most important technique was x-ray diffraction analysis, by which information was obtained about the arrangement and dimensions of various parts of the molecule. The most significant observations were that the molecule is helical and that the bases of the nucleotides are stacked with their planes separated by a spacing of 3.4 Å.

Chemical analysis of the molar content of the bases (generally called the **base composition**) adenine, thymine, guanine, and cytosine in DNA molecules isolated from many organisms provided the important fact that [A] = [T] and [G] = [C], in which [ ] denotes molar concentration, from which followed the corollary [A + G] = [T + C] or [purines] = [pyrimidines].

James Watson and Francis Crick combined chemical and physical data for DNA with a feature of the x-ray diffraction diagram that suggested that two helical strands are present in DNA and showed that

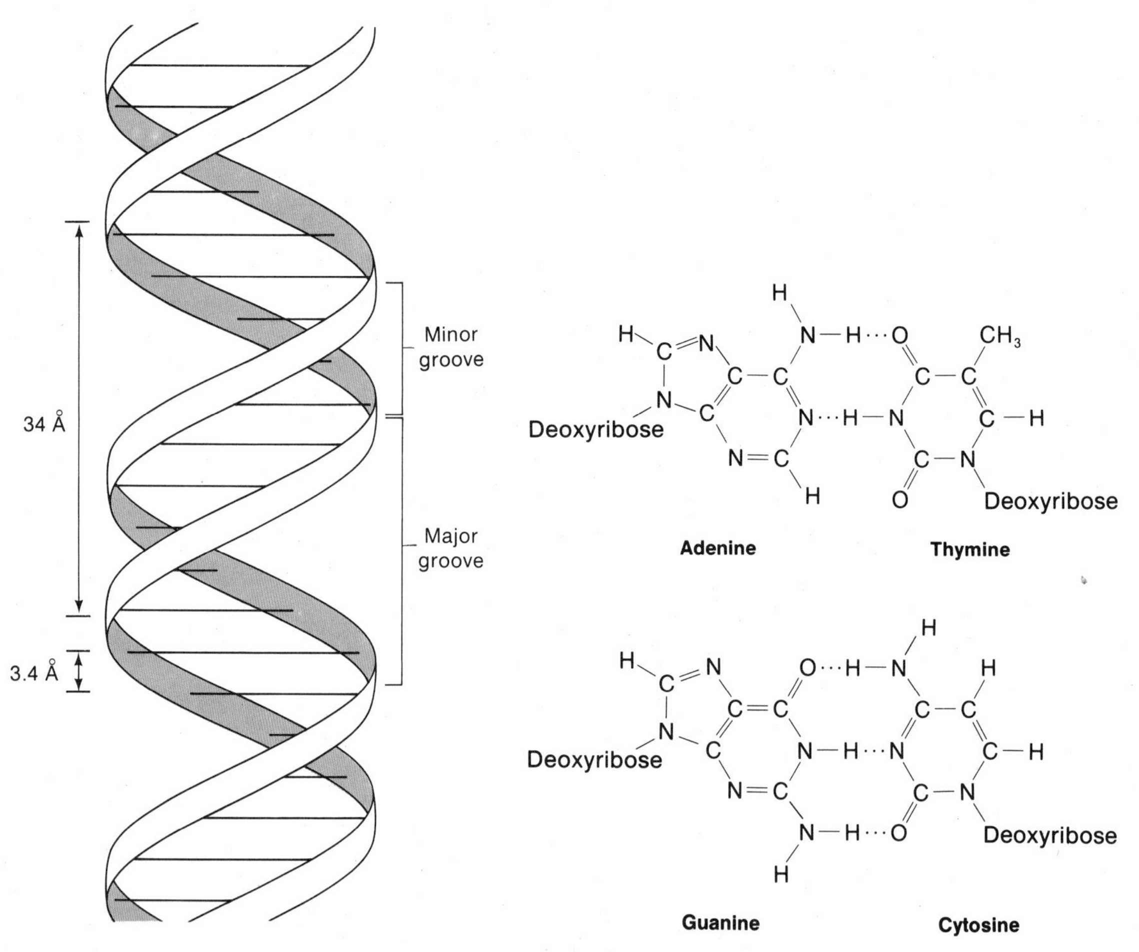

Figure 3-1
Diagrammatic model of the DNA double helix in the common B form. In the A form, which may not exist within cells, the bases are tilted and there are 10 bases per turn.

Figure 3-2
The two common base pairs of DNA. Note that hydrogen bonds (dotted lines) are between the weakly charged groups noted in Figure 2-6.

the two strands are coiled about one another to form a double-stranded helix (Figure 3-1). In this model the sugar-phosphate backbones follow a helical path at the outer edge of the molecule and the bases are in a helical array in the central core. The bases of one strand are hydrogen-bonded to those of the other strand to form the purine-to-pyrimidine base pairs A · T and G · C (Figure 3-2).

The two bases in each base pair lie in the same plane and the plane of each pair is perpendicular to the helix axis. The base pairs are rotated 36° with respect to each adjacent pair, so that there are ten pairs per helical turn. The diameter of the double helix is 20 Å and the molecular weight per unit length of the helix is approximately 2×10^6 per micrometer.

The helix has two external helical grooves, a deep wide one (the **major groove**) and a shallow narrow one (the **minor groove**); both of these grooves are large enough to allow protein molecules to come in contact with the bases.

The double helix shown in Figure 3-1 is a right-handed helix. This means that when an observer looks down the axis of the helix in either direction, each strand follows a clockwise path as it moves away from the observer. If the path were counterclockwise, the helix would be left-handed.

Naturally occurring DNA molecules are generally right-handed helices.

In a later section we will examine a synthetic DNA molecule, which in certain conditions forms a left-handed helix, and discuss the recent observation that some naturally occurring DNA molecules apparently have regions that are left-handed helices.

ATGGTCAACTG
| | | | | | | | | | |
TACCAGTTGAC

Figure 3-3
A diagram of a DNA molecule showing complementary base-pairing of the individual strands. The light lines represent hydrogen bonds.

Base-pairing is one of the most important features of the DNA structure because it means that the base sequences of the two strands are complementary (Figure 3-3); that is, if one strand has the base sequence AATGCT, the other strand has the sequence TTACGA, reading in the same direction. This has deep implications for the mechanism of DNA replication because, in this way, the replica of each strand is given the base sequence of its complementary strand.

The two polynucleotide strands of the DNA double helix are antiparallel—that is, the 3′-OH terminus of one strand is adjacent to the 5′-P (5′-phosphate) terminus of the other (Figure 3-4). The significance of this is twofold. First, in a linear double helix there is one 3′-OH and one 5′-P terminus at each end of the helix. The second and generally more significant point is that the orientations of the two strands are different. That is, if two nucleotides are paired, the sugar portion of one nucleotide lies upward along the chain, whereas the sugar of the other nucleotide lies downward. We will see in Chapter 7 that this structural

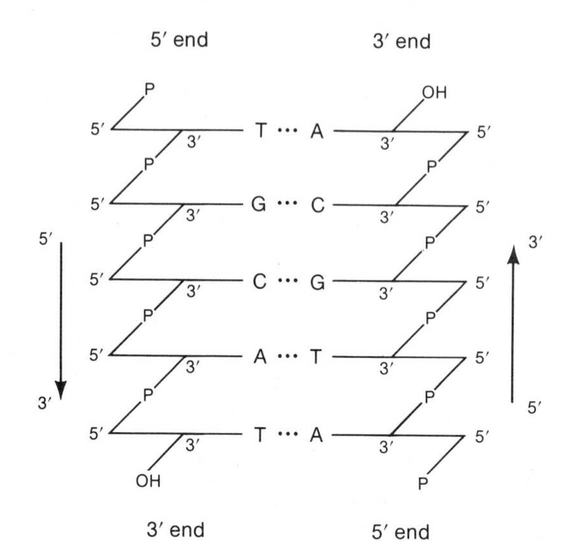

Figure 3-4
A stylized drawing of a segment of a DNA duplex showing the antiparallel orientation of the complementary chains. The arrows indicate the 5′ to 3′ direction of each strand. The phosphates (P) join the 3′ carbon of one deoxyribose (horizontal line) to the 5′ carbon of the adjacent deoxyribose.

THE DOTS BETWEEN THE PAIRS INDICATE THE NUMBER OF HYDROGEN BONDS

feature poses an interesting constraint on the mechanism of DNA replication.

Because the strands are antiparallel, a convention is needed for stating the sequence of bases of a single chain. The convention is to write a sequence with the 5′-P terminus at the left; for example, ATC denotes the trinucleotide P-5′—ATC—3′-OH.

It is always true that [A] = [T] and [G] = [C], but there is no rule governing the ratio of total concentrations of guanine and cytosine to adenine and thymine in a DNA molecule—that is, ([G] + [C])/([A] + [T]). In fact, there is enormous variation in this ratio, ranging from 0.37 to 3.16 for different species of bacteria. The usual way that the base composition of the DNA molecule from a particular organism is expressed is not by the ratio just given but by the fraction of all bases that are G • C pairs; that is, ([G] + [C])/[all bases]; this is termed the G+C content, or percent G+C. The base composition of hundreds of organisms has been determined. Generally speaking, the value of the G+C content is near 0.50 for the higher organisms and has a very small range from one species to the next (0.49 – 0.51 for the primates); for the lower organisms the value of the G+C content varies widely from one genus to another. For example, for bacteria the extremes are 0.27 for the genus *Clostridium* and 0.76 for the genus *Sarcina*; *E. coli* DNA has the value 0.50.

THE FACTORS THAT DETERMINE THE STRUCTURE OF DNA

The helical structure of nucleic acids is determined by stacking between adjacent bases in the same strand, and the double-stranded helical structure of DNA is maintained by hydrogen-bonding between the bases in the base pairs. This conclusion, as well as features of the structure that have not yet been described, has come from studies of denaturation, analyzed in this section.

The free energies of the noncovalent interactions described in Chapter 2 are not much greater than the energy of thermal motion at room temperature, so that at elevated temperatures the three-dimensional structures of both proteins and nucleic acids are disrupted. A macromolecule in a disrupted state, in which the molecules are in a nearly random coil configuration, is said to be **denatured**; the ordered state, which is presumably that originally present in nature, is called **native.** A transition from the native to the denatured state is called **denaturation.** When double-stranded DNA or native DNA is heated, the bonding forces between the strands are disrupted and the two strands separate; thus, *denatured DNA is single-stranded.*

A great deal of information about structure and stabilizing interactions has been obtained by studying nucleic acid denaturation. This is done by measuring some property of the molecule that changes as denaturation proceeds—for example, the absorption of ultraviolet light. Originally, denaturation was always accomplished by heating a DNA solution, so that a graph of a varying property as a function of temperature is called a **melting curve.** For DNA, denaturation can be detected by observing the increase in the ability of a DNA solution to absorb ultraviolet light at a wavelength of 260 nanometers (nm). A standard measure of such absorption is a logarithmic ratio of transmitted light and incident light called the **absorbance** $\boldsymbol{A}$; at 260 nm, this is written A_{260}. Nucleic acid bases absorb 260-nm light strongly, and the amount of light absorbed by a given number of bases depends on their proximity. When the bases are highly ordered (very near one another), as in double-stranded DNA, A_{260} is lower than that for the less ordered state in single-stranded DNA. In particular, if a solution of double-stranded DNA has a value of $A_{260} = 1.00$, a solution of single-stranded DNA at the same concentration has a value of $A_{260} = 1.37$. This relation is often described by stating either that double-stranded DNA is **hypochromic** or that single-stranded DNA is **hyperchromic.** (The corresponding value for free bases is 1.60, so that absorbance can also be used to measure degradation of nucleic acids.)

If a DNA solution is slowly heated and the A_{260} is measured at various temperatures, a melting curve such as that shown in Figure 3-5 is obtained. The following features of this curve should be noted:

1. The A_{260} remains constant up to temperatures well above those encountered by living cells in nature.
2. The rise in A_{260} occurs over a range of 6–8°C.
3. The maximum A_{260} is about 37 percent higher than the starting value.

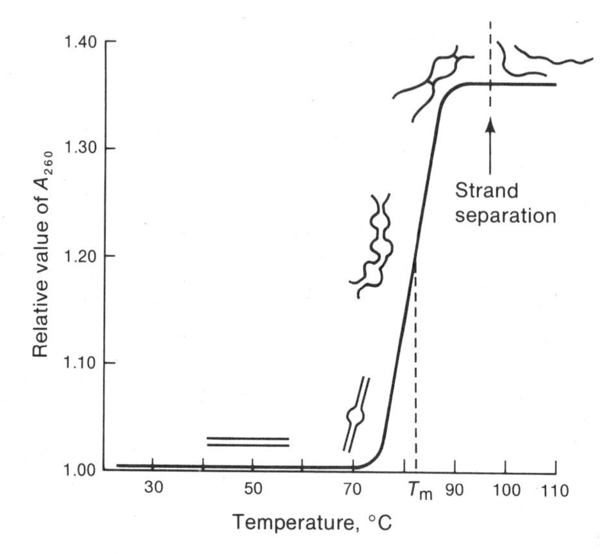

Figure 3-5
A melting curve of DNA showing T_m and possible molecular configurations for various degrees of melting.

The state of a DNA molecule in different regions of the melting curve is also shown in the figure. Before the rise begins, the molecule is fully double-stranded. In the rise region, base pairs in various segments of the molecule are broken; the number of broken base pairs increases with temperature. In the initial part of the upper plateau a few base pairs remain to hold the two strands together until a critical temperature is reached at which the last base pair breaks and the strands separate completely.

A convenient parameter to characterize a melting transition is the temperature at which the rise in A_{260} is half complete. This temperature is called the **melting temperature**; it is designated $\boldsymbol{T_m}$.

A great deal of information is obtained by observing how T_m varies with base composition and experimental conditions. For example, it has been observed that T_m increases with percent G+C, which has been interpreted in terms of the relative number of hydrogen bonds in a G · C pair (three) and an A · T pair (two). That is, a higher temperature is required to disrupt a G · C pair than an A · T pair, because the double-stranded structure is stabilized, at least in part, by hydrogen bonds. In addition, reagents that increase the solubility of the bases and therefore decrease the hydrophobic interaction also lower T_m. These results suggest that a hydrophobic interaction is present in DNA. Furthermore, the results imply that a polynucleotide would tend to have a three-dimensional structure that maximizes the contact of the highly soluble phosphate group with water and minimizes contact between bases and water. This explains why, in DNA, the sugar-phosphate chain is on the outside and the bases are on the inside, in a stacked array. Clearly, stacked bases are more easily hydrogen-bonded, and correspondingly, hydrogen-bonded bases, which are oriented by the bonding, stack more easily. Thus, the two interactions act cooperatively to form a very stable structure. If one of the interactions is eliminated, the other is weakened; this explains why T_m drops so markedly after the addition of a reagent that destroys either type of interaction.

A useful denaturant is high pH, for the charge of several groups engaged in hydrogen-bonding is changed and a base bearing such a group loses its ability to form these bonds. At a pH greater than 11.3 all hydrogen bonds are eliminated and DNA is completely denatured. Since DNA is quite resistant to alkaline hydrolysis, this procedure is the method of choice for denaturing DNA, as heating breaks phosphodiester bonds.

A repulsive force is also present in DNA, namely, that between the negatively charged phosphate groups in complementary strands. These charges are neutralized by bound ions, such as Na^+ and Mg^{2+}. In distilled water, the repulsion is so great that strand separation occurs.

Although DNA is stable, it is a dynamic structure in which double-stranded regions frequently open to become single-stranded bubbles. This important phenomenon, called **breathing,** is thought to enable specialized proteins to interact with the DNA molecule and to "read" its encoded information. Note that since a G · C pair has three hydrogen bonds and an A · T pair has only two, breathing occurs more often in regions rich in A · T pairs than in regions rich in G · C pairs.

RENATURATION

A solution of denatured DNA can be treated in such a way that native DNA re-forms. The process is called **renaturation** or **reannealing** and the re-formed DNA is called **renatured DNA.**

Renaturation has proved to be a valuable tool in molecular biology since it can be used to demonstrate genetic relatedness between different organisms, to detect particular species of RNA, to determine whether certain sequences occur more than once in the DNA of a particular organism, and to locate specific base sequences in a DNA molecule.

Two requirements must be met for renaturation to occur:

1. The salt concentration must be high enough that electrostatic

repulsion between the phosphates in the two strands is eliminated—usually 0.15 to 0.50 *M* NaCl is used.

2. The temperature must be high enough to disrupt the random, intrastrand hydrogen bonds present in single-stranded DNA. However, the temperature cannot be too high, or stable interstrand base-pairing will not occur and be maintained. The optimal temperature for renaturation is 20–25° below the value of T_m.

The molecular mechanism of renaturation is explained in *MB*, pages 121–122.

An important point is that a renatured DNA molecule does not contain its own original single strands—*renaturation is a random-mixing process*. For example, if DNA labeled with the nonradioactive isotope ^{15}N is mixed with DNA containing the normal isotope ^{14}N and then denatured and renatured, three kinds of double-stranded DNA molecules result—25 percent with ^{14}N in both strands, 25 percent with ^{15}N in both strands, and 50 percent with one ^{14}N strand and one ^{15}N strand. When renatured DNA is formed from distinguishable molecules, renaturation is called **hybridization.**

Thin filters **(membrane filters)** made of nitrocellulose are commercially available. These filters bind single-stranded DNA very tightly but fail to bind either double-stranded DNA or RNA. They provide an important technique for measuring hybridization.

Renaturation can be carried out on a filter in the following way (Figure 3-6). A sample of denatured DNA is filtered. The single strands bind tightly to the filter along the sugar-phosphate backbone, but the bases remain free. The filter is then placed in a vial with a solution

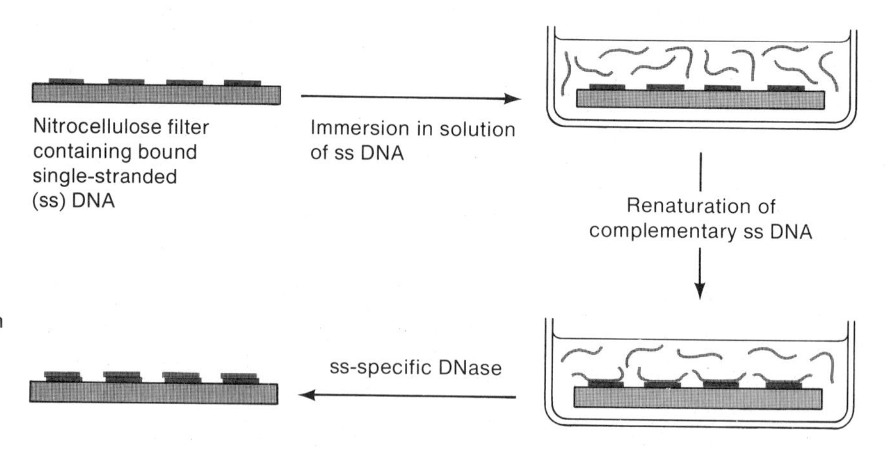

Figure 3-6
Method of hybridizing DNA to nitrocellulose filters containing bound, single-stranded DNA. In the final step the filter is treated with a single-strand-specific DNase, an enzyme that depolymerizes single-stranded, but not double-stranded, DNA.

containing a small amount of radioactive denatured DNA and a reagent that prevents additional binding of single-stranded DNA to the filter. After a period of renaturing, the filter is washed. Radioactivity is found on the filter only if renaturation has occurred.

The most important use of filter hybridization is the detection of sequence homology between a single strand of DNA and an RNA molecule—this is called **DNA-RNA hybridization,** and it is the method of choice to detect an RNA molecule that has been copied from a particular DNA molecule. In this procedure, a filter to which single strands of DNA have been bound, as above, is placed in a solution containing radioactive RNA. After renaturation the filter is washed and hybridization is detected by the presence of radioactive RNA on the filter.

The initial event in renaturation is a collision; thus, the renaturation rate obeys the law of mass action and increases with DNA concentration. This concentration dependence has been used to discover some surprising properties of the DNA of eukaryotes and prokaryotes.

Consider two DNA solutions having equal concentrations in terms of g/ml (not molar concentration). If the molecular weights of the molecules in each solution differ, the smaller molecules will renature faster because their molar concentration will be higher and more collisions will occur per unit time. This fact is the basic principle of renaturation kinetics.

Application of this principle to a collection of identical molecules broken into fragments gives information about repeated based sequences within the molecule. Consider a very long molecule having a repeating base sequence as shown in Figure 3-7(a). This sequence contains 3 percent of the total number of bases of the DNA, and there are five copies of the repeated sequence in the DNA. If this large molecule is broken into fragments much smaller than 1 percent of the total length of the molecule, the molar concentration of this sequence becomes five times larger than that of the remainder, but the weight fraction remains $(5 \times 3)/100 = 0.15$. Such a DNA would have a renaturation curve such as that shown in Figure 3-7(b). Thus, this curve, which is obtained with DNA from a single organism, contains two components. The more rapidly renaturing component accounts for 15 percent of A_{260}, from which one can conclude that 15 percent of the DNA (on a weight base) contains a repeating sequence. In the following we show that from the kinetics one can determine the number of bases in the repeating sequence (called its **complexity**), and thus the number of copies of this sequence and the number of bases in a single nonrepeating sequence.

The following equation describes the kinetics of renaturation in terms of the initial DNA concentrations C_0 (expressed in moles of bases

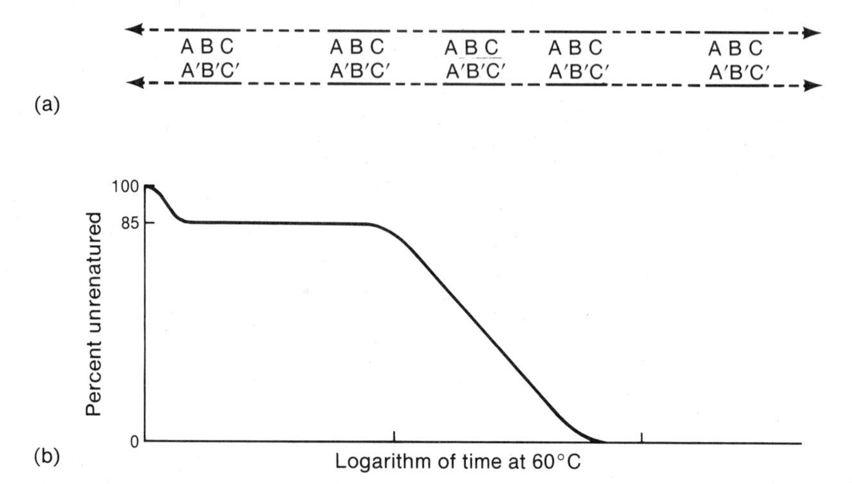

Figure 3-7
(a) A hypothetical DNA molecule having a base sequence that is 3 percent of the total length of the DNA and is repeated five times. The dashed lines represent the non-repetitive sequences; they account for 85 percent of the total length. (b) A renaturation curve for the DNA in part (a). Time is logarithmic in order to keep the curve on the page. The *y* axis—percent unrenatured DNA—is equivalent to the unit, relative value of A_{260}, used in previous figures with $A_{260} = 1.37$ equal to 100 percent.

per liter), the concentration C of unrenatured DNA at times t (in minutes), and k_2, a rate constant that depends on the temperature and the size of the DNA fragments:

$$C/C_0 = 1/(1 + k_2C_0t).$$

Experimentally, one chooses several values of C_0 and then measures C as a function of time; the data obtained are plotted as C/C_0 versus log C_0t. When $C/C_0 = 1/2$, then $C_0t_{1/2} = 1/k_2$. This is a significant expression because it can be shown that k_2 is inversely proportional to the number N of bases per repeating unit. Thus, the value of $C_0t_{1/2}$ is directly proportional to N. Note that if there are no repeating sequences, so that the DNA molecule itself represents a unique sequence, then N is the number of base pairs in the complete DNA molecule of the organism. Figure 3-8 shows the kind of graph obtained. This graph is the sequence complexity of the DNA, one first notes that 53 percent of the sequences have a $C_0t_{1/2}$ of 10^2, 27 percent have a $C_0t_{1/2}$ of 1, and 20 percent have a $C_0t_{1/2}$ of 10^{-3}. These data are then used to determine the number of copies and the sizes of each sequence. From analysis of pure samples of molecules having a unique sequence of known length, a size scale, such as that shown in Figure 3-8(b), can be obtained. However, molecular size cannot be read directly from the observed $C_0t_{1/2}$ values,

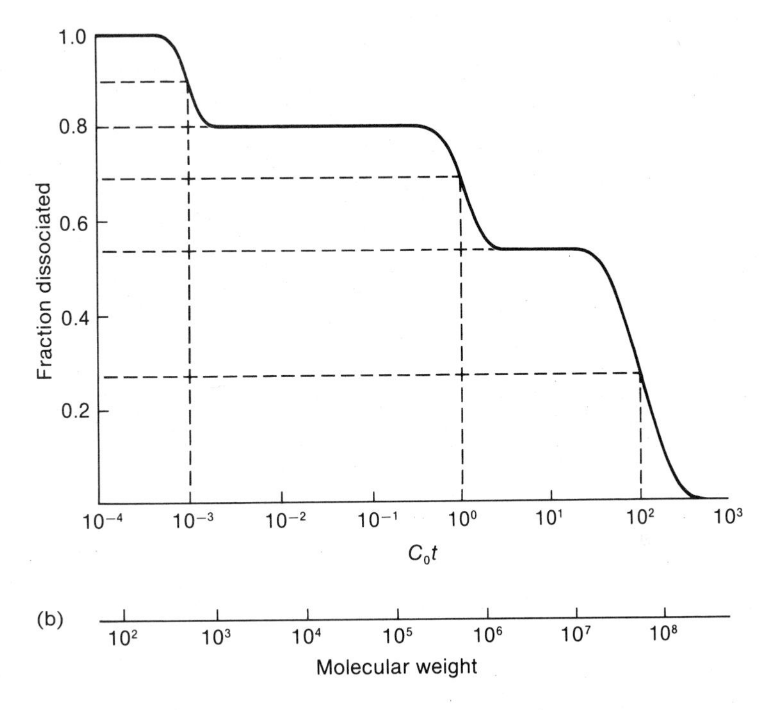

Figure 3-8
C_0t analysis. (a) The C_0t curve analyzed in the text. (b) A standard scale relating the number of nucleotide pairs in a unique sequence and $C_0t_{1/2}$. The horizontal axis in (a) and the scale in (b) are aligned vertically so values can be compared directly.

because the values of C_0 used in plotting the horizontal axis in panel (a) is the *total* DNA concentration in the renaturation mixture, rather than the concentration of each component. To utilize the size scale, one must calculate the concentration of each component; this is done by simply multiplying each observed $C_0t_{1/2}$ value by the fraction of the total DNA that it represents. Thus, the "real" $C_0t_{1/2}$ values are 0.53×10^2, 0.27×1, and 0.20×10^{-3}; the corresponding sizes are 4.2×10^7, 2.2×10^5, and 160 base pairs; respectively. The number of copies of each sequence is inversely proportional to $t_{1/2}$ and hence to the observed (uncorrected) $C_0t_{1/2}$ value. Thus, if one assumes that the genome contains one copy of the longest sequence (possible only in a haploid organism), then the cell from which the DNA has been isolated would contain 1, 10^2, and 10^5 copies of sequences having 4.2×10^7, 2.2×10^5, and 160 base pairs, respectively. The total number of base pairs per genome would be

$$4.2 \times 10^7 + 100(2.2 \times 10^5) + 10^5(160) = 8 \times 10^7,$$

and the molecular weight of the total cellular DNA would be

$$(8 \times 10^7)(660) = 5.3 \times 10^{10}.$$

Note that if an independent measurement of the total molecular weight of the DNA were to yield the value $1 \times 10^{11} \cong 2(5.3 \times 10^{10})$, then there would be 2, 200, and 2×10^5 copies of the 4.2×10^7-, 2.2×10^5-, and 160-base-pair sequences, respectively.

Bacterial DNA contains a few repeating units such as the genes for transfer RNA but these account for such a small fraction (0.1 percent) of the DNA that they are not readily detected. However, C_0t analysis has shown that in eukaryotic DNA a significant fraction of the DNA consists of repeated sequences and in some cases there are a million or more copies of a particular repeated sequence. This will be discussed shortly.

DNA Heteroduplexes

Renaturation has been combined with the technique of electron microscopy in a procedure that allows the localization of common, distinct, and deleted sequences in DNA. This procedure is called **heteroduplexing.** Consider the DNA molecules #1 and #2 shown in Figure 3-9(a). These molecules differ in sequence only in one region. If a mixture of the two molecules is denatured and renatured, hybrid molecules having unpaired single strands are produced, as shown in Figure 3-9(b). Figure 3-10 shows an actual electron micrograph of a heteroduplex. Measurement of the lengths of the single- and double-stranded regions yields the endpoints of the regions of nonhomology. Note that the two single strands of the bubble always have the same length.

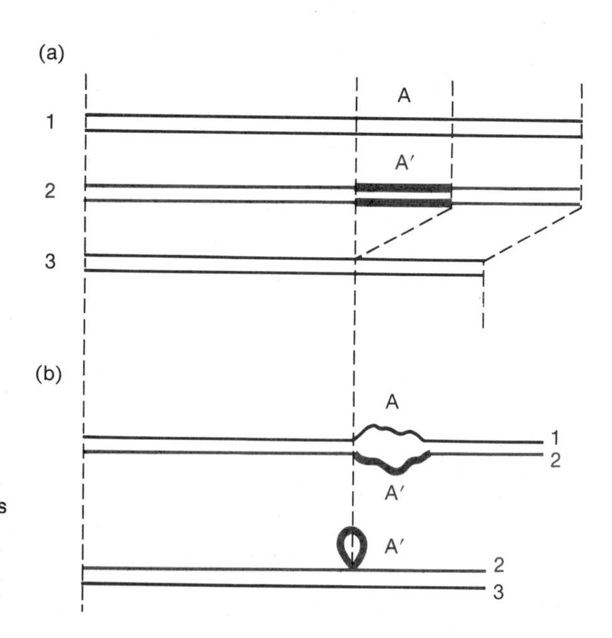

Figure 3-9
(a) Three DNA molecules to be heteroduplexed. Sequences A and A′ of molecules 1 and 2 differ. Neither sequence is present in the deletion molecule 3. The dashed lines indicate reference points. (b) Heteroduplexes resulting from renaturing molecules 1 and 2 or 2 and 3.

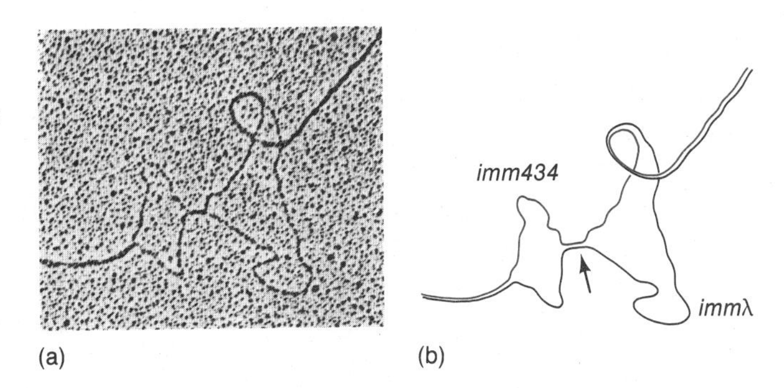

Figure 3-10
An electron micrograph of a portion of a heteroduplex between two *E. coli* phage λ molecules λ*imm434* and λ*imm*λ, which have nonhomologous segments with a short common sequence. (a) The micrograph. (b) An interpretive drawing. The arrow points to the common sequence. (Courtesy of W. Szybalski and B. Westmoreland.)

Consider now molecule #3, shown in Figure 3-9(a). In this molecule, region A is deleted. If a hybrid is made between this molecule and molecule #2, the result is a molecule with a single loop, as shown in Figure 3-9 (b).

CIRCULAR AND SUPERHELICAL DNA

The DNA molecules of prokaryotes and most viruses are circular. A circular molecule may be a **covalently closed circle**, which consists of two unbroken complementary single strands, or it may be a **nicked circle**, which has one or more interruptions (**nicks**) in one or both strands, as shown in Figure 3-11. With few exceptions, covalently closed circles are twisted, as shown in Figure 3-12. Such a circle is said to be a **superhelix** or a **supercoil.** Let us now examine what is meant by a circular DNA molecule being twisted.

The ends of a linear DNA helix can be brought together and joined in such a way that each strand is continuous. If in so doing one of the ends is rotated 360° with respect to the other to produce some unwinding of the double helix, and then the ends are joined, the resulting covalent circle will, if the hydrogen bonds re-form, twist in the opposite sense to form a twisted circle, in order to relieve strain. Such a molecule will look like a figure 8 (that is, have one crossover point). If it is instead twisted 720° prior to joining, the resulting superhelical molecule will have two crossover points. The reason for the twisting is the following. In the case of a 720° unwinding of the helix, 20 base pairs must be broken (because the linear molecule has 10 base pairs per turn of the helix). However, a DNA molecule has such a propensity for maintaining a right-handed (positive) helical structure with 10 base pairs per turn that it will deform itself in such a way that the underwinding is rewound and compensated for by negative (left-handed) twisting of the circle. Similarly, the initial rotation might

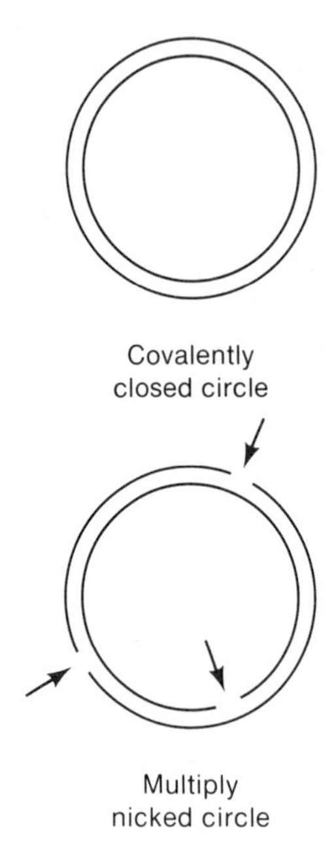

Figure 3-11
A covalently closed circle and a nicked circle. Arrows indicate the nicks.

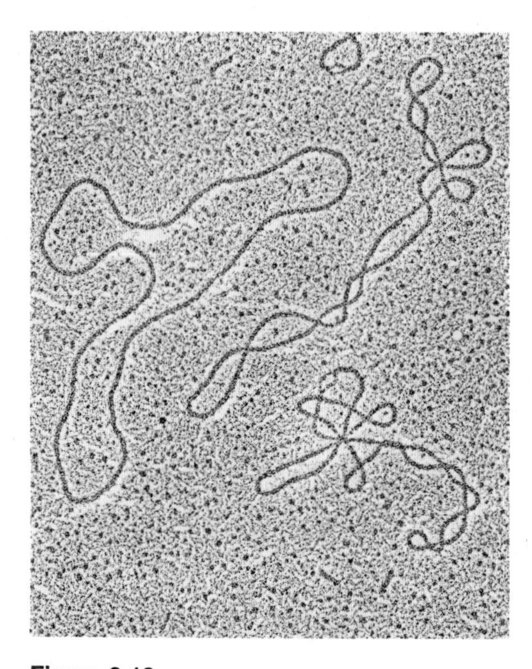

Figure 3-12
Nicked circular and supercoiled DNA of phage PM2. (Courtesy of K. G. Murti.)

instead be in the sense of overwinding, in which case the joined circle will twist in the opposite sense; this is called a positive superhelix. *All naturally occurring superhelical DNA molecules are initially underwound and hence form negative superhelices.* Furthermore, the degree of twisting—that is, the **superhelix density**—is about the same for all molecules; namely, one negative twist is produced per 200 base pairs, or, 0.05 twists per turn of the helix.

In addition to supercoiling of a covalent circle there are three other possible arrangements that would counteract the strain of underwinding: (1) The number of base pairs per turn of the helix could change. This does not happen, though, for thermodynamic reasons. (2) All of the underwinding could be taken up by having one or more large single-stranded regions (Figure 3-13). (3) The underwinding could be taken up in part by superhelicity and in part by bubbles. The real situation is alternative 3—that is, a state intermediate between forms (b) and (c) of Figure 3-13—in which the molecule is partly single-stranded and partly supercoiled. Furthermore, owing to breathing, the regions that are either unwound, unaffected, or twisted vary with time.

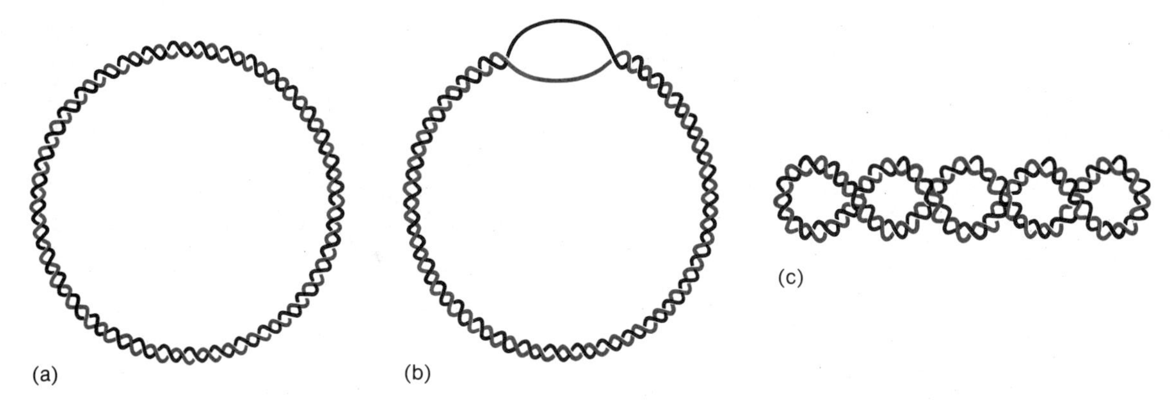

Figure 3-13
Different states of a covalent circle. (a) A nonsupercoiled covalent circle having 36 turns of the helix. (b) An underwound covalent circle having only 32 turns of the helix. (c) The molecule in part (b), but with four superhelical turns to eliminate the underwinding. In solution, (b) and (c) would be in equilibrium; the equilibrium would shift toward (b) with increasing temperature.

In bacteria, the underwinding of superhelical DNA is not due to unwinding prior to end-joining but is introduced into preexisting circles by an enzyme called **DNA gyrase,** which will be described in Chapter 8, when replication is examined. In eukaryotes, the underwinding is a result of the structure of chromatin, a DNA-protein complex of which chromosomes are composed. In this complex, DNA is wound about specific protein molecules in a direction that introduces underwinding. Chromatin will be discussed in Chapter 6.

A variety of techniques enable one to determine whether a particular covalent circle is supercoiled. The most direct method is electron microscopy. Other techniques in common use—sedimentation through alkali, gel electrophoresis, and density gradient centrifugation in the presence of ethidium bromide (which binds to DNA, reducing the density of DNA less with a supercoil than with other forms of DNA) —are described in *MB*, pages 133–136.

THE STRUCTURE OF RNA

A typical cell contains about ten times as much RNA as DNA, yet we have said little about RNA up to this point. With the exception of the RNA of one phage and a few viruses, RNA is a single-stranded polynucleotide. There are three primary types of RNA—ribosomal RNA (of which there are three or four forms), transfer RNA (of which there are about fifty different structures), and messenger RNA (of which there are almost as many different molecules as there are genes). These molecules superficially resemble single-stranded DNA in that single-stranded regions are interspersed with intramolecular double-stranded regions. Between 1/2 and 2/3 of the bases are paired. In single-stranded DNA the pairing is random and the paired regions tend to contain six or fewer pairs. Furthermore, if a sample of identical DNA molecules is denatured and intramolecular hydrogen bonds are allowed to form, the base-pairing pattern may differ from one molecule to the next. On the contrary, in RNA the double-stranded regions may contain up to twenty base pairs, and a particular molecule has a definite base-pairing pattern.

The structures of the different classes of RNA molecules will be discussed in detail in Chapters 9 and 10.

HYDROLYSIS OF NUCLEIC ACIDS

Both DNA and RNA can be hydrolyzed to free nucleotides either chemically or enzymatically.

At very low pH (1 or less), phosphodiester hydrolysis of both DNA and RNA occurs. This is accompanied by breakage of the *N*-glycosylic bond between the base and the sugar so that free bases are produced. At high pH the behavior of DNA is strikingly different from that of RNA. DNA remains very resistant to hydrolysis at pH 13 (about 0.2 phosphodiester bonds broken per million bonds per minute at 37°C), whereas at pH 11 a typical RNA molecule is totally hydrolyzed to ribonucleotides in a few minutes at 37°.

A variety of enzymes hydrolyze nucleic acids. They are called **nucleases.** They usually show chemical specificity and are either deoxyribonucleases (DNase) or ribonucleases (RNase). Many DNases act on only single-stranded or only double-stranded DNA, although some act on either kind. Furthermore, some nucleases act only at the end of a nucleic acid, removing a single nucleotide at a time; these are called **exonucleases,** and they may also be specific for the 3′ or 5′ end of the strand. Most nucleases act within the strand and are **endonucleases**; some of these are specific in that they cleave only between particular bases. Nucleases serve a variety of biological functions and have been useful in the laboratory for removing unwanted nucleic acids and as an analytical tool. Site-specific nucleases, called restriction enzymes, are the basis of genetic engineering and are discussed in Chapter 13.

SEQUENCING NUCLEIC ACIDS

Throughout this book, base sequences within DNA molecules will be described as a means of understanding the kinds of physical and chemical signals that are used by cells. This section describes how these sequences are determined.

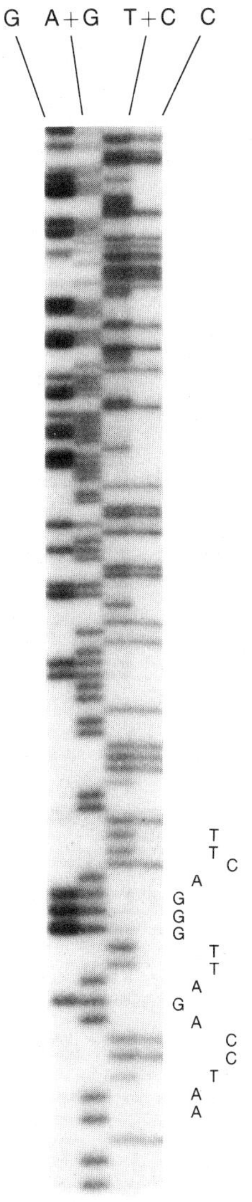

Figure 3-14
A portion of a sequencing gel. The sequence is read from the bottom to the top. Each horizontal row represents a single base. Each vertical column represents a sample treated to indicate the position of G, A or G, T or C, and C, respectively. The sequence from a portion of the gel is shown.

There are two methods for determining the base sequence of a DNA molecule—the Maxam-Gilbert procedure and the Sanger procedure. Although both methods are equally effective, we shall describe only the Maxam-Gilbert procedure, as it is used more frequently at present. The procedure will be given only in outline; details can be found in *MB*, pages 141–145.

A sample containing a particular fragment of DNA is first denatured, and then the individual single strands are isolated. Each of these strands is treated separately but identically in order to determine the base sequence of each strand. Since the sequences are complementary, one serves as a check on the other. The procedure begins by making the single strands radioactive by putting a radioactive phosphorus atom at their 5′ termini; since only one terminus (the same for all strands) has a 5′ group, only one terminus is labeled.Then, the sample is divided into four portions and each portion is treated by a separate chemical procedure that (1) removes either one adenine, guanine, cytosine, or thymine and (2) cleaves the strand at the site of removal of the base. Thus, if a guanine were at position 36 from the 5′ end, a radioactive fragment containing 35 nucleotides would be produced in the sample that we will call G-only. Similarly, if there were a cytosine at position 23, a 22-base radioactive fragment would be produced in the sample called C-only. Note that there would be no 34-base fragment in sample C-only and no 22-base fragment in sample G-only. However, each fragment size is not unique to one of the four portions, because the procedure that produces a cut at an adenine also occasionally makes a cut at a guanine, and the procedure that produces a cut at a cytosine also sometimes makes a cut at a thymine. Thus, an adenine is identified as a base determined by a cut in sample A + G, and a guanine is at a position determined by a cut in both the A + G and the G-only classes. Similarly, a thymine is at the position indicated by the T + C class, and a cytosine is at the position observed in the C-only and T + C classes. The four samples are analyzed by gel electrophoresis, a procedure that separates the fragments by size (Chapter 2). The DNA is detected by autoradiography (a photographic procedure that detects radioactivity), so that only radioactive fragments are observed. The kind of data obtained is shown in Figure 3-14; the base sequence can be read directly from the gel.

Although the gels can resolve two fragments differing in length by one nucleotide, they are not capable of distinguishing one hundred different fragments simultaneously, unless the gel is exceedingly long. To avoid this problem, after all of the chemical treatments one usually divides all samples into three aliquots and electrophoreses each set of four samples for different times. For example, run 1 might give the sequence of bases 3 through 40, run 2 that of bases 35 through 75, and run 3 that of bases 65 through 100. By observing the common sequences (for example, bases 35–40 of runs 1 and 2), the complete sequence can be worked out.

This method, as well as other procedures, has been used to determine the base sequence of a DNA molecule having 40,000 base pairs.

LEFT-HANDED DNA HELICES

The Watson-Crick form of DNA is a right-handed helix. In concentrated salt solutions, short synthetic double-stranded polynucleotides in which purines and pyrimidines alternate (for example, AGAGAG...) form a left-handed helix. The structure differs from the regular Watson-Crick form in several ways; one feature, a zig-zag arrangement of the strands, has led to the term **Z DNA** for such molecules. Alternating purines and pyrimidines present in long natural DNA molecules, for example,

···TGATCCGCGCGCGAGTCTT···
···ACTAGGCGCGCGCTCAGAA···

can also assume the Z configuration. Some evidence exists for the presence of Z-DNA regions in the chromosomes of animal cells. The significance is unknown, but hypotheses about the role of Z DNA in regulating gene expression have been considered. Details about the structure of Z DNA and its potential function can be found in *MB*, pages 145–148.

SEQUENCE ORGANIZATION OF EUKARYOTIC DNA

The mean base composition of DNA varies widely throughout the microbiological world. The extremes are 23 percent G+C for the *Clostridium* genus of bacteria and 76 percent G+C for *Sarcina* and the *Micrococci*. *E. coli*, the bacterium that we will study most often in this book, is 50 percent G+C. Proceeding upward along the evolutionary scale, the variation becomes smaller. For instance, for most plants and animals, the extremes are 48 percent and 52 percent.

A convenient way to measure the mean base composition is by density gradient centrifugation, because the density of DNA is linearly related to base composition. When DNA of prokaryotes is broken

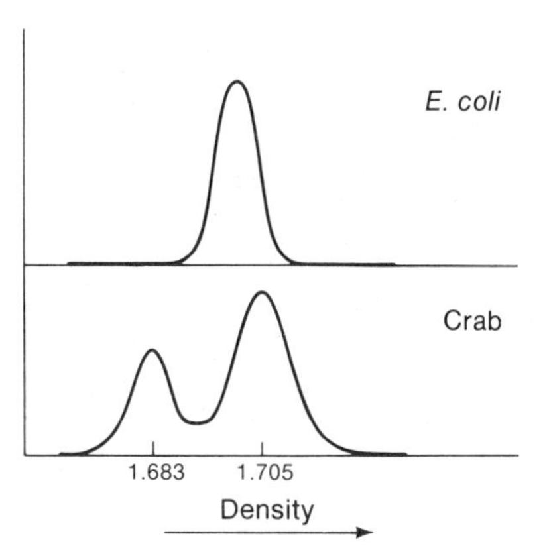

Figure 3-15
The concentration distribution of the DNA of the bacterium *E. coli* and the crab *Cancer borealis* after density-gradient centrifugation. As is common with bacteria, the *E. coli* DNA has a narrow range of densities. The crab DNA consists of two discrete fractions, one of very low density. This minor band is called *satellite DNA*. (From N. Sueoka, 1961. *J. Mol. Biol.*, 3:31–40. Copyright: Academic Press Inc. [London] Ltd.)

randomly into about 1000 fragments, the bands that result are usually very narrow. This indicates that the mean base composition does not vary very much from fragment to fragment—that is, the DNA of an organism is not usually 25 percent G + C in one region and 75 percent G + C in another.

However, when the DNA of the crab was examined, two discrete bands were observed (Figure 3-15). The main band accounted for 70 percent of the total DNA and was 52 percent G+C. The smaller band, which was 3 percent G+C, was termed **satellite DNA.** This phenomenon has been observed often in eukaryotes in which the satellite DNA may have a higher or lower G+C content than the main fraction. Some organisms have two satellites. C_0t analysis has shown that satellite DNA is usually highly repetitive, consisting sometimes of 10^6 copies of a single sequence.

Repeated sequences are found in all eukaryotes but unicellular organisms; however, these sequences do not always show up as satellite bands because their base compositions may not be very different from the average base composition of the bulk of the DNA. The fraction of the DNA that is repetitive and the size of the repeated sequences can be determined by C_0t analysis. Figure 3-16 shows a C_0t curve for bacterial and mouse DNA. Close examination of C_0t curves from many species shows that there are four classes of sequences—**unique** (single copy), **slightly repetitive** (1 to 10 copies), **middle repetitive** (10 to several hundred copies), and **highly repetitive** (several hundred to several million copies).* The bounds of these classes are arbitrary, so one must be careful when making generalizations about the classes.

*Note that since one usually studies a diploid cell, unless the DNA is isolated from eggs, there are actually two copies of a unique sequence.

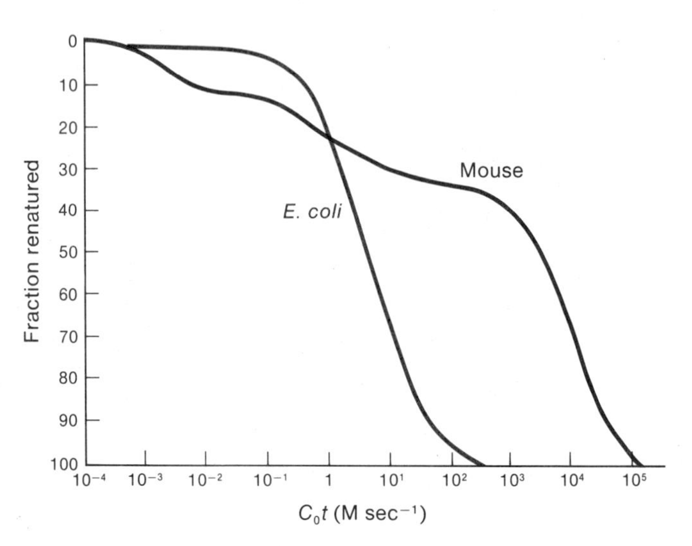

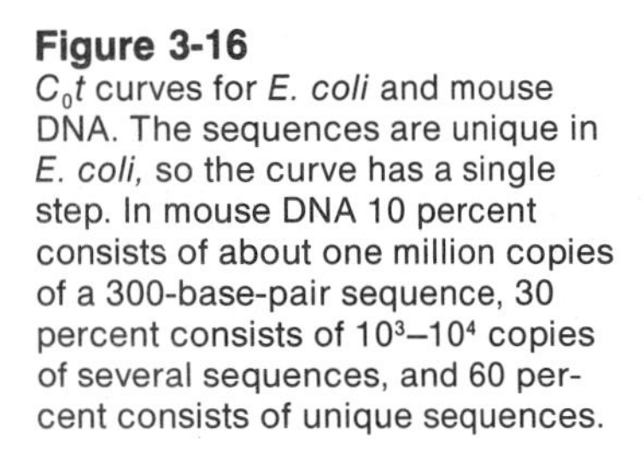

Figure 3-16
C_0t curves for *E. coli* and mouse DNA. The sequences are unique in *E. coli,* so the curve has a single step. In mouse DNA 10 percent consists of about one million copies of a 300-base-pair sequence, 30 percent consists of 10^3–10^4 copies of several sequences, and 60 percent consists of unique sequences.

The unique sequences account for most of the genes of the cell but often for only a few percent of the DNA. The slightly repetitive class consists of genes that exist in a few copies; in some cases the sequences are not perfectly repetitive: the gene products may have slight differences in amino acid composition—for example, the various forms of hemoglobin. The middle-repetitive sequences are of two types—clustered and dispersed. The clustered sequences usually represent genes that exist in multiple copies in order to increase the amount or the rate of synthesis of the gene product. Some of the dispersed sequences may have regulatory function. However, the major fraction consists of 20–60 copies of sequences, containing 2000–7000 nucleotide pairs, that are able to move from one location to another in a chromosome and between chromosomes. These are called transposable elements and are discussed in detail in Chapter 12. The highly repetitive sequences include the short sequences in satellite DNA and sequences of normal length in certain genes that exist in very large numbers, such as those making ribosomal RNA, which is needed for all protein synthesis. There are in fact very few different highly repetitive sequences, but the number of copies is so great that these sequences may account for 20 percent or more of the mass of the DNA.

4 The Physical Structure of Protein Molecules

All proteins are polymers of amino acids, yet each species of protein molecule has a unique three-dimensional structure determined principally by the amino acid sequence of that protein, in contrast with DNA, which has a universal structure—the double helix. This makes the study of proteins very complex but, on the other hand, the diversity of protein structures enables these molecules to carry out the thousands of different processes required by a cell and makes proteins fascinating objects to study.

The study of the detailed structure of proteins is beyond the scope of this book, being heavily dependent on a variety of optical techniques, especially the mathematically complex technique of x-ray diffraction. For this reason the discussion of proteins in this chapter will be a survey.

BASIC FEATURES OF PROTEIN MOLECULES

In Chapter 2 the basic chemical features of protein molecules were described. That is, a protein is a polymer of amino acids in which carbon atoms and peptide groups alternate to form a linear polypeptide chain and specific groups—the amino-acid side chains—project from the α-carbon atom (Figure 2-1). The term "linear" requires careful consideration, for, as will be seen in the following sections, a polypeptide chain is highly folded and can assume a variety of three-dimensional shapes; each shape in turn consists of several standard elementary three-dimensional configurations and other configurations which may be unique to that molecule.

In Chapter 3, it was seen that nucleic acid molecules are very large, having molecular weights often as high as 10^{10}. Protein molecules are much smaller; in fact, the molecular weight of a typical protein molecule is comparable to that of the smallest nucleic acid molecules, the transfer RNA molecules. The molecular weights of hundreds of different proteins have been measured. Typical polypeptide chains have molecular weights ranging from 15,000 to 70,000. The average molecular weight of an amino acid is 110, which means that typical polypeptide chains contain some 135 to 635 amino acids.

The length of a typical polypeptide chain, if it were fully extended, would be 1000 to 5000 Å. A few of the longer fibrous proteins, such as myosin (from muscle) and tropocollagen (from tendon) are in this range—with lengths of 1600 Å and 2800 Å, respectively. However, most proteins are highly folded and their longest dimension usually ranges from 40 to 80 Å. This folding is described in the next section.

THE FOLDING OF A POLYPEPTIDE CHAIN

A fully extended polypeptide chain, if it were to exist, would have the configuration shown in Figure 4-1. (The chain is not perfectly straight because the C—N and C—C bonds in which the α-carbon atom participates are not colinear.) Such an extended zig-zagged molecule could not exist without stabilizing interactions to maintain the extension. In fact, a single polypeptide is never completely extended but is folded in a complex way. If there were free rotation about every bond in the chain and no interaction between different parts of the chain, the folding would be random. Instead, three rules govern the manner of folding:

1. The peptide bond has a partial double-bond character (Figure 4-2) and hence is constrained to be planar. Free rotation occurs only between the α carbon and the peptide unit. Thus, the polypeptide chain is flexible but is not as flexible as would be the case if there were free rotation about all of the bonds.
2. The side chains of the amino acids cannot overlap. Thus, the folding can never be truly random because certain orientations are forbidden.
3. Two charged groups having the same sign will not be very near one another. Thus, like charges tend to cause extension of the chain.

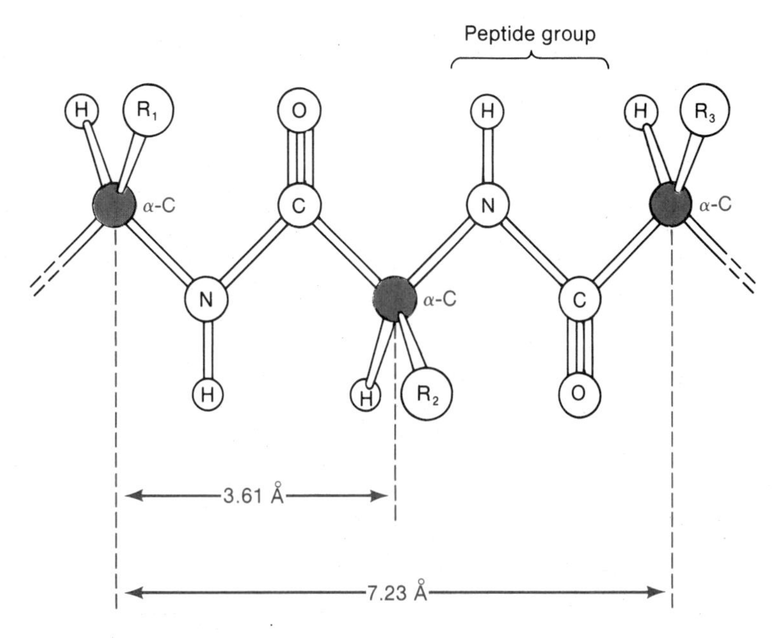

Figure 4-1
The configuration of a hypothetical fully extended polypeptide chain. The length of each amino acid residue is 3.61 Å; the repeat distance is 7.23 Å. The α-carbon atoms are shown in red. Side chains are denoted by R.

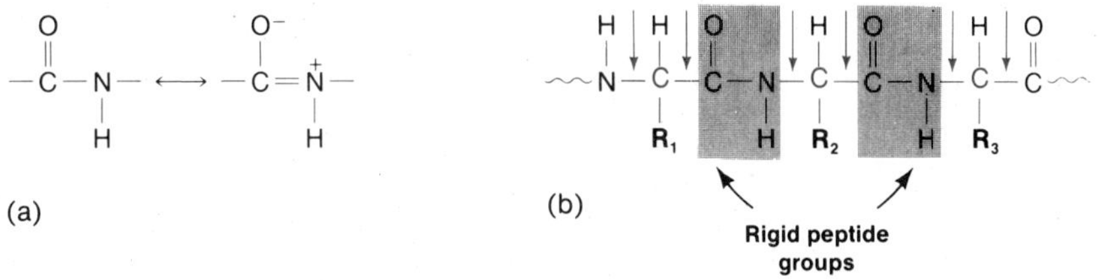

(a)

(b)

Rigid peptide groups

Figure 4-2
The planarity of the peptide group. (a) Two forms of the peptide bond in equilibrium. In the form at the right the double bond creates a rigid unit. Owing to the equilibrium, the peptide group has a partial double-bond character and is rigid. (b) The peptide groups in a protein molecule. They are planar and rigid, but there is freedom of rotation about the bonds (red arrows) that join peptide groups to α-carbon atoms (red).

However, in addition to these rules, folding behaves according to the following general tendencies:

1. Amino acids with polar side chains tend to be on the surface of the protein in contact with water.
2. Amino acids with nonpolar side chains tend to be internal. Very hydrophobic side chains tend to cluster.

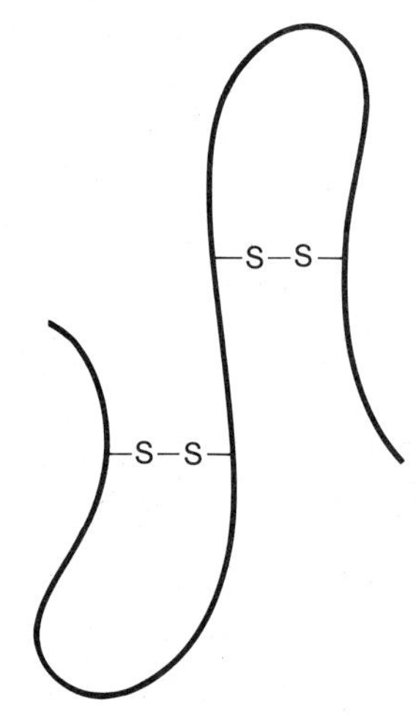

Figure 4-3
A polypeptide chain in which four cysteines are engaged in two disulfide bonds.

3. Hydrogen bonds tend to form between the carbonyl oxygen of one peptide bond and the hydrogen attached to a nitrogen atom in another peptide bond.
4. The sulfhydryl group of the amino acid cysteine tends to react with an —SH group of a second cysteine to form a covalent S—S (**disulfide**) bond. Such bonds pose powerful constraints on the structure of a protein (Figure 4-3).

These tendencies should not be considered invariable because there are many exceptions. Nevertheless they indicate that the structure of a protein may be changed markedly by a single amino acid substitution—for example, a polar amino acid for a nonpolar one; similarly, the change might be minimal if one nonpolar amino acid replaces another nonpolar one. This notion will be encountered again in Chapter 11 when mutations are considered.

The three-dimensional shape of a polypeptide chain is a result of a balance between all of the rules and tendencies just described and can be very complex. However, in examining many polypeptide chains, it has become apparent that certain geometrically regular arrays of the chain are found repeatedly in different polypeptide chains and in different regions of the same chain. These are the arrays resulting from hydrogen-bonding between different peptide groups. These arrays are described in the next section.

THE α HELIX AND β STRUCTURES

In the absence of any interactions between different parts of a polypeptide chain, a random coil would be the expected configuration. However, hydrogen bonds easily form between the H of the N—H group and the O of the carbonyl group. A variety of structures can result from this hydrogen-bonding, the most common being the α helix and the β structures.

In an α helix the polypeptide chain follows a helical path that is stabilized by hydrogen-bonding between peptide groups. Each peptide group is hydrogen-bonded to two other peptide groups, one three units ahead and one three units behind in the chain direction (Figure 4-4). The helix has a pitch of 5.4 Å, which is the repeat distance, a diameter of 2.3 Å, and contains 3.6 amino acids per turn. Thus, it is a much tighter helix than the DNA helix. The side chains, including those that are ionized, do not participate in forming the α helix.

In the absence of all interactions other than the hydrogen-bonding just described, the α helix is the preferred form of a polypeptide chain

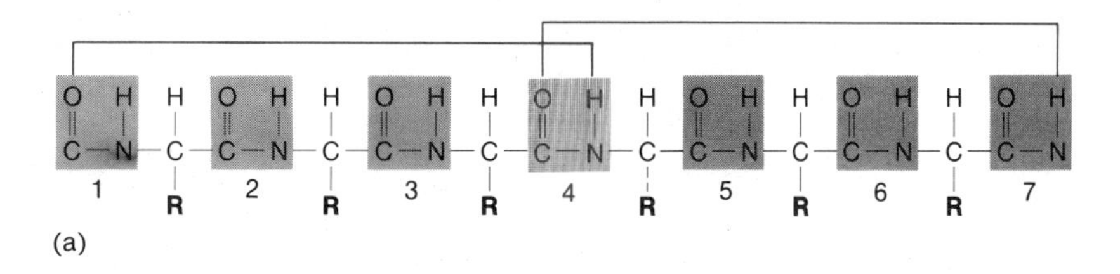

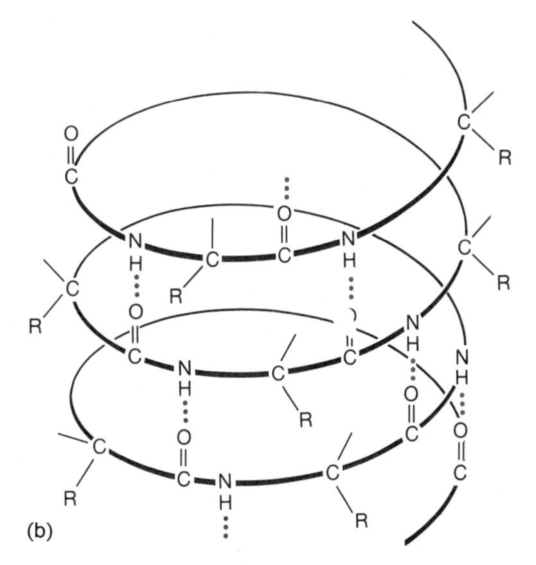

Figure 4-4
Properties of an α helix. (a) The two hydrogen bonds in which peptide group 4 (red) is engaged. The peptide groups are numbered below the chain. (b) An α helix drawn in three dimensions, showing how the hydrogen bonds stabilize the structure. The red dots represent the hydrogen bonds. The hydrogen atoms that are not in hydrogen bonds are omitted for the sake of clarity.

because, in this structure, all monomers are in identical orientation and each forms the same hydrogen bonds as any other monomer. Thus, polyglycine, which lacks side chains and hence cannot participate in any interactions other than those just described, is an α helix.

If all monomers are not identical or if there are secondary interactions that are not equivalent, the α helix is not necessarily the most stable structure. Then, not only is it true that the amino-acid side chains do not participate in forming the α helix, but also that the side chains are responsible for preventing α-helicity. A striking example of the disruptive effect of a side chain on an α helix is evident with the synthetic polypeptide polyglutamic acid, a polypeptide containing only glutamic acid. At a pH below about 5, the carboxyl group of the side chain is not ionized and the molecule is almost purely an α helix. However, above pH 6, when the side chains are ionized, electrostatic repulsion totally destroys the helical structure.

If the amino acid composition of a real protein is such that the helical structure is extended a great distance along the polypeptide backbone, the protein will be somewhat rigid and fibrous (not all rigid fibrous proteins are α helical, though). This structure is common in many structural proteins, such as the α-keratin in hair.

Another hydrogen-bonded configuration is the **β structure**. In this form, the molecule is almost completely extended (repeat distance = 7 Å) and hydrogen bonds form between peptide groups of polypeptide segments lying adjacent and parallel with one another (Figure 4-5(a)). The side chains lie alternately above and below the main chain.

Two segments of a polypeptide chain (or two chains) can form two types of β structure, which depend on the relative orientations of the segments. If both segments are aligned in the N-terminal-to-C-terminal direction or in the C-terminal-to-N-terminal direction, the β structure is **parallel.** If one segment is N-terminal to C-terminal and the other is C-terminal to N-terminal, the β structure is **antiparallel.** Figure 4-5(b) shows how both parallel and antiparallel β structures can occur within a single polypeptide chain.

When many polypeptides interact in the way just described, a pleated structure results called the **β-pleated sheet** (Figure 4-5(c)). These sheets can be stacked and held together in rather large arrays by van der Waals forces and are often found in fibrous structures such as silk.

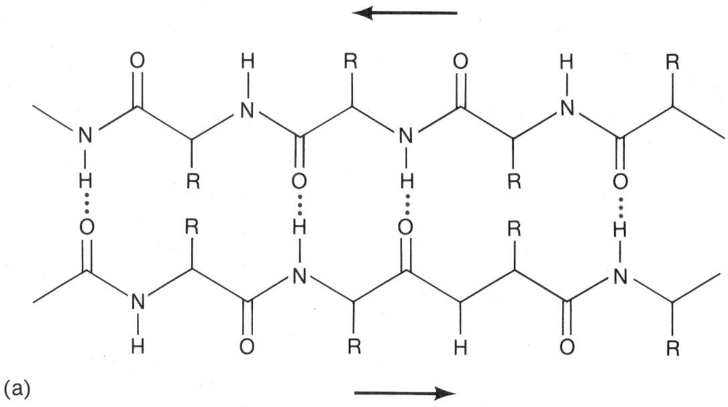

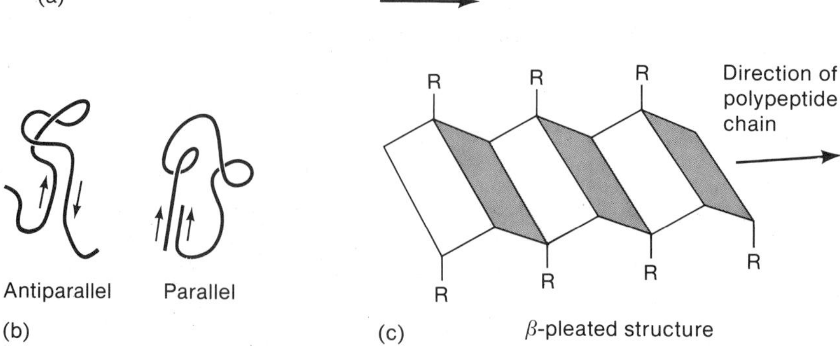

Figure 4-5
β structures. (a) Two regions of nearly extended chains are hydrogen-bonded (red dots) in an antiparallel array (arrows). The side chains (R) are alternately up and down. (b) Antiparallel and parallel β structures in a single molecule. (c) A large number of adjacent chains forming a β pleat.

PROTEIN STRUCTURE

Few proteins are pure α helix or β structure; usually regions having each structure are found within a protein. Since these configurations are rigid, a protein in which most of the chain has one of these forms is usually long and thin and is called a **fibrous protein.** In contrast are the quasi-spherical proteins called the **globular proteins** in which α helices and β structures are short and interspersed with randomly coiled regions and compact structures.

The fibrous proteins are typically responsible for the structure of cells, tissues, and organisms. Some examples of structural proteins are collagen (the protein of tendon, cartilage, and bone), elastin (a skin protein), tubulin (a protein that maintains the shape of nerve cells), and actin (a ubiquitous protein that contributes to the shape of almost all animal cells). Some of the fibrous proteins are not soluble in water—examples are the proteins of hair and silk.

The catalytic and regulatory functions of cells are performed by proteins that have a well defined but deformable structure. These are the globular proteins, of which the catalytic proteins, or enzymes, are the most widely studied. Globular proteins are compact molecules having a generally spherical or ellipsoidal shape. Large segments of the polypeptide backbone of a typical globular protein are α-helical. However, the molecule is extensively bent and folded. Usually, the stiffer α-helical segments alternate with very flexible randomly coiled regions, which permit bending of the chain without excessive mechanical strain. Numerous segments of the chain, which might be quite distant along the backbone, form short parallel and antiparallel β structures; these also are responsible in part for the folding of the backbone (Figure 4-5(c)). The α helix and β structures are called the **secondary structure** of the molecule. The extensive folding of the backbone is usually called the **tertiary structure** or **tertiary folding.**

A very important distinction can be made between secondary and tertiary structure; namely,

> Secondary structure results from hydrogen-bonding between peptide groups whereas tertiary structure is formed from β structures and several different side-chain interactions.

The most prevalent interactions responsible for tertiary structure are the following:

1. Ionic bonds between oppositely charged groups in acidic and basic amino acids.

2. Hydrogen bonds between the hydroxyl group in tyrosine and a carboxyl group of aspartic or glutamic acids.
3. Hydrophobic clustering between the hydrocarbon side chains in phenylalanine, leucine, isoleucine, and valine.
4. Metal-ion coordination complexes between amino, hydroxyl, and carboxyl groups, ring nitrogens, and pairs of SH groups.

Hydrophobic clustering (item 3 above) is the most important stabilizing feature.

Figure 4-6 shows a schematic diagram of a hypothetical protein (in two dimensions) in which several of these interactions determine the structure. This figure should be examined carefully, for it indicates the

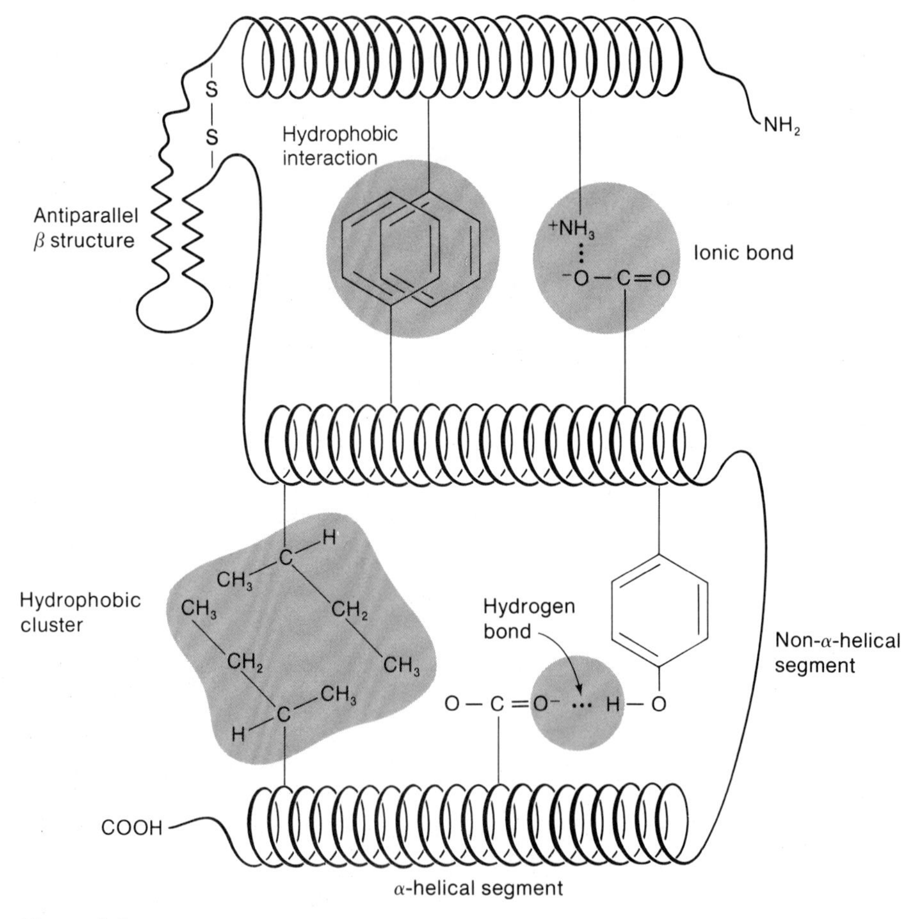

Figure 4-6
A hypothetical globular protein having several types of side-chain interactions.

role of different features of a protein molecule in determining the overall conformation of the molecule. One can see the following:

1. Disulfide bonds bring distant amino acids together.
2. Hydrophobic interactions bring distant amino acids together.
3. Hydrogen bonds sometimes bring distant amino acids together, but usually a single hydrogen bond makes a more subtle change in position.
4. Electrostatic interactions bring amino acids together or keep them apart, depending on the signs of the charges.
5. A β structure brings distant segments of the polypeptide backbone together and creates rigidity.
6. An α helix makes adjacent regions of a polypeptide backbone stiff and linear.
7. Van der Waals forces produce specific interactions between clusters of amino acids that may or may not be nearby in the polypeptide chain.

Figure 4-7 is an idealized drawing of the three-dimensional structure of the enzyme carbonic anhydrase, showing a β structure, a

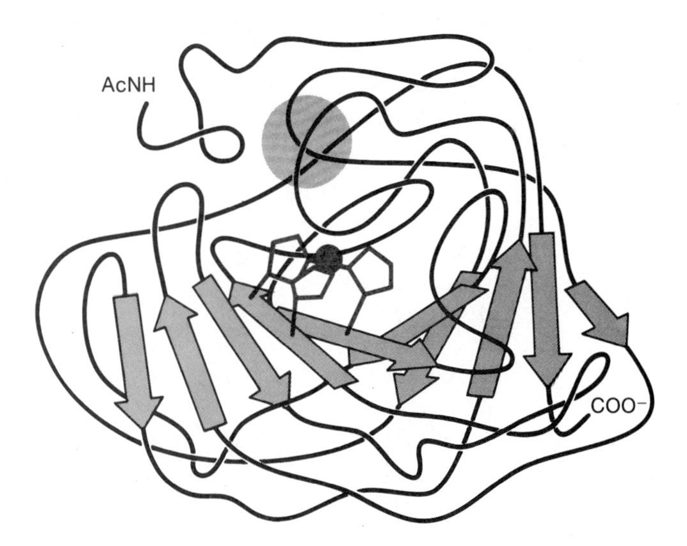

Figure 4-7
An idealized drawing of the tertiary structure of human carbonic anhydrase. Shown are the peptide chain and the three histidines (red) that coordinate to a zinc ion at the active site. Individual β-sheet strands are drawn as arrows from the amino to carboxyl ends. Note the twist of the β sheets. A hydrophobic cluster is shaded in red. (After K. K. Kannan, et al. 1971. *Cold Spring Harbor Symp. Quant. Biol.,* 36: 221).

hydrophobic cluster, and three amino acids in a coordination bond with a metal ion.

PROTEINS WITH SUBUNITS

A polypeptide chain usually folds such that nonpolar side chains are internal—that is, isolated from water. However, it is rarely possible for a polypeptide chain to fold in such a way that *all* nonpolar groups are internal. Thus, it is often the case that nonpolar amino acids on the surface form clusters in an effort to minimize contact with water. A protein molecule having a large hydrophobic patch can further reduce contact with water by pairing with a hydrophobic patch on another protein molecule. Similarly, if a molecule has several distantly located hydrophobic patches, a structure consisting of several protein molecules in contact effectively minimizes contact with water. The protein is then said to consist of identical **subunits**; this is in fact a very common phenomenon, with two, three, four, and six subunits occurring most frequently. A multisubunit protein may also contain unlike subunits, and this is quite common. For example, hemoglobin, the oxygen carrier of blood, consists of four subunits, two each of two different types; likewise, RNA polymerase, which catalyzes synthesis of RNA, has five subunits of which four are different; and DNA polymerase III, which synthesizes DNA in *E. coli,* consists of ten different subunits.

Multisubunit proteins have certain advantages, particularly with respect to economy of synthesis and regulation of enzymatic activity; a detailed presentation of the value of subunits can be found in *MB*, pages 164–165. An example of a multisubunit protein, whose physical and biological properties are well known, is immunoglobulin G. It will be described in some detail in the following paragraphs; its synthesis is discussed in Chapter 16.

The immunoglobulins (**antibodies**) are the proteins of the immune system. Their function is to interact with specific foreign molecules (**antigens**) and thereby render them inactive. This interaction is called the **antigen-antibody reaction.** The best-understood immunoglobulin is immunoglobulin G (**IgG**); other classes of immunoglobulins are IgA, IgM, IgD, and IgE. In this section, we shall examine only the IgG class, itself comprising several subclasses defined by slight structural differences.

Cleavage of the disulfide bonds of IgG yields two polypeptide chains whose molecular weights are about 25,000 and 50,000. The lighter polypeptide is called an **L chain;** the heavier one is an **H chain.** IgG is a tetramer containing two L chains and two H chains. A schematic structure of IgG is shown in Figure 4-8. Experimentally, IgG can be

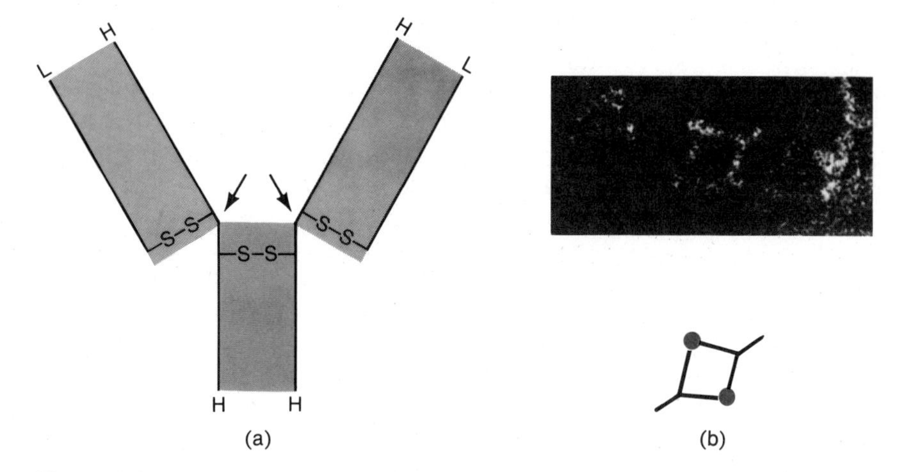

Figure 4-8
(a) The subunit structure of immunoglobulin G. There are two L subunits and two H subunits. Disulfide bonds join each L strand to an H strand and join the two H strands to one another. Treatment with papain cleaves the H strands at the arrows, yielding two F_{ab} units (shaded black) and one F_c unit (shaded red). (b) An electron micrograph showing two immunoglobulin G molecules joined at the antigen binding site. The joining is a result of binding two antigen molecules not visible in the micrograph but shown as red dots in the interpretive drawing. (Courtesy of R. C. Valentine.)

cleaved with the proteolytic enzyme papain, which causes each of the H chains to break, as shown in the figure, thus producing three separate subunits. The two units that consist of an L chain and a fragment of the H chain equal in mass to the L chain are called F_{ab} fragments (the subscript stands for "antigen-binding"). The third unit, consisting of two equal segments of the H chain, is called the F_c fragment.

Two sites on an IgG molecule can bind antigen. Each site is at the end of an F_{ab} segment, as shown in Figure 4-9. The F_c segment is not

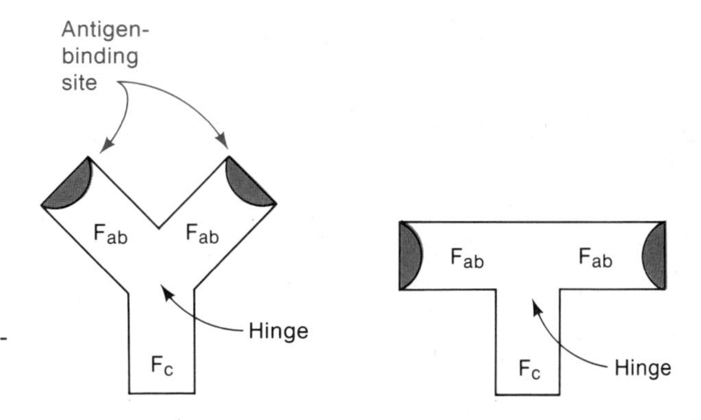

Figure 4-9
Immunoglobulin G is a Y-shaped molecule. It contains a hinge that gives it segmental flexibility.

involved in antigen-antibody binding but is used in later processes needed to destroy the antigen.

The amino acid sequences of many subclasses of IgG molecules, each capable of combining with only a single kind of antigen molecule. have been determined. All immunoglobulins are structurally similar inasmuch as L and H chains both have what is called a **variable (V)** and a **constant (C)** region.

The V and C regions have separate functions in IgG. The V regions confer antigen-binding specificity, whereas the C regions are responsible for the overall structure of the molecule and for its recognition by other components of the immune system.

The C regions of the L chains (C_L) of all types of IgG molecules have an identical amino acid sequence. Likewise, the C regions of all H chains (C_H) are identical for all IgG, though different from the C_L sequences. The V regions differ from one type of IgG to the next, however. The comparative sizes of the C_L, V_L, C_H, and V_H regions are shown in Figure 4-10.

Some of the similarities between amino acid sequences in different parts of an IgG molecule are quite striking. For example, the C region of the H chain can be divided into three segments C_H1, C_H2, and C_H3, whose amino acid sequences are quite similar though not identical to

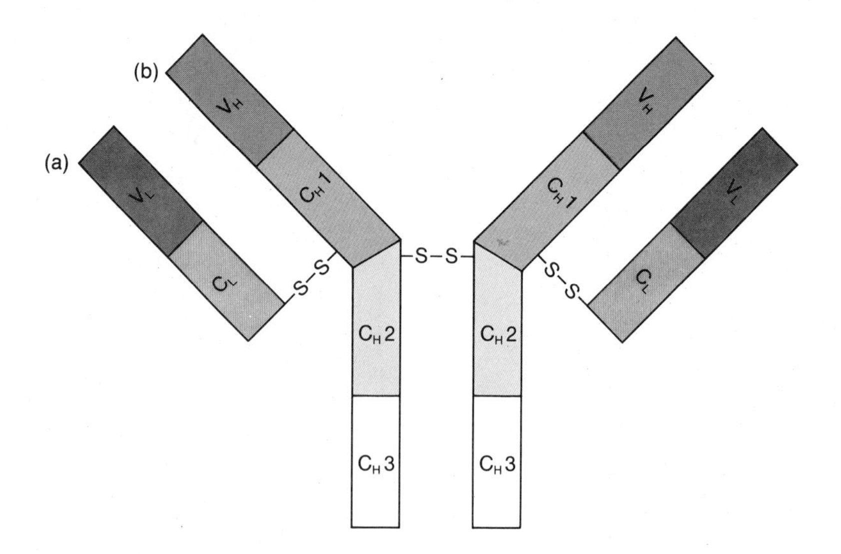

Figure 4-10
The arrangement and relative sizes of the variable and constant regions of the IgG (a) L chain and (b) H chain. The numbers refer to amino acids counted from the amino terminus. Note that the lengths of the variable regions are the same.

one another. Furthermore, the sequence of C_L resembles the C_H sequences, though generally they are different. The V_L and V_H sequences also are nearly the same. This means that the IgG molecule, which already has twofold symmetry, consists of four domains, each of which has twofold symmetry, as shown in Figure 4-10, in which the intensity of the shading indicates two homologous regions. The symmetry of the molecules is made especially evident in Figure 4-11, which shows the three-dimensional structure of IgG. A feature of IgG, not clearly shown in the figure, is that the antigen-binding site is formed by joining parts of different subunits, namely, the V_H and V_L regions. Formation of a binding site by a subunit interaction is a property of many multisubunit proteins.

The structures of some multisubunit proteins are quite complex. For example, collagen, the protein of tendon, cartilage, bone, and skin, is formed by the mutual binding of three polypeptide chains as a triple helix and the interaction of numerous triple helices to form a tough fiber. Details about collagen structure are given in *MB*, pages 190–195.

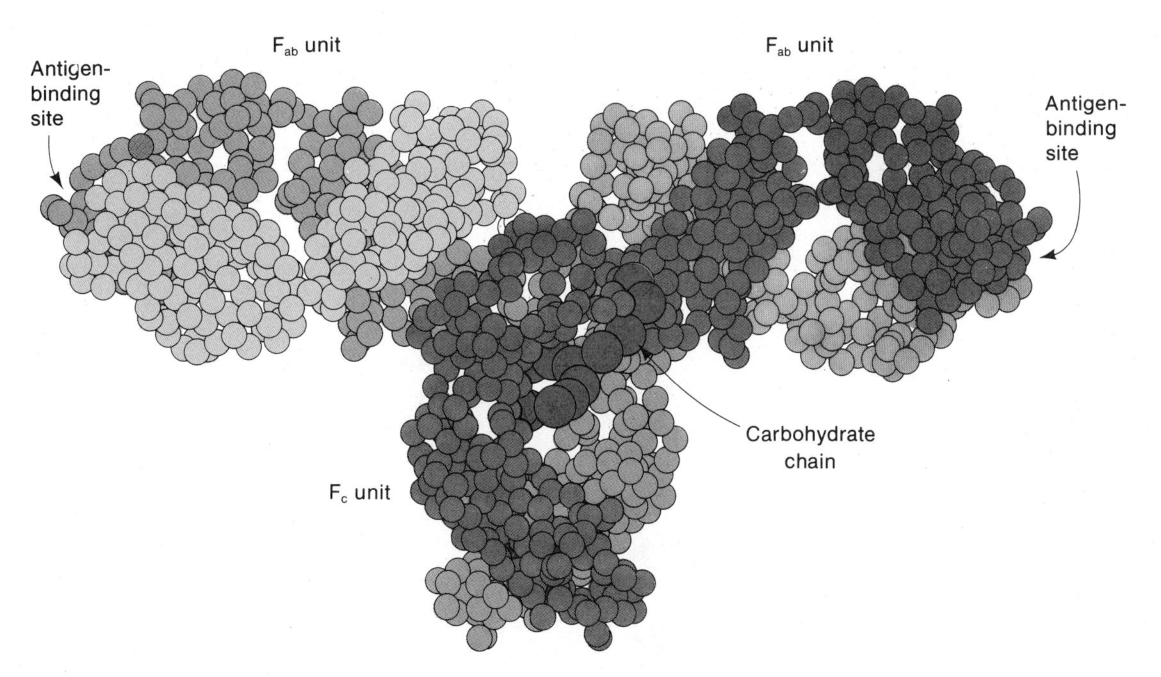

Figure 4-11
Schematic drawing of the three-dimensional structure of an IgG molecule. One of the H chains is shown in dark pink, the other in dark gray. One of the L chains is shown in light pink, the other in light gray. (After E. W. Silverton, M. A. Navia, and D. R. Davies. 1977. *Proc. Nat. Acad. Sci.*, 74: 5142.)

ENZYMES

Enzymes are special proteins able to catalyze chemical reactions. Their catalytic power exceeds all manmade catalysts. A typical enzyme accelerates a reaction 10^8- to 10^{10}-fold, though there are enzymes that increase the reaction rate by a factor of 10^{15}. Enzymes are also highly specific in that each catalyzes only a single reaction or set of closely related reactions. Furthermore, only a small number of reactants, often only one, can participate in a single catalyzed reaction. Since nearly every biological reaction is catalyzed by an enzyme, these clearly require a very large number of distinct enzyme molecules.

The detailed mechanism of catalysis by particular enzymes is beyond the scope of this book. However, all enzymes have certain general features, the knowledge of which is important for understanding molecular biological phenomena. These features are described in this unit.

The Enzyme-Substrate Complex

In any reaction that is enzymatically catalyzed, one reactant always forms a tight complex with the enzyme. This reactant is called the **substrate** of the enzyme; in descriptive formulas, it is denoted S. The complex between the enzyme E and the substrate is called the enzyme-substrate or **ES complex.**

Initially the enzyme and the substrate are bound by a variety of weak bonds, though in a few cases a covalent bond forms. The site on the enzyme at which the substrate binds is the **active site.** The extraordinary selectivity in enzyme catalysis is almost entirely a result of specificity of enzyme-substrate binding. After the ES complex forms, the substrate is usually altered in some way that facilitates further reaction. When the substrate is in the altered state, the ES-complex is said to be active and is usually denoted **(ES)***. The (ES)* complex then engages in one or a series of transformations, which result in conversion of the substrate to the product and dissociation of the product from the enzyme. The extent to which ES forms is determined by the strength of binding between E and S; this is called the **affinity** of E and S.

Theories of Formation of the Enzyme-Substrate Complex

Catalysis by an enzyme occurs in several steps—binding of the substrate, conversion to the product, and release of the product. The initial step, formation of the ES complex, is in principle easy to understand

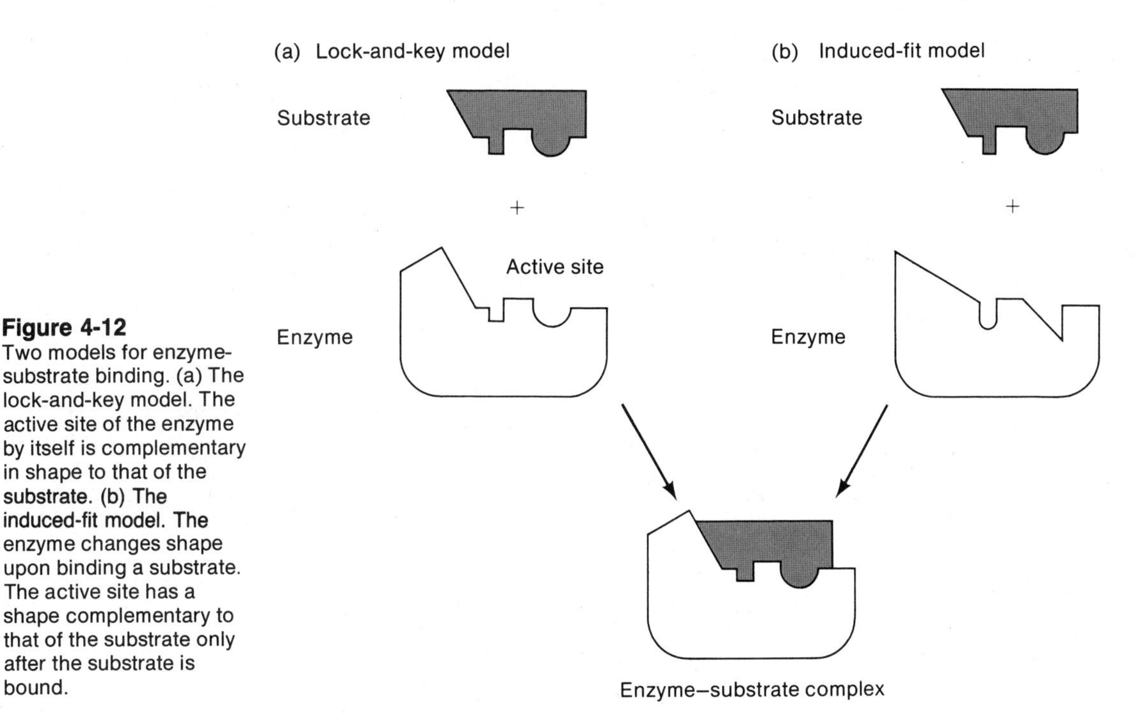

Figure 4-12
Two models for enzyme-substrate binding. (a) The lock-and-key model. The active site of the enzyme by itself is complementary in shape to that of the substrate. (b) The induced-fit model. The enzyme changes shape upon binding a substrate. The active site has a shape complementary to that of the substrate only after the substrate is bound.

and is often considered in thinking about molecules. The subsequent rearrangments are chemical phenomena and will not be discussed in this book.

There are two major theories of enzyme binding, the **lock-and-key** and **induced-fit** models (Figure 4-12). In the lock-and-key model, the shape of the active site of the enzyme is complementary to the shape of the substrate. In the induced-fit model, the enzyme changes shape upon binding the substrate and the active site has a shape that is complementary to that of the substrate only *after* the substrate is bound. For every enzyme-substrate interaction examined to date, one of these two models applies. It is often the case, though, that the substrate itself undergoes a small change in shape; in fact, the strain to which the substrate is subjected is often the principle mechanism of catalysis—that is, the substrate is held in an enormously reactive configuration.

Molecular Details of an Enzyme-Substrate Complex

The first detailed analysis of enzyme-substrate binding was carried out by using hen egg-white lysozyme. This enzyme cleaves certain bonds between sugar residues in some of the polysaccharide components of bacterial cell walls and is responsible for maintaining sterility within eggs. The amino acid sequence of lysozyme is shown in Figure 4-13. The 19 amino acids that are part of the active site are printed in red; it should be noticed that they form widely separated clusters along the chain. Only when the chain is folded do they come into proximity and form the active site. The folding of the chain is shown in Figure 4-14. The deep cleft indicated by the arrow is the active site. This is seen more clearly in the space-filling model shown in Figure 4-15. The substrate is a hexasaccharide segment that fits into the cleft and is distorted upon binding. The enzyme itself changes shape when the substrate is bound.

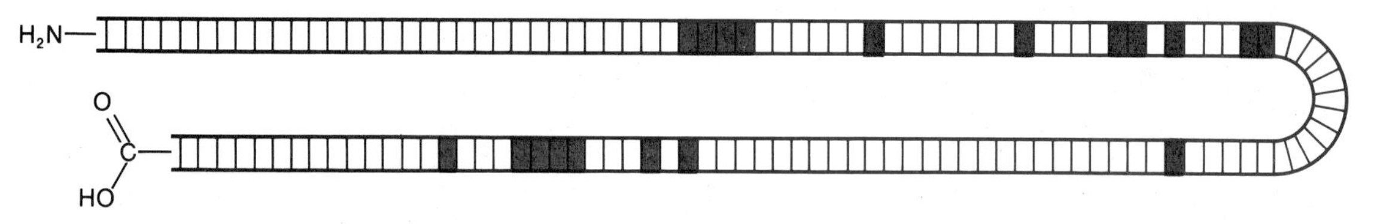

Figure 4-13
Schematic diagram of the amino acid sequence of lysozyme showing that the amino acids (red) in the active site are separated along the chain. Folding of the chain brings these amino acids together.

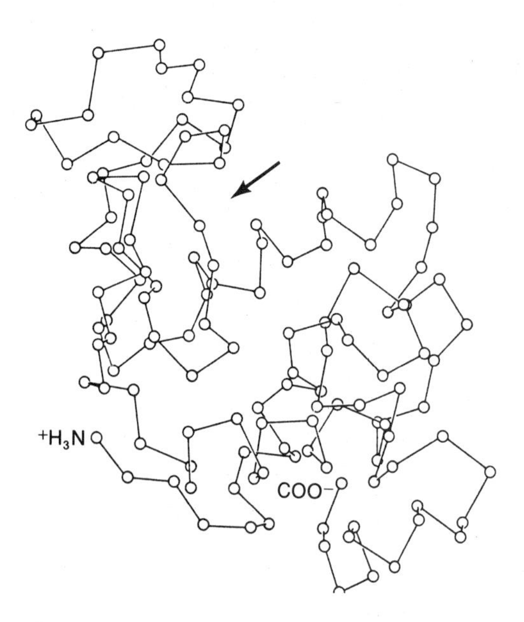

Figure 4-14
Three-dimensional structure of lysozyme. Only the α-carbon atoms are shown. The active site is in the cleft indicated by the arrow. (Courtesy of Dr. David Phillips.)

Figure 4-15
A space-filling molecular model of the enzyme lysozyme. The arrow points to the cleft that accepts the polysaccharide substrate. (C atoms are black; H, white; N, gray; O, gray with slots.) (Courtesy of John A. Rupley.)

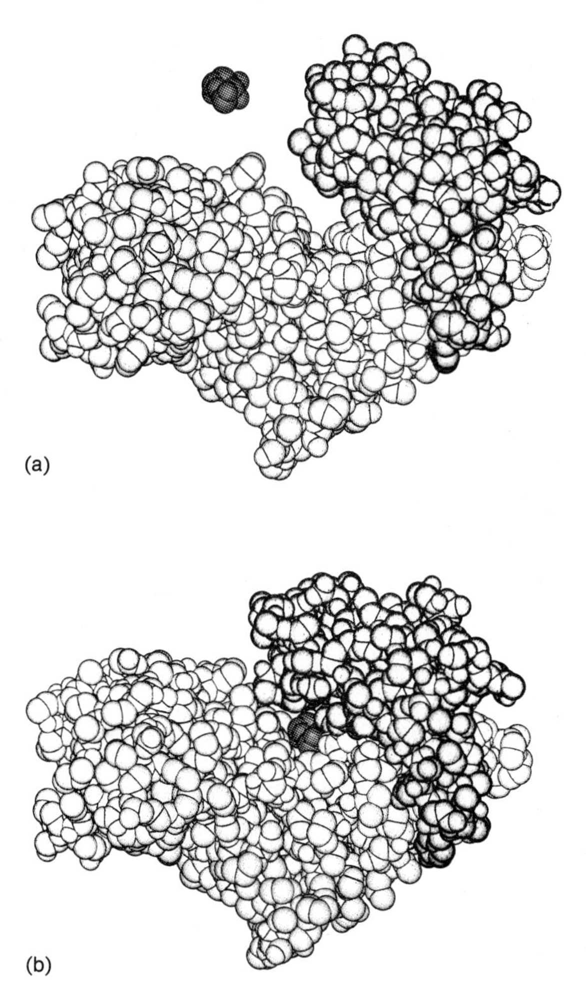

Figure 4-16
Structure of yeast hexokinase A. All atoms except hydrogen are shown separately. (a) Free hexokinase and its substrate, glucose. (b) Hexokinase complexed with glucose. Note the marked change in enzyme structure that accompanies glucose binding. (With permission, from J. Bennett and T. A. Steitz. 1980. *J. Mol. Biol.,* 140: 211. Copyright: Academic Press Inc. [London] Ltd.)

Another enzyme, yeast hexokinase A, has been studied in order to examine what structural changes occur on substrate binding; these changes are now well documented. This is shown in the pair of space-filling models shown in Figure 4-16.

5 Macromolecular Interactions and the Structure of Complex Aggregates

Interactions between macromolecules underly most biological phenomena, and structural components both within individual cells and extracellular in organisms are invariably assemblies of macromolecules. For example, nucleic acids are often associated with proteins, as in chromosomes (which are DNA-protein complexes), extracellular viral nucleic acids are encased in protein shells, and bone and cartilage are complex assemblies of proteins and other macromolecules. Proteins can also interact with lipids to produce membranes such as those which separate the contents of a cell from the environment and which separate different intracellular components from one another. Finally, polysaccharides form extraordinarily complex structures such as the cell walls of bacteria and of plants.

The study of such complex structures and how they are formed is called structural biology. A few structures are almost completely understood and many more are actively being studied. In this chapter only a few structures will be described. These have been selected by two criteria—they are generally important or illustrate general principles, and their structures are reasonably well understood.

A COMPLEX DNA STRUCTURE: THE *E. COLI* CHROMOSOME

All genes of *E. coli*, and presumably of most bacteria, are contained on a single supercoiled circular DNA molecule. The total length of the circle

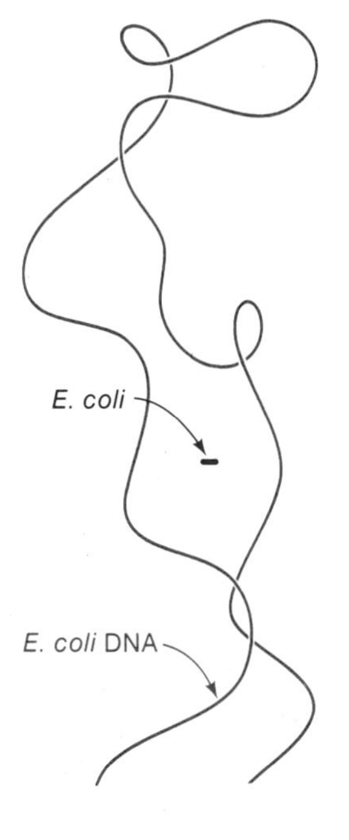

(a)

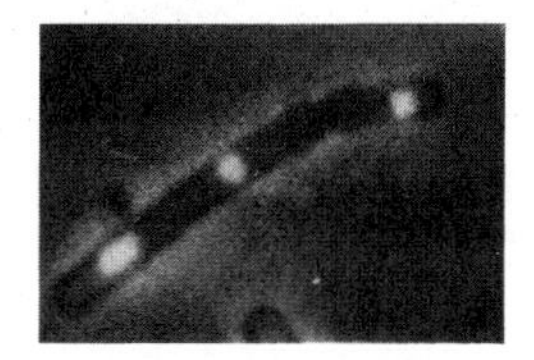

(b)

Figure 5-1
(a) Schematic diagram showing the relative sizes of *E. coli* and its DNA molecule, drawn to the same scale except for the width of the DNA molecule, which is enlarged approximately 10^6 times. (b) An *E. coli* cell, whose DNA has bound an added fluorescent dye. (Courtesy of Todd Steck and Karl Drlica.)

is about 1300 μm. A cylindrical bacterium has a diameter and a length of about 1 and 3 μm respectively (Figure 5-1); clearly the bacterial DNA must be highly folded when it is in a cell.

When *E. coli* DNA is isolated by a technique that avoids both breakage of the molecule and denaturation of proteins, a highly compact structure known as a bacterial chromosome, or **nucleoid,** is found. This structure contains protein and a single supercoiled DNA molecule. Some RNA is also present in the isolated nucleoid, but it is probably not an essential component, possibly becoming associated with the structure during the isolation procedure. An electron micrograph of the *E. coli* chromosome is shown in Figure 5-2. The particular feature of the structure to note is that the DNA is in the form of loops, which are supercoiled and emerge from a dense protein-containing structure, sometimes called the **scaffold.** The physical dimensions of the isolated chromosome are affected by a variety of factors, and some controversy exists about the state of the nucleoid within the cell. If the nucleoid is treated with an enzyme that degrades proteins or various detergents that break down protein-protein interactions, the chromosome expands markedly, though it remains much more compact than a

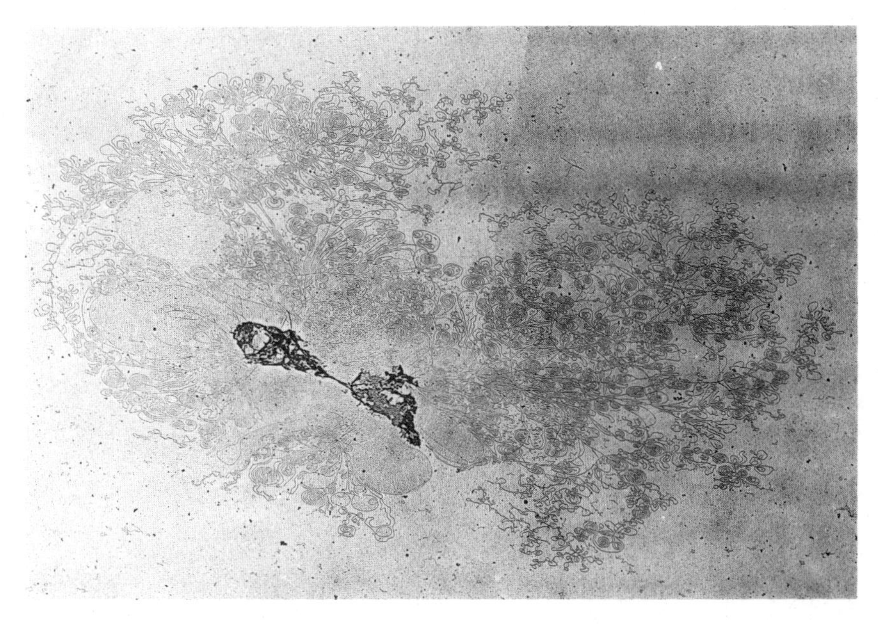

Figure 5-2
Electron micrograph of the chromosome of *E. coli* attached to two fragments of a proteinaceous substance, which may be the cell membrane. This single molecule of double-helical DNA is intact and supercoiled. (From H. Delius and A. Worcel. 1974. *J. Mol. Biol.*, 82: 108. Copyright: Academic Press, Inc. [London] Ltd.)

free DNA molecule. As proteins are removed, the scaffold is disrupted and the chromosome goes through a series of transitions between forms having different degrees of compactness. The conclusion from these and other observations is that the chromosome is held in compact form by the binding of different regions of the DNA molecule to the scaffold. Whether the scaffold has a well-defined organization and is a true "structure" (as opposed to a disorganized aggregate) is not known.

Treatment of a chromosome with very tiny amounts of a DNase, producing single-strand breaks, followed by sedimentation of the treated DNA, gives some insight into the physical structure of the chromosome. In Chapter 3 it was explained that if a supercoiled DNA molecule receives one single-strand break, the strain of underwinding is immediately removed by free rotation about the opposing sugar-phosphate bond and all supercoiling is lost. Since a nicked circle is much less compact than a supercoil of the same molecular weight, the nicked circle sediments much more slowly. Thus, one single-strand break causes an abrupt decrease (by about 30 percent) in the s value. However, if one single-strand break is introduced by a nuclease into the *E. coli* chromosome, the s value decreases by only a few percent. Furthermore, a second break causes a second decrease in the s value, and subsequent breaks cause additional stepwise changes. After about forty-five breaks the form of the chromosome remains constant. This clearly indicates that free rotation of the entire DNA molecule does not occur when a single-strand break is introduced. The structure of the *E. coli* chromosome that has been deduced from these data is shown in Figure 5-3. The DNA is assumed to be fixed to the scaffold at 45 ± 10 positions, each of which prevents free rotation. Thus, there would be about 45 ± 10

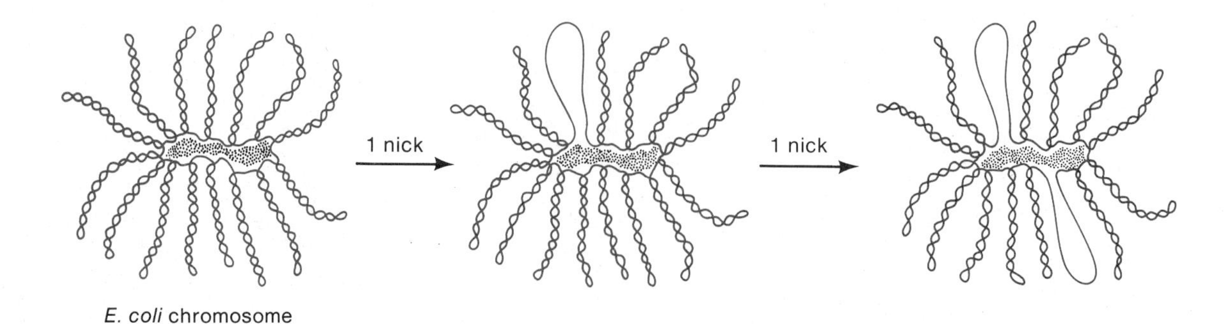

Figure 5-3
A schematic drawing of the highly folded supercoiled *E. coli* chromosome, showing only 15 of the 46 loops attached to the scaffold and the opening of a loop by a single-strand break (nick).

supercoiled loops of DNA. One single-strand break would then cause one supercoiled loop to become open, and each subsequent break would, on the average, open another loop. This notion has been confirmed in electron micrographs of chromosomes in which there are one or two nicks; structures such as that in Figure 5-2, but with one or two open loops, are observed.

The enzyme DNA gyrase, which plays an important role in DNA replication (Chapter 7), is responsible for the supercoiling. If coumermycin, an inhibitor of *E. coli* DNA gyrase, is added to a culture of *E. coli* cells, the chromosome loses its supercoiling in about one generation time. Indirect measurements of the number of binding sites for gyrase on the chromosome indicate that there are roughly forty-five sites. The spatial distribution of these sites is not known but it is tantalizing to think that there may be one binding site in each supercoiled loop.

In the next section it will be seen that DNA is arranged in a much more complex way in eukaryotic cells.

CHROMOSOMES AND CHROMATIN

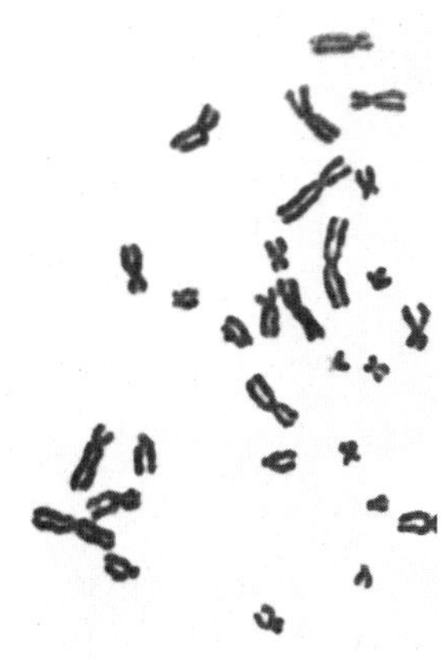

Figure 5-4
Human chromosomes from a cell at metaphase. Each chromosome is partially separated along its long axis prior to separation of the two daughter chromosomes. The constriction is the site of attachment to the mitotic spindle. (Courtesy of Theodore Puck.)

The DNA of all eukaryotes is organized into morphologically distinct units called **chromosomes** (Figure 5-4). Each chromosome contains only a single enormous DNA molecule. For example, the DNA molecule in a single chromosome of the fruit fly *Drosophila* has a molecular weight greater than 10^{10} and a length of 1.2 cm. (Since the width of all DNA molecules is 20 Å or 2×10^{-7} cm, the ratio of length to width of *Drosophila* DNA has the extraordinary value of 6×10^6.) These molecules are much too long to be seen in their entirety by electron microscopy because, at the minimum magnification needed to see a DNA molecule, the field of view is only about 0.01 cm across.

A chromosome is much more compact than a DNA molecule and in fact a DNA molecule cannot spontaneously fold to form such a compact structure because the molecule would be strained enormously. Instead, DNA is made compact by a hierarchy of different types of folding, each of which is mediated by one or more protein molecules.

The DNA molecule in a eukaryotic chromosome is bound to basic proteins called **histones.** The complex comprising DNA and histones is called **chromatin.** There are five major classes of histones—H1, H2A, H2B, H3, and H4—whose properties are listed in Table 5-1. Histones have an unusual amino acid composition in that they are extremely rich in the positively charged amino acids lysine and arginine. The lysine-to-arginine ratio differs in each type of histone. (In the older literature H1 was called the lysine-rich histone and H3 and H4 were called

Table 5-1
Types of histones

Type	*Lys/Arg ratio*	*Location*
H1	20.0	Linker
H2A	1.25	Core
H2B	2.5	Core
H3	0.72	Core
H4	0.79	Core

arginine-rich.) The positive charge of the histones is one of the major features of the molecules, enabling them to bind to the negatively charged phosphates of the DNA. This electrostatic attraction is apparently the most important stabilizing force in chromatin since, if chromatin is placed in solutions of high salt concentration (e.g., 0.5 *M* NaCl), which breaks down electrostatic interactions, chromatin dissociates to yield free histones and free DNA. Chromatin can also be reconstituted by mixing purified histones and DNA in a concentrated salt solution and gradually lowering the salt concentration by dialysis. The result shows that no other components are needed to form chromatin.

Reconstitution experiments have been carried out in which histones from different organisms are mixed. Usually, almost any combination of histones works because, except for H1, the histones from different organisms are very much alike. In fact, the amino acid sequences of both H3 and H4 are nearly identical (sometimes one or two of the amino acids differ) from one organism to the next. Histone H4 from the cow differs by only two amino acids from H4 from peas—arginine for lysine and isoleucine for valine—which shows that the structure of histones has not changed in the 10^9 years since plants and animals diverged. Clearly histones are very special proteins indeed.

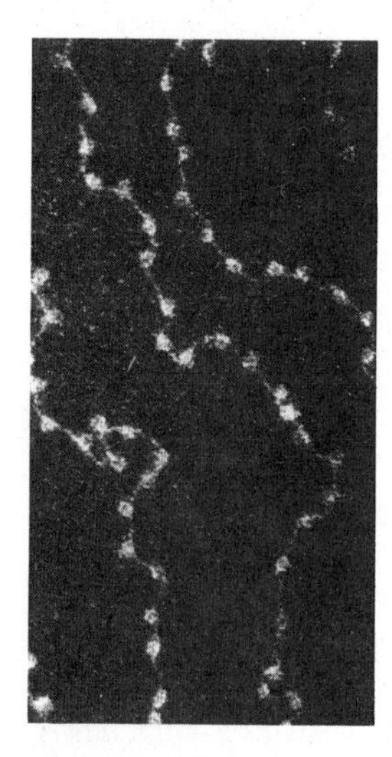

Figure 5-5
Electron micrograph of chromatin. The beadlike particles have diameters of nearly 100 Å. (Courtesy of Ada Olins.)

As a cell passes through its growth cycle, chromatin structure changes. In a resting cell the chromatin is dispersed and fills the entire nucleus. Later, after DNA replication has occurred, the chromatin condenses about 100-fold and chromosomes form. Chromosomes have been isolated and gradually dissociated, and chromosomes at various degrees of dissociation have been observed by electron microscopy. The most elementary structural unit is a fiber 10 nm wide, which appears like a string of beads (Figure 5-5). The beadlike structure, which is chromatin, is also seen when chromatin is isolated from resting cells. The structural hierarchy of a chromosome, which has been deduced from a variety of studies, is shown in Figure 5-6. The higher orders of structure are still somewhat speculative but a great deal of information

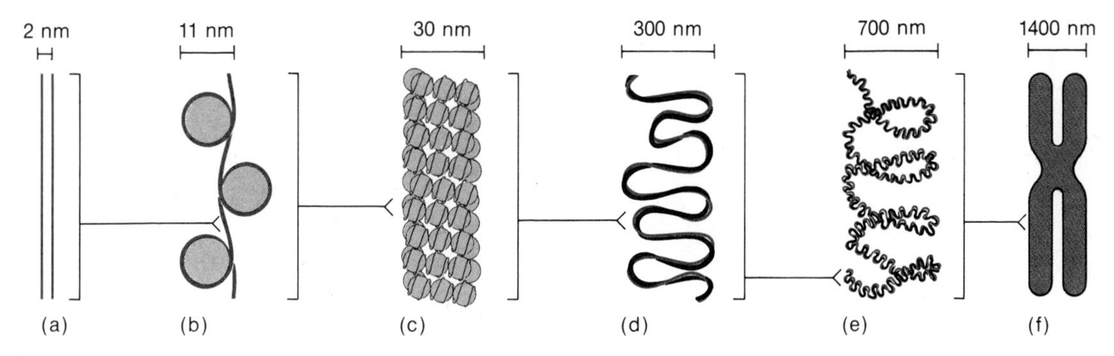

Figure 5-6
Various stages in the condensation of (a) DNA and (b–e) chromatin in forming (f) a metaphase chromosome. The dimensions indicate known sizes of intermediates, but the detailed structures are hypothetical.

is known about the beads themselves, each of which is an ordered aggregate of DNA and histones.

Prior to electron microscopic studies, the effect of various DNA endonucleases on chromatin also suggested that chromatin contains repeating units. Treatment of chromatin with pancreatic DNase I (which cannot attack DNA that is in contact with protein) yields a collection of small particles containing DNA and histones (Figure 5-7). If, after enzymatic digestion, the histones are removed, DNA fragments having roughly 200 base pairs, or a multiple of 200, are found. After a long period of digestion the multiple-size units are not found and all of the DNA has the 200-base-pair-unit size. Fragments have also been isolated before removal of the histones and have been examined by electron microscopy; it has been found that a fragment containing 200*n* base pairs consists of *n* connected beads, indicating that there is a fundamental bead unit containing about 200 base pairs.

The beadlike particles are called **nucleosomes.** Each nucleosome is found to consist of one molecule of histone H1, two molecules each of histones H2A, H2B, H3, and H4, and a DNA fragment. Treatment of the nucleosomes (which have been obtained by digestion of chromatin with pancreatic DNase) with the enzyme micrococcal nuclease gradually removes additional amounts of DNA. All histones remain associated with the DNA until the number of base pairs is less than 160, at which point H1 is lost. More bases can be removed but the number cannot be reduced to less than 140 base pairs. The structure that remains is called the **core particle.** It contains an octameric protein disc consisting of two

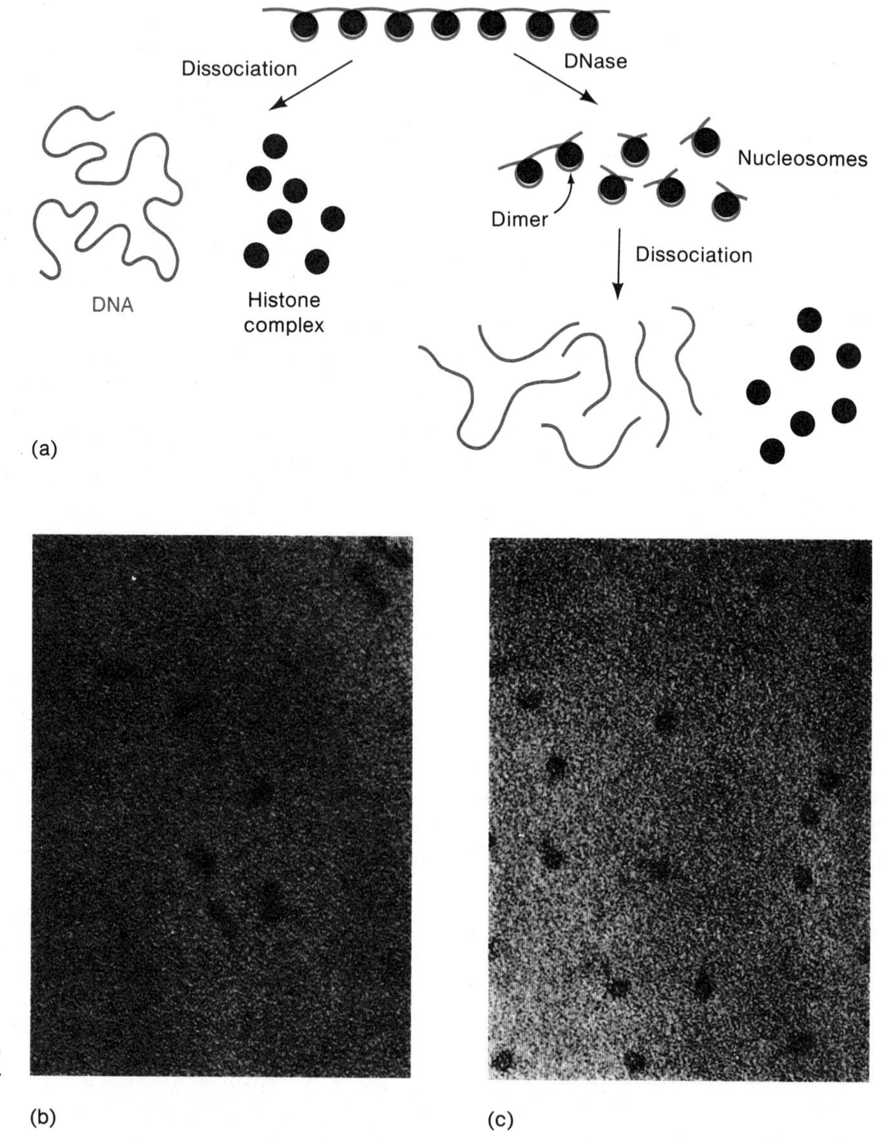

Figure 5-7
(a) The DNase-digestion method for production of individual nucleosomes and 200-base-pair fragments (b,c). (b) Electron micrographs of monomers and dimers. (c) Nucleosomes produced as described in part (a). (From J. T. Finch, M. Noll, and R. D. Kornberg. 1975. *Proc. Nat. Acad. Sci.,* 72: 3321.)

copies each of H2A, H2B, H3, and H4, around which the 140-base-pair segment is wrapped like a ribbon (Figure 5-8). Thus a nucleosome consists of a core particle, additional DNA that links adjacent core particles (**linker DNA**), and one molecule of H1. The H1 binds to the histone octamer and to the linker DNA, causing the linkers extending from both sides of the core particle to cross and draw nearer to the octamer, though some of the linker DNA does not come into contact

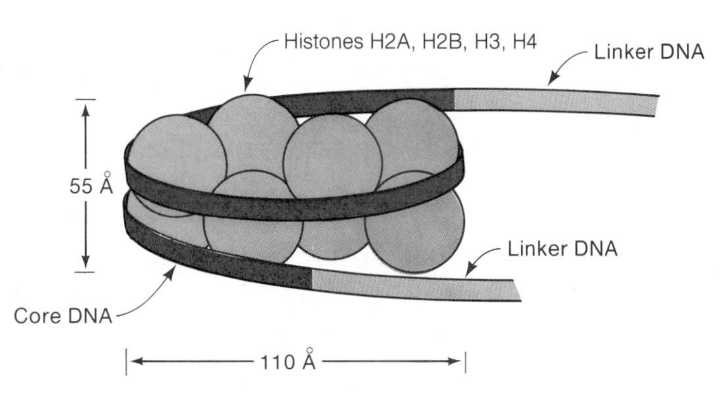

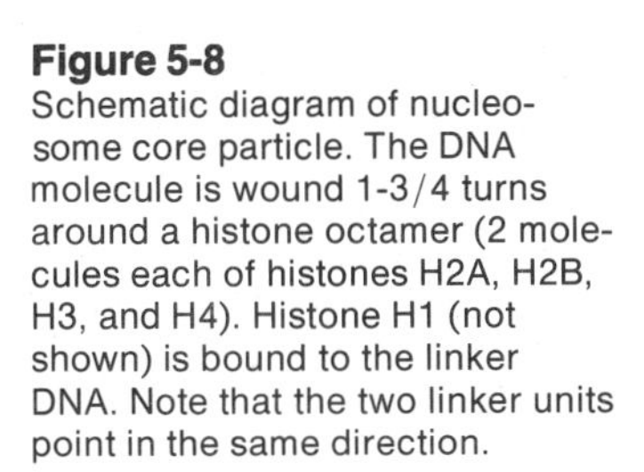

Figure 5-8
Schematic diagram of nucleosome core particle. The DNA molecule is wound 1-3/4 turns around a histone octamer (2 molecules each of histones H2A, H2B, H3, and H4). Histone H1 (not shown) is bound to the linker DNA. Note that the two linker units point in the same direction.

with any histones. The term **chromatosome** is coming into use for the nucleosome containing H1 in which the DNA makes almost two full turns around the octamer. The overall structure of the chromatin fibril is probably a zigzag, as shown in panels (a) and (b) of Figure 5-6. Assembly of DNA and histones is the first stage of shortening of the DNA strand in a chromosome—namely, a sevenfold reduction in length of the DNA and the formation of a beaded flexible fiber 110 Å (11 nm) wide, roughly five times the width of free DNA.

The binding of DNA and histone has been examined. About 80 percent of the amino acids in the histones are in α-helical regions. Preliminary data suggest that many of the extended α-helical regions lie in the larger groove of the DNA helix and that the complex is stabilized by an electrostatic attraction between the positively charged lysines and arginines of the histones and the negatively charged phosphates of the DNA.

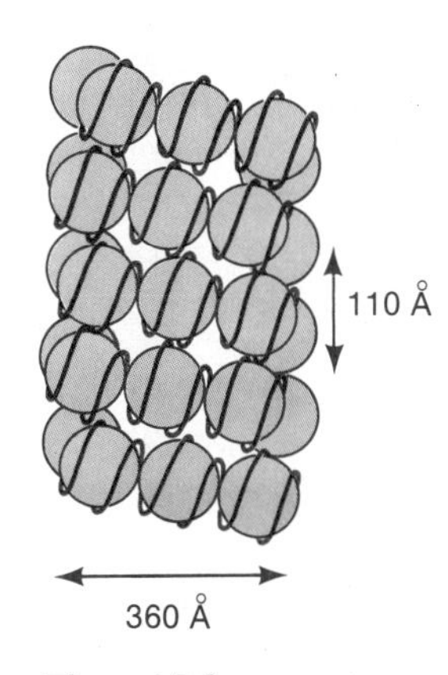

Figure 5-9
A proposed solenoidal model of chromatin. Six nucleosomes (shaded) form one turn of the helix. The DNA double helix (shown in red) is wound around each nucleosome. (After J. T. Finch and A. Klug. 1976. *Proc. Nat. Acad. Sci.*, 73: 1900.)

The DNA content of nucleosomes varies from one organism to the next, ranging from 150 to 240 base pairs per unit. The core particles of all organisms have the same DNA content (140 base pairs), so that the observed variation results from different sizes of the linker DNA between the nucleosomes (namely, 10–140 nucleotide pairs). Little is known about the structure of linker DNA or whether it has a special function, and the cause of variation of its length is unknown.

In forming a compact chromosome, the DNA molecule is folded and refolded in several ways, as shown in Figure 5-6. The second level of folding is the shortening of the 11-nm fiber to form a solenoidal supercoil with six nucleosomes per turn, called the **30-nm fiber** (Figures 5-6(c) and 5-9); this form has been isolated and well characterized. The remaining levels of organization—folding of the 30-nm fiber and further folding of the folded structure—shown in Figure 5-6(d–f), are less well understood. Electron micrographs of isolated metaphase chromosomes from which histones have been removed indicate that the

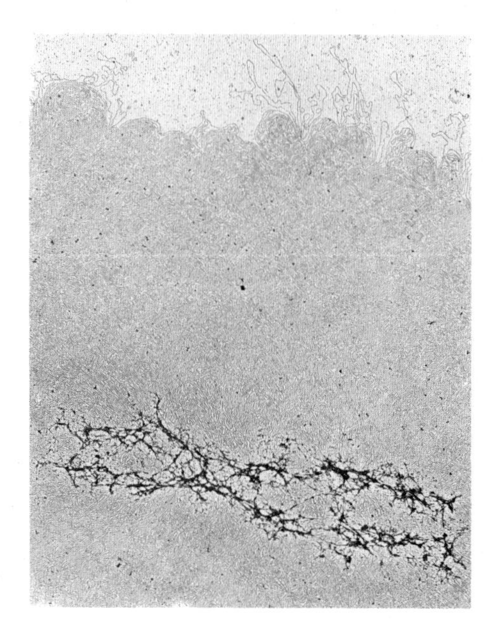

Figure 5-10
Electron micrograph of a segment of a human metaphase chromosome from which the histones have been chemically removed. The dense network near the bottom of the figure is the protein scaffold on which the chromosome is assembled. (Courtesy of Ulrich Laemmli.)

partially unfolded DNA has the form of an enormous number of loops that seem to extend from a central core or scaffold, composed of nonhistone chromosomal proteins, as was seen for the *E. coli* chromosome (Figure 5-10).

The compaction of DNA and protein into chromatin and ultimately into the chromosome greatly facilitates the distribution of the genetic material during nuclear division. We will see in later chapters that certain variations in chromatin structure affect gene expression.

INTERACTION OF DNA AND A PROTEIN THAT RECOGNIZES A SPECIFIC BASE SEQUENCE

In the formation of chromatin the histones do not bind to specific base sequences but recognize only the general DNA structure. There are, however, numerous proteins that bind only to particular sequences of bases, and we will encounter many such proteins in this book. These proteins are of three types: (1) polymerases, which initiate synthesis of DNA and RNA from particular base sequences; (2) regulatory proteins, which turn on and off the activity of particular genes; and (3) certain nucleases, which cut phosphodiester bonds between a single pair of adjacent nucleotides contained in unique sequences of four to six nucleotides. Although the structures of only a few of the binding regions in DNA-protein complexes have been elucidated, a general pattern seems to be emerging, one which applies both to sequence-specific binding and to nonspecific binding. This pattern is the following:

1. The base sequence of the DNA has some sort of twofold symmetry, allowing binding of the protein to two sites.
2. The protein-DNA contact region is on only one side of the DNA molecule.
3. The protein usually has one or more α-helical regions and these regions bind to the DNA within two segments of the major groove of the DNA.
4. The protein usually is in the form of a dimer having twofold symmetry; the monomers are arranged so that the α helices are in a symmetric array that matches the symmetry of the base sequence.
5. Amino acids of the α-helical regions are in contact with bases in particular positions; in sequence-specific binding the correct bases must be present at those positions.
6. Positively charged amino acids often form ionic bonds with the phosphate groups, which are negatively charged. In sequence-specific binding these do not confer specificity but stabilize the interaction.

A detailed analysis of the structure of a DNA-protein complex is beyond the scope of this book (for more information, see *MB*, pages 206–212). However, some of the features just listed will be illustrated by an examination of the complex between the Cro protein of *E. coli* phage λ and the operator sequence in λ DNA. Cro is a small protein, consisting of 66 amino acids; it forms a dimer and binds to a DNA sequence containing 17 nucleotide pairs (Figure 5-11). Each monomer of the Cro protein contains three α-helical regions, one of which is in direct contact with DNA bases in the Cro-DNA complex. Figure 5-12 shows the binding region of the DNA molecule, with the bases and phosphate groups that contact the Cro protein shown. These sites have been identified by treating the complex with reagents that attack the sites in free DNA but that cannot alter a region of the DNA that is in contact with Cro. Analysis of the chemical structure of the DNA after exposure of the complex to these reagents showed that certain bases and phosphates were unaltered and, hence, are in the contact region. Figure 5-13 shows the Cro-DNA complex. The symmetry of the Cro dimer is apparent; the relevant α helices are in the upper and lower parts of the dimer and can be seen to fit nicely into the major groove of the DNA. This kind of symmetric binding array has been observed in repressor-operator binding (Chapters 14 and 15), in binding of the *E.*

TATCACCGCAAGGGATA
ATAGTGGCGTTCCCTAT

Figure 5-11
The base sequence of one of the λ Cro protein binding sites. This sequence is called *oR3*. The adenines and guanines shown in Figure 5-12 are printed in red.

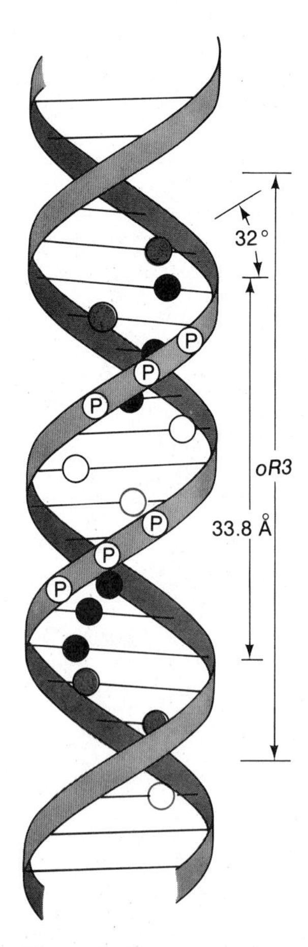

Figure 5-12
Points at which the *oR3* region of λ DNA makes contact with Cro. Phosphates are labeled P; guanines and adenines in contact are solid black and red circles; guanines and adenines not in contact are open black and red circles, respectively. Note that the bases in contact are only in the wider (major) groove of the DNA. (After W. F. Anderson *et al.*, 1981. *Nature*, 290: 754.)

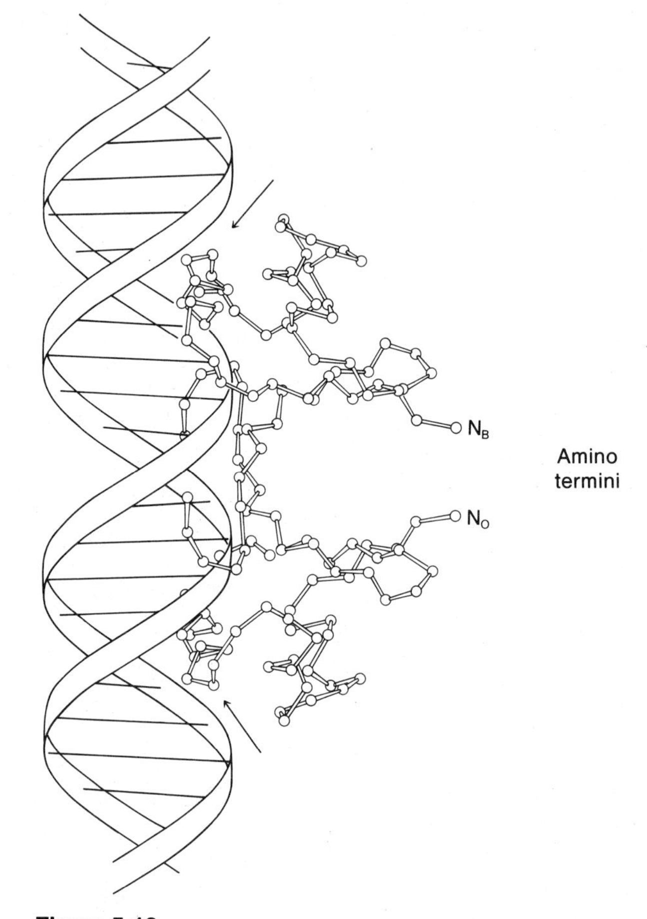

Figure 5-13
Presumed structure of the Cro-*oR3* complex. The DNA is rotated 90° relative to that in Figure 5-12 so that the contact points are on the right side of the molecule. The amino termini of each Cro monomer are labeled N_O and N_B. A pair of α helices, related by two-fold symmetry, occupy successive major grooves of the DNA (arrows), and the two extended chains run parallel to the axis of the DNA. (After W. F. Anderson *et al.*, 1981. *Nature*, 290: 754.)

coli CAP protein to the relevant sequence in the *lac* operon (Chapter 14), and in chromatin.

A close look at the complex (for which there is not enough detail in the figures) would show that amino acids in each Cro monomer are situated in a way that they can form hydrogen bonds with the DNA bases, which confers binding specificity, and also interact electrostatically with the phosphate groups, which confers binding strength.

BIOLOGICAL MEMBRANES

Biological membranes are organized assemblies consisting mainly of proteins and lipids. The structures of all biological membranes have many common features but small differences exist to accommodate the varied functions of different membranes.

A basic function of membranes is to separate the contents of a cell from the environment. However, it is always necessary that cells take up nutrients from the surroundings; therefore, the enclosing membrane of a cell must be permeable. The permeability of the membrane must be selective though, so that the concentrations of compounds within a cell can be controlled; otherwise, intracellular concentrations would be inseparable from extracellular concentrations. Selective permeability is attained by means of restricting free passage of most intra- and extracellular substances and allowing transport of specific substances by molecular pumps, channels, and gates. External membranes also have a signal function. For instance, they often contain receptors for signal molecules such as hormones. The external membrane is also responsible for transmitting electrical impulses, as in nerve cells.

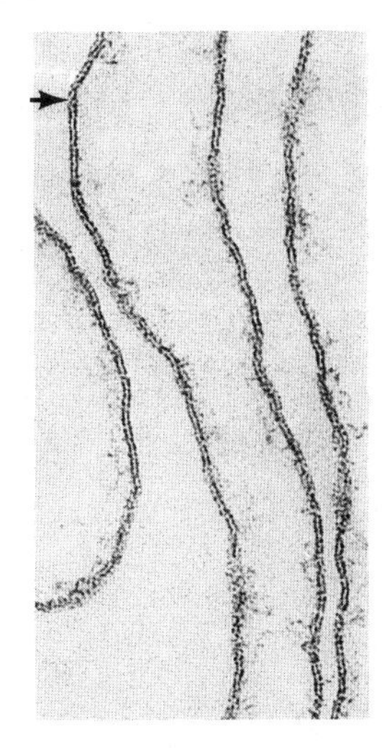

Figure 5-14
Electron micrograph of a preparation of plasma membranes from red blood cells. A single membrane is denoted by an arrow. (Courtesy of Vincent Marchesi.)

There are also numerous types of intracellular membranes. These membranes serve to compartmentalize various intracellular components and to provide surfaces on which certain molecules can be adsorbed prior to chemical reaction with other adsorbed molecules.

Figure 5-14 shows an electron micrograph in which the membrane enclosing a red blood cell is seen in cross-section. The most notable feature of this membrane is that it consists of two layers—it is called a **membrane bilayer.** This structure is a consequence of the chemical nature of the lipids that form the membrane. There are many membrane lipids, of which the most prevalent are the phospholipids, molecules consisting of a polar phosphate-containing group and two long nonpolar hydrocarbon chains. Figure 5-15 shows a space-filling model of one phospholipid and a schematic drawing. The essential feature of the molecule is a polar "head" group to which hydrocarbon "tails" are attached. Such a molecule, which has distinct polar and nonpolar segments, is called **amphipathic.** When placed in water, amphipathic molecules tend to aggregate. This is because only the polar head is capable of interaction with the polar water molecules. The hydrocarbon tails are brought together by hydrophobic interactions—that is, they cluster because they are unable to interact with water. If the length of

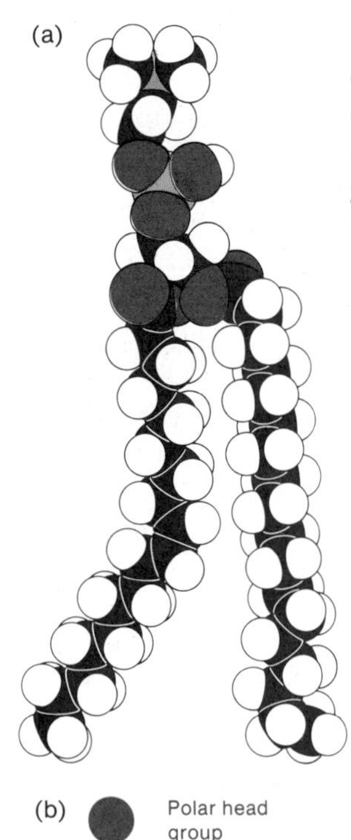

Figure 5-15
(a) Space-filling model of phosphatidyl choline, a phospholipid. (b) The essential features of a phospholipid or glycolipid molecule.

the hydrocarbon tail satisfies certain geometric constraints, a collection of amphipathic molecules forms a **micelle,** a spherical array of molecules in which the nonpolar tails form a hydrocarbon microdroplet enclosed in a shell composed of the polar heads (Figure 5-16(a)). The geometric constraints limit the size of a stable micelle, and as the size of the hydrocarbon tail increases, the array becomes ellipsoidal (Figure 5-16(b)) with the ratio of the lengths of the major axis to the minor axis of the ellipsoid increasing with length of the tail. Above a certain tail length, the only stable configuration is an ellipsoid whose major axis is "infinitely" long—that is, an extended lipid bilayer sheet, as shown in part (c) of the figure. The lipid bilayer can also be stabilized by van der Waals attractive forces between the hydrocarbon tails.

The ends of a lipid bilayer are unstable, as the hydrocarbon chains are exposed to water. Thus lipid bilayers tend to close upon themselves to form hollow bilayer spheres known as **vesicles** (Figure 5-17). If either natural membranes or surface films of individual lipid molecules are physically disrupted, many fragments will form vesicles. If such vesicles are formed from a solution of a small molecule, some will contain these molecules. Studies of synthetic vesicles formed from a single type of phospholipid and containing a variety of small molecules have shown that the lipid bilayer is impermeable to all ions and highly polar molecules, raising the question of how these molecules get in and out of cells. The answer is that naturally occurring biological membranes contain certain proteins that are responsible for transport of all molecules having polar regions and for most nonpolar molecules as well.

There are two classes of membrane proteins—the **integral membrane proteins,** which are contained wholly or in part within the membrane, and the **peripheral proteins,** which lie on the membrane surface and are bound to the integral proteins. An electron micrograph

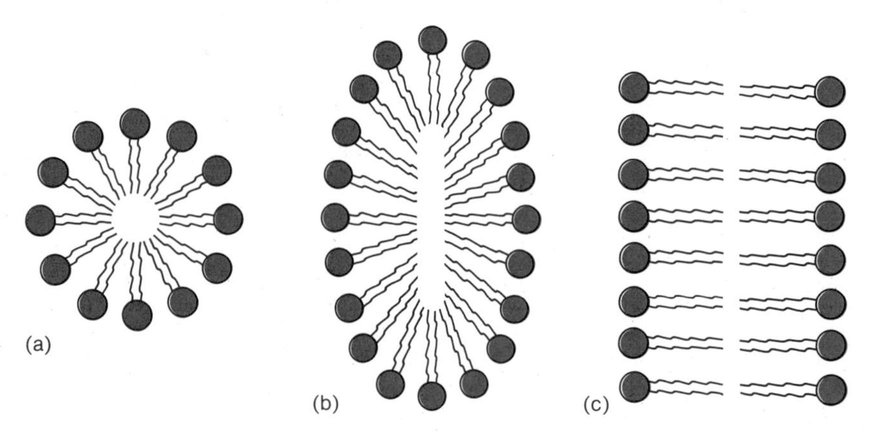

Figure 5-16
(a) Diagram of a section of a micelle formed from phospholipid molecules. (b) An ellipsoidal micelle. (c) Diagram of a section of a membrane bilayer formed from phospholipid molecules.

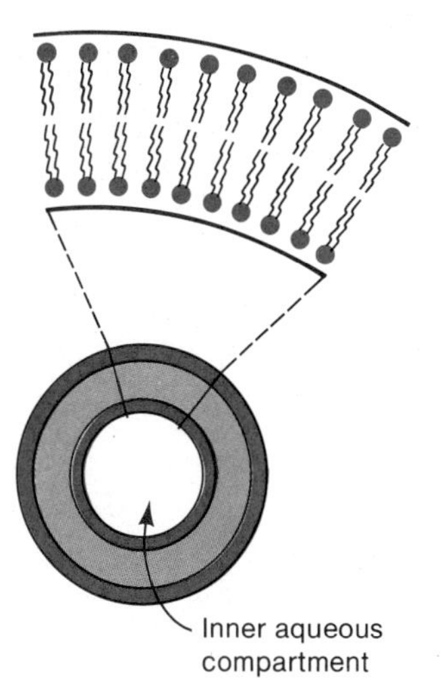

Figure 5-17
Schematic diagram of a lipid vesicle. The membrane is enlarged to show the double layer.

of the plasma membrane of a red blood cell in which integral membrane proteins can be seen is shown in Figure 5-18. The outer surface of the membrane is smooth but the inner surface is covered with many globular protein molecules.

Numerous physical measurements have shown that the integral membrane proteins can freely diffuse laterally throughout the bilayer.

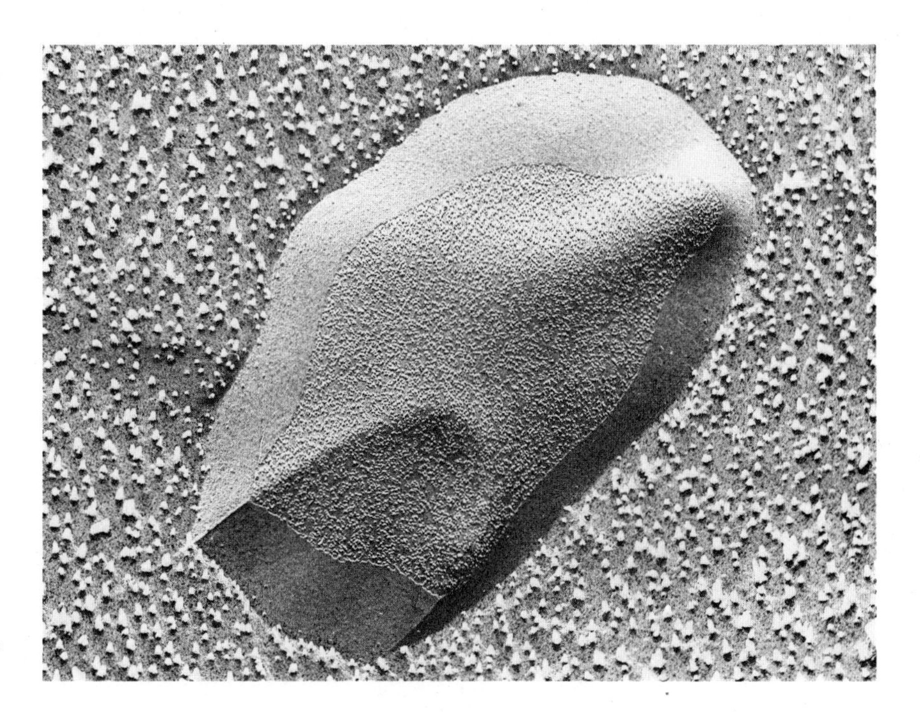

Figure 5-18
An electron micrograph of the plasma membrane of red blood cells. The interior of the membrane, which has been exposed by fracture of the membrane, contains numerous globular particles having a diameter of about 75 Å. These particles are termed integral membrane proteins. (Courtesy of Vincent Marchesi.)

The degree of motion is determined by the fluidity of the lipid layer, which is in turn a function of the van der Waals interactions between the hydrocarbon tails of the particular phospholipids in the membrane. This phenomenon was first noted by S. Jonathan Singer and Garth Nicholson and led to the **fluid mosaic model** of membrane structure (Figure 5-19). It was further shown that the proteins do not spontaneously rotate (flip-flop) from one side of the membrane to another. This is because the integral proteins also have polar and nonpolar regions as shown in the figure, and inasmuch as the external region is polar and the internal region is nonpolar, rotation would require passage of the polar region through the nonpolar center of the bilayer.

Since different proteins protrude from the two sides of the membrane, the membrane is an asymmetric structure having, if one were to take the point of view of a cell, an inside and an outside. It is this asymmetry that determines the direction of movement of molecules entering and leaving a cell.

There are many modes of transport of molecules across the bilayer and only a few will be mentioned. One mode makes use of channels through the membrane. These channels usually are passages through the integral membrane proteins; the passages will have an abundance of polar amino acids if their function is to allow transit of polar substances. Often the channels can be opened and closed by means of conformational changes of the membrane proteins. Other transport systems utilize chemical reactions that convert the substance to be transported to a molecule that can enter the membrane and then, after transit of the modified molecule, restore the original molecule at the other side of the membrane; these chemical mechanisms are usually very complex, consume a great deal of energy, and are poorly understood.

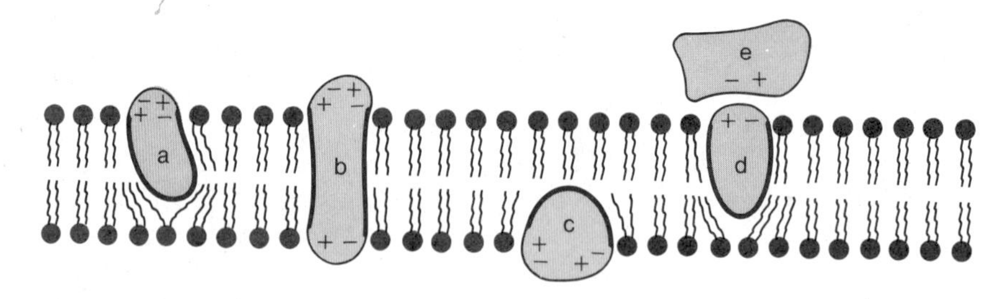

Figure 5-19
The structure of a membrane according to the fluid mosaic model. Four integral proteins (a)—(d) are embedded to various degrees into a lipid bilayer such that the hydrophobic surface of each protein (heavy lines) is in the membrane and the polar region (indicated by + and −) are external. Peripheral proteins (e.g., (e)) are on the surface and bound to a polar region of an integral protein. Integral proteins can drift laterally but cannot flip-flop.

The Genetic Material

The genetic material of any organism is the substance that carries the information determining the properties of that organism. Furthermore, it is responsible for transferring the genetic information from parent to progeny. In almost all organisms the genetic material is DNA; the exceptions are a few bacteriophages and numerous plant and animal viruses in which the genetic material is RNA. Two classic experiments led to the identification of DNA as the genetic material and, in so doing, laid the foundation for molecular genetics. These experiments are described in the following section.

IDENTIFICATION OF DNA AS THE GENETIC MATERIAL

The Transformation Experiments

The development of the current idea that DNA is the genetic material began with an observation in 1928 by Fred Griffith, who was studying the bacterium responsible for human pneumonia—i.e., *Streptococcus pneumoniae* or *Pneumococcus*. The virulence of this bacterium was known to be dependent on a surrounding polysaccharide capsule that protects it from the defense systems of the body. This capsule also causes the bacterium to produce smooth-edged (S) colonies on an agar surface. It was known that mice were normally killed by S bacteria (Figure 6-1(a)). Griffith then isolated a rough-edged (R) colony mutant, which proved to be both nonencapsulated and nonlethal (Figure 7-4(b)). He subsequently made a significant observation—namely, whereas both R and heat-killed S were nonlethal, a mixture of live R and heat-killed S was lethal (Figure 6-1(c,d)). Furthermore, the bacteria isolated from a mouse that had died from such a mixed infection were only S—i.e., the live R had somehow been replaced by or **transformed** to S bacteria. Several years later it was shown that the mouse itself was not needed to mediate this transformation because when a mixture of R and heat-killed S was grown in a culture fluid, living S cells were produced. A possible explanation for this surprising phenomenon was that the R cells restored the viability of the dead S cells; but this idea was eliminated by the observation that living S cells grew even when the heat-killed S culture in the mixture was replaced by a cell *extract*

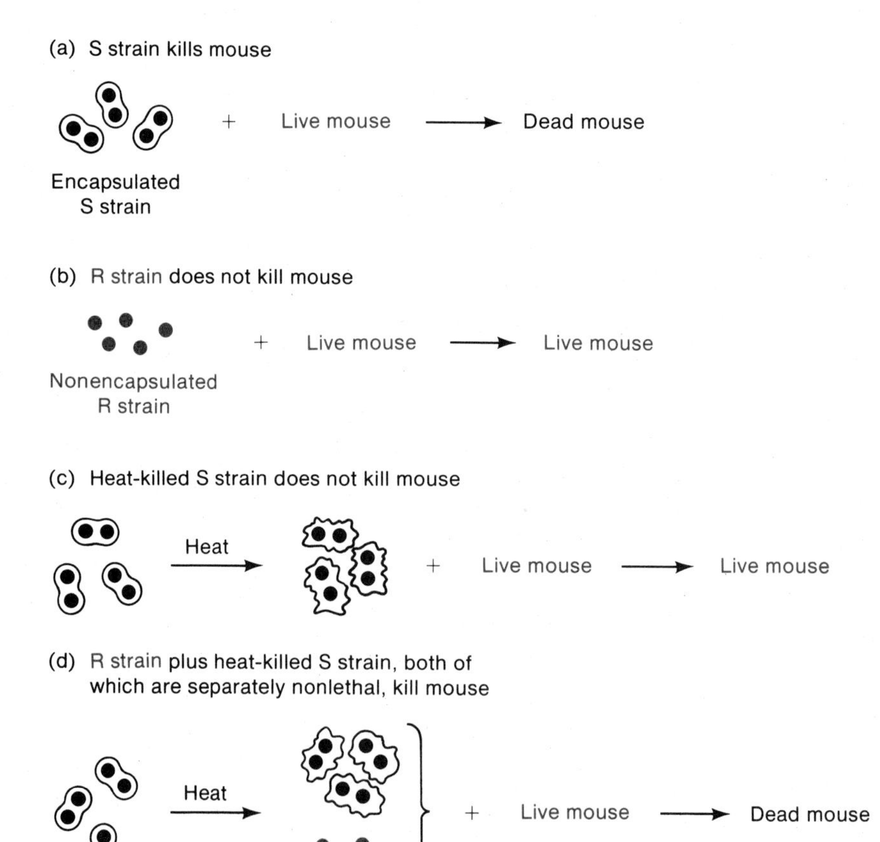

Figure 6-1
The Griffith experiment showing conversion of a nonlethal bacterial strain to a lethal form by a cell extract.

prepared from broken S cells, which had been freed from both intact cells and the capsular polysaccharide by centrifugation (Figure 6-2). Hence, it was concluded that the cell extract contained a **transforming principle,** the nature of which was unknown.

The next development occurred some 15 years later when Oswald Avery, Colin MacLeod, and Maclyn McCarty partially purified the transforming principle from the cell extract and demonstrated that it was DNA. These workers modified known schemes for isolating DNA and prepared samples of DNA from S bacteria. They added this DNA to a live R bacterial culture; after a period of time they placed a sample of the S-containing R bacterial culture on an agar surface and allowed it to grow to form colonies. Some of the colonies (about 1 in 10^4) that grew were S type (Figure 6-2). To show that this was a permanent genetic change, they dispersed many of the newly formed S colonies and placed them on a second agar surface. The resulting colonies were again S type. If an R colony arising from the original mixture was dispersed, only R bacteria grew in subsequent generations. Hence the R colonies retained the R character, whereas the transformed S colonies bred true as S. Because S and R colonies differed by a polysaccharide coat around each S bacterium, the ability of purified polysaccharide to transform was also tested, but no transformation was observed. Since the procedures for isolating DNA then in use produced DNA containing many impurities, it was necessary to provide evidence that the transformation was actually caused by the DNA alone.

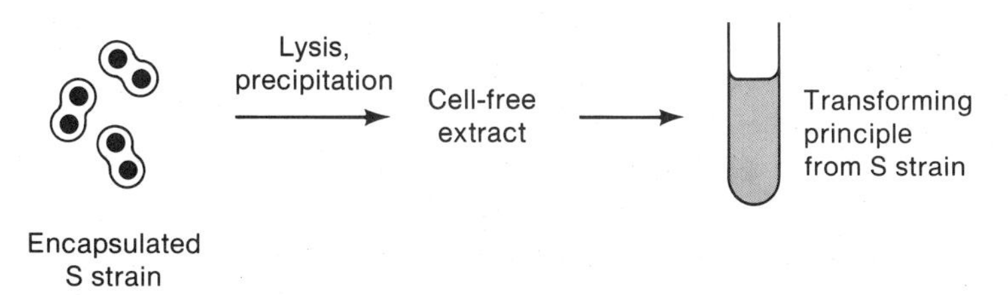

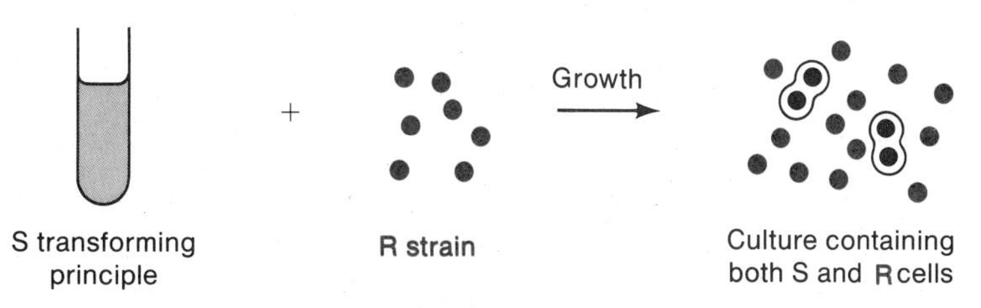

Figure 6-2
The transformation experiment.

This evidence was provided by the following four procedures.

1. Chemical analysis showed that the major component was a deoxyribose-containing nucleic acid.
2. Physical measurements showed that the sample contained a highly viscous substance having the properties of DNA.
3. Experiments demonstrated that transforming activity is not lost by reaction with either (a) purified proteolytic (protein-hydrolyzing) enzymes—trypsin, chymotrypsin, or a mixture of both—or (b) ribonuclease (an enzyme that depolymerizes RNA).
4. It was demonstrated that treatment with materials known to contain DNA-depolymerizing activity (DNase) inactivated the transforming principle.

The transformation experiment was not accepted by the scientific community as proof that DNA is the genetic material, because it was widely believed that DNA was a simple tetranucleotide incapable of carrying the information required of a genetic substance. The tetranucleotide hypothesis was based on chemical analyses that indicated that DNA consists of equimolar amounts of the four bases; this conclusion was based on inadequate chemical procedures for analyzing base composition and on the use of higher organisms as sources of DNA; in these organisms the base composition is not far from being equimolar. However, as techniques improved and a greater range of organisms was examined, it was found that the base composition varies widely from one species to the next and that DNA is not a simple small molecule. With these results, the idea that DNA is the genetic material became acceptable. The following experiment provided the final proof.

The Blendor Experiment

An elegant confirmation of the genetic nature of DNA came from an experiment with *E. coli* phage T2. This experiment, known as the **blendor experiment** because a kitchen blendor was used as a major piece of apparatus, was performed by Alfred Hershey and Martha Chase, who demonstrated that the DNA injected by a phage particle into a bacterium contains all of the information required to synthesize progeny phage particles.

A single particle of phage T2 consists of DNA (now known to be a single molecule) encased in a protein shell (Figure 6-3(a)). The DNA is the only phosphorus-containing substance in the phage particle; the proteins of the shell, which contain the amino acids methionine and cysteine, have the only sulfur atoms. Thus, by growing phage in a nutrient medium in which radioactive phosphate ($^{32}PO_4^{3-}$) is the sole source of phosphorus, phage containing radioactive DNA can be

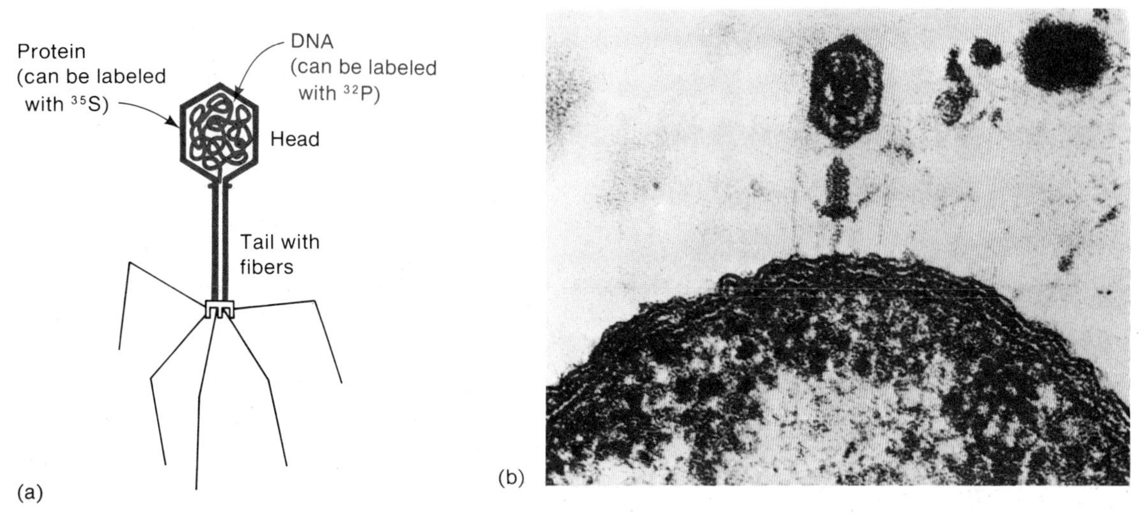

Figure 6-3
(a) A diagram of T2 phage. (b) Phage T2 adsorbed to the surface of *E. coli.* (Courtesy of Lee Simon and Thomas Anderson.)

prepared. If instead the growth medium contains radioactive sulfur as $^{35}SO_4^{3-}$, phage containing radioactive proteins are obtained. If these two kinds of labeled phage are used in an infection of a bacterial host, the phage DNA and the protein molecules can always be located by their radioactivity. Hershey and Chase used these phage to show that ^{32}P but not ^{35}S is injected into the bacterium.

Each phage T2 particle has a long tail by which it attaches to sensitive bacteria (Figure 6-3(b)). Hershey and Chase showed that an attached phage can be torn from a bacterial cell wall by violent agitation of the infected cells in a kitchen blendor. Thus, it was possible to separate an adsorbed phage from a bacterium and determine the component(s) of the phage that could not be shaken free by agitation—presumably, those components had been injected into the bacterium.

In the first experiment ^{35}S-labeled phage particles were adsorbed to bacteria for a few minutes. The bacteria were separated from unadsorbed phage and phage fragments by centrifuging the mixture and collecting the sediment (the pellet), which consisted of the phage-bacterium complexes. These complexes were resuspended in liquid and blended. The suspension was again centrifuged, and the pellet, which now consisted almost entirely of bacteria, and the supernatant were collected. It was found that 80 percent of the ^{35}S label was in the supernatant and 20 percent was in the pellet. The 20 percent of the ^{35}S that remains associated with the bacteria was shown some years later to consist mostly of phage tail fragments that adhered too tightly to the bacterial surface to be removed by the blending. A very different result

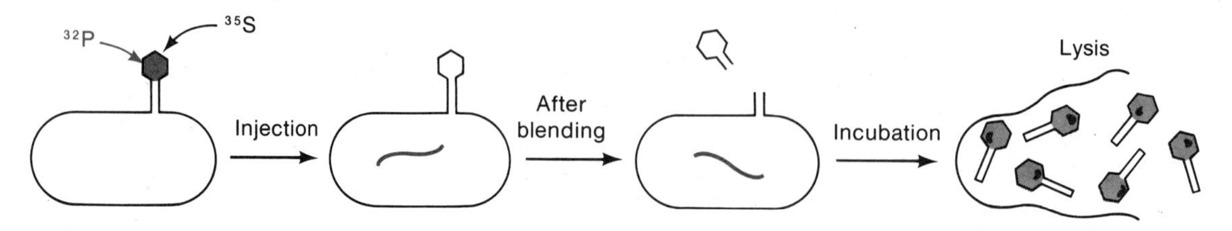

Figure 6-4
The blendor experiment. The parental DNA molecule is drawn in red. During incubation the DNA replicates and progeny phage are produced. Some, but not all, of the parental DNA appears in these progeny. Although not known to Hershey and Chase, progeny DNA molecules engage in genetic recombination resulting in dispersal of parental DNA among the progeny phage, as shown.

was observed when the phage population was labeled with ^{32}P. In this case 70 percent of the ^{32}P remained associated with the bacteria in the pellet after blending and only 30 percent was in the supernatant. Of the radioactivity in the supernatant roughly one-third could be accounted for by breakage of the bacteria during the blending. (The remainder was shown some years later to be a result of defective phage particles that could not inject their DNA.) When the pellet material was resuspended in growth medium and reincubated, it was found to be capable of phage production. Thus, the ability of a bacterium to synthesize progeny phage is associated with transfer of ^{32}P, and hence of DNA, from parental phage to the bacteria (Figure 6-4).

Another series of experiments, known as **transfer experiments,** supported the interpretation that genetic material contains ^{32}P but not ^{35}S. In these experiments progeny phage were isolated, after blending, from cells that had been infected with either ^{35}S- or ^{32}P-containing phage and the progeny were then assayed for radioactivity, the idea being that some parental genetic material should be found in the progeny. It was found that no ^{35}S but about half of the injected ^{32}P was transferred to the progeny. This result indicated that though ^{35}S might be residually associated with the phage-infected bacteria, it was not part of the phage genetic material. The interpretation (now known to be correct) of the transfer of only half of the ^{32}P was that progeny DNA is selected at random for packaging into protein coats and that all progeny DNA is not successfully packaged.

PROPERTIES OF THE GENETIC MATERIAL

The genetic material must have the following properties:

1. Ability to store genetic information and to transmit it to the cell as needed.

2. Ability to transfer its information to daughter cells with minimal error.
3. Physical and chemical stability, so that information is not lost.
4. Capability for genetic change, though without major loss of parental information.

We will see in the following sections that DNA is particularly suited to be the genetic material.

Storage and Transmission of Genetic Information by DNA

The information possessed and conveyed by the DNA of a cell is of several types:

1. The sequence of amino acids in every protein synthesized by the cell.
2. A start and a stop signal for the synthesis of each protein.
3. A set of signals that determine whether a particular protein is to be made and how many molecules are to be made per unit time.

This information is contained in the sequence of DNA bases. The amino acid sequence and the start and stop signals are not obtained directly from the DNA base sequence but via RNA intermediates. That is, DNA serves as a template for synthesis of specific RNA molecules called messenger RNA. The base sequence of the mRNA is complementary to that of one of the DNA strands and it is from the mRNA sequence that the amino acid sequence is translated. In particular, the base sequence is "read" in order in groups of three bases called **triplets** or **codons** and each group corresponds to a particular amino acid or to a start or stop codon. This two-stage process has the advantage that the DNA molecule neither has to be used very often nor has to enter the protein-synthetic mechanism. Since protein synthesis occurs continually, one can appreciate why DNA evolved with bases having groups capable of forming hydrogen bonds. First, by specific patterns of hydrogen-bonding—that is, base-pairing—the base sequence of DNA is easily transcribed into an RNA molecule by means of an enzymatic system that adds a base to the polymerizing end of an RNA molecule only if that base can hydrogen-bond with the base being copied. This system of transcription is discussed in detail in Chapter 9. The use of hydrogen-bonding enables the cell to use less genetic material to hold information than if van der Waals forces had been selected in the course of evolution; this is because van der Waals forces are so weak that the error frequency in transmitting information would be much greater than with hydrogen bonds unless more bases (or other molecules) were used for complementary binding.

Transmission of Information from Parent to Progeny

When a cell divides, each daughter cell must contain identical genetic information; that is, each DNA molecule must become two identical molecules, each carrying the information that was contained in the parent molecule. This duplication process is called **replication.** Once again, the value of bases capable of hydrogen-bonding is apparent. That is, to make a replica, the replication system need only require that the base being added to the growing end of a new chain be capable of hydrogen-bonding to the base being copied.

Chemical Stability of DNA and of Its Information Content

In long-lived organisms a single DNA molecule may have to last one hundred years or more. Furthermore, the information contained in the molecule is passed on to successive generations for millions of years with only small changes. Thus, DNA molecules must have great stability.

The sugar-phosphate backbone of DNA is extremely stable. The C—C bonds in the sugar are resistant to chemical attack under all conditions other than strong acid at very high temperatures. The phosphodiester bond is a little less stable and can be hydrolyzed at room temperature at pH 2, but this is not a physiological condition. In considering the stability of the phosphodiester bond, one can see immediately why 2′-deoxyribose rather than ribose is the constituent sugar in DNA. The phosphodiester bond in RNA is rapidly broken in alkali. The chemical mechanism for this breakage requires the OH group on the 2′ carbon of the sugar. In DNA, because deoxyribose is used, there is no 2′–OH group, so the molecule is exceedingly resistant to alkaline hydrolysis. The N–glycosylic bond is also very stable, though it would not be so if the bases were not rings.

An alteration in the chemical structure of a base definitely means loss of genetic information. In a cell there are certainly a large number of chemical compounds that can attack a free base. In considering this problem one can see immediately the value of the double helix. One aspect of this is that the molecule is redundant in the sense that identical information is contained in both strands; that is, the base sequence of one strand is complementary to that of the other strand. In fact, there exist in cells elegant repair systems that can remove an altered base and then by reading the sequence on the complementary strand, can replace the correct base. (Repair will be discussed in detail in Chapter 8.) The more important aspect of this double-helical structure is that the nature of the bases and the duplex structure of DNA provides extreme protection against chemical attack. The bases are hydrophobic rings having charged groups that contain the genetic information. It is therefore the charged groups that need protection. The hydrophobic nature of the bases causes the bases to stack so extensively that water is almost completely excluded from the stacked array. This has the effect that water-soluble compounds are often unable to come into close contact with the "dry" stack of bases. Certainly the likelihood of reaching the hydrogen-bonded charged groups is small.

The bases themselves, with the exception of cytosine, are very stable. At a very low rate, however, cytosine is deaminated to form uracil:

$$\text{Cytosine} + H_2O \longrightarrow \text{Uracil} + NH_3$$

Cytosine **Uracil**

Deamination is a disastrous change because the deamination product, uracil, pairs with adenine rather than with guanine. This has two effects: (1) an incorrect base will appear in mRNA and (2) an adenine instead of a guanine will occur in newly replicated DNA strands (Figure 6-5). There exists an intracellular system, which will be described in Chapter 7, for removing uracil from DNA and replacing it with thymine.

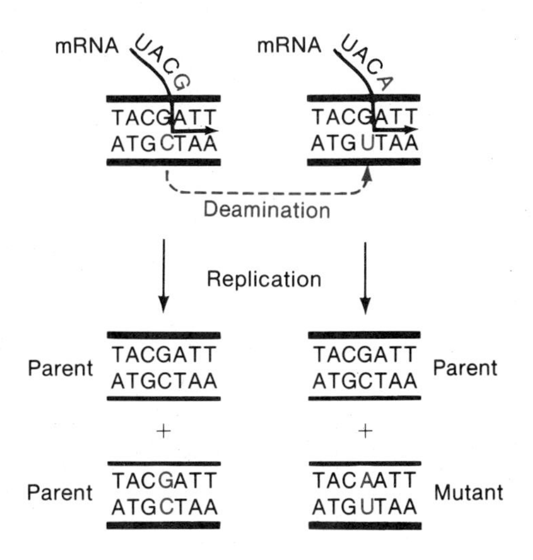

Figure 6-5
The effect of deamination of cytosine to form uracil in the base sequences of mRNA and of two daughter DNA molecules. The C→U transition is shown in red in the uppermost panel. "Parent" and "mutant" refer to base sequences before and after deamination. Newly replicated DNA is indicated by a thin line.

The necessity for eliminating a uracil formed by deamination explains why DNA utilizes thymine and not uracil as the base complementary to adenine. If uracil were the normal DNA base, there would be no way to distinguish a correct uracil from an incorrect uracil produced by deamination of cytosine. By using thymine, the cell follows the rule:

> Always remove a uracil from DNA because it is unwanted.

It is not obvious why RNA uses uracil and not thymine nor why DNA evolved with cytosine rather than with some base that would not deaminate. This may have been an evolutionary accident. It is possible, though, that the original DNA and RNA molecules both contained cytosine and uracil because these were the only pyrimidines available in the primordial sea. Cells then gained the ability to methylate uracil to form thymine because this provides a cell with a criterion for eliminating the result of a cytosine→uracil conversion. It is significant that the final step in the synthesis of thymidylic acid is the methylation of deoxyuridylic acid. The thymidylic acid is then converted to the triphosphate needed for incorporation into DNA.

The Ability of DNA to Change: Mutation

All of the information contained in a cell resides in its DNA. Thus, if a cell is to be able to improve through time—that is, to evolve—then the base sequence of its DNA must be capable of change. Furthermore, the new sequence must persist so that progeny cells will have the new property. The process by which a base sequence changes is called **mutation.** There are two main mechanisms of mutation:

1. A **chemical alteration** of the base that gives it new hydrogen-bonding properties and causes a new base to be present in a newly replicated daughter molecule.
2. A **replication error** by which an incorrect base or an extra base is accidentally inserted in the daughter molecules.

On the average, mutational changes are deleterious and lead to cell death. Therefore, it is important that not too many mutations occur in a single DNA molecule because otherwise the rare advantageous alteration would always occur in a cell destined to die by virtue of lethal mutations. Thus, the mutation rate must be low or controlled. This is accomplished in two ways. First, the hydrophobic, water-free core of the DNA molecule reduces the accessibility of the DNA to attacking molecules, as we discussed earlier. Second, the cell has evolved several repair mechanisms (discussed in Chapter 8) for correcting alterations and replication errors. These repair systems are not entirely efficient though and allow mutations to occur at a rate that is very low but useful in the long run. These mutations are, as we have said, usually deleterious, so that it is important that the parental information is not lost. Such loss is prevented in two ways: (1) The other members of the species retain the parental base sequence, and (2) a double-stranded molecule is redundant. Normally only one strand is altered, and DNA replicates in such a way that after cell division the DNA molecule in each daughter cell receives only one of the parental single strands. Thus, it is possible for one of the daughter DNA molecules to be normal and the other to be mutant; the mutant may not be able to survive but the cell with the parental base sequence will. If the mutant is better equipped to survive and multiply than the parent organisms or any other member of the species, then after a great many generations, Darwin's principle of survival of the fittest will lead to ultimate replacement of the parental genotype by the mutant phenotype in nature.

Occasional replication errors provide the second mechanism of mutation. These errors are usually corrected by systems described in Chapters 7 and 8. However, a small fraction of the errors persist and these give rise to mutations.

RNA AS THE GENETIC MATERIAL

In the preceding sections the properties of DNA that suit it to be the genetic material have been described. RNA is clearly not as suitable but nonetheless has survived in some organisms even though it lacks beneficial features of the base protection and the redundancy associated with a double helix (though one virus and one phage are known to have double-stranded RNA). It is noteworthy that RNA is not present as the genetic material in cellular organisms, for which chemical stability is required, but only in phages and viruses, in which the RNA is protected from the environment by a protein coat. Furthermore, RNA molecules spend most of their time as inert particles, replicating only infrequently in host organisms. When they do replicate, they do so very rapidly, so that an enormous number of progeny particles are produced in a short period of time and sheer numbers compensate for the lesser chemical stability of RNA. Thus the RNA phages and viruses have evolved special compensatory features that enable them to survive despite their deficiencies relative to DNA-containing organisms.

7 DNA Replication

Genetic information is transferred from parent to progeny organisms by a faithful replication of the parental DNA molecules. Usually the information resides in one or more double-stranded DNA molecules. Some bacteriophage species contain single-stranded instead of double-stranded DNA. In these systems replication consists of several stages in which single-stranded DNA is first converted to a double-stranded molecule, which then serves as a template for synthesis of single strands identical to the parent molecule.

Replication of double-stranded DNA is a complicated process that is not completely understood. This complexity is, at least in part, an acknowledgement of the importance of the following facts: (1) replication requires a supply of energy to unwind the helix; (2) single-stranded DNA tends to form intrastrand base pairs; (3) a single enzyme can catalyze only a limited number of physical and chemical reactions; (4) several safeguards have evolved that are designed both to prevent replication errors and to eliminate the rare errors that do occur; and (5) both circularity and the enormous size of DNA molecules impose geometric constraints on the replicative system, and how these fit into the system has to be understood. To add to the difficulty for the researcher, there is not a unique mode of replication common to all organisms having double-stranded DNA.

All genetically relevant information contained in a nucleic acid molecule resides in its base sequence, so the prime role of any mode of replication is to duplicate the base sequence of the parent molecule. The specificity of base pairing—adenine with thymine and guanine with cytosine—provides the mechanism used by all replication systems. Furthermore,

1. Nucleotide monomers are added one by one to the end of a growing strand by an enzyme called a **DNA polymerase.**
2. The sequence of bases in each new or **daughter strand** is complementary to the base sequence in the old or **parent strand** being copied—that is, if there is an adenine in the parent strand, a thymine will be added to the end of the growing daughter strand when the adenine is being copied.

In the following section we consider how the two strands of a daughter molecule are physically related to the two strands of the parent molecule.

SEMICONSERVATIVE REPLICATION OF DOUBLE-STRANDED DNA

The purpose of DNA replication is to create daughter DNA molecules that are identical to the parental molecule. This is accomplished by **semiconservative replication.** In this replication mode each parental single strand is a template for the synthesis of one new or daughter strand and as each new strand is formed, it is hydrogen-bonded to its parental template (Figure 7-1). Thus, as replication proceeds, the parental double helix unwinds and then rewinds again into two new double helices, each of which contains one originally parental strand and one newly formed daughter strand.

Unwinding a double helix in semiconservative replication is a mechanical problem. Either the two daughter branches at the Y-fork shown in Figure 7-1 must revolve around one another or the unreplicated portion must rotate. If the molecule were fully extended in solution, there would be no problem and rotation of the unreplicated portion would be the simpler motion, as there would be less friction with the solvent. However, since the molecule is 600 times longer than the cell that contains it (so that it must be repeatedly folded, as was shown in Chapter 5), such rotation is unlikely. A simple solution would be to make a single-strand break in one parental strand ahead of the growing fork; this would enable a small segment of the unreplicated region to rotate, thereby eliminating the geometric problem (Figure 7-2). The only requirement would then be to re-form the broken bond. However, this repair would have to occur before the replication fork

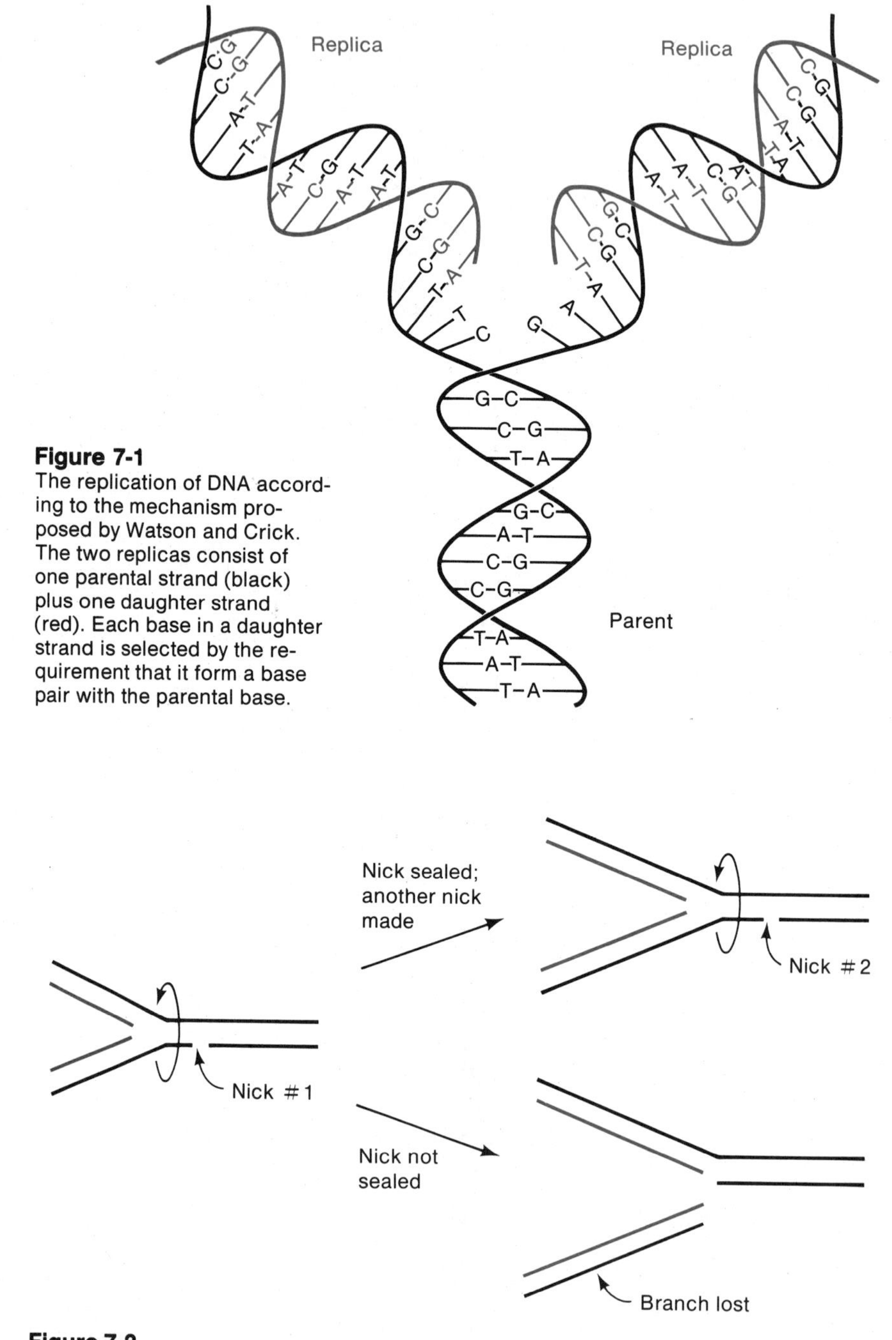

Figure 7-1
The replication of DNA according to the mechanism proposed by Watson and Crick. The two replicas consist of one parental strand (black) plus one daughter strand (red). Each base in a daughter strand is selected by the requirement that it form a base pair with the parental base.

Figure 7-2
Mechanism by which a nick ahead of a replication fork allows rotation. If the nick is not sealed, a newly formed branch is lost.

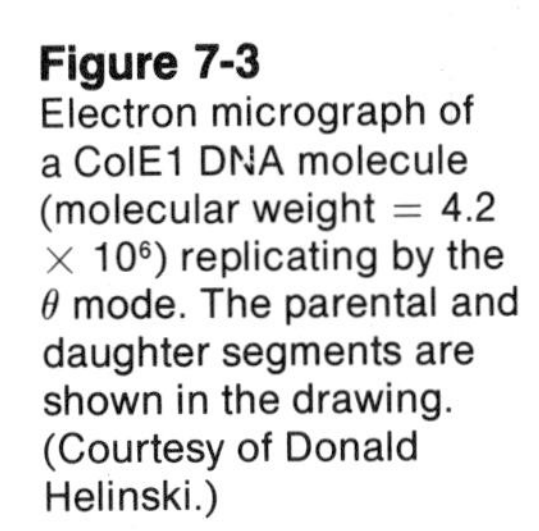

Figure 7-3
Electron micrograph of a ColE1 DNA molecule (molecular weight = 4.2 × 10^6) replicating by the θ mode. The parental and daughter segments are shown in the drawing. (Courtesy of Donald Helinski.)

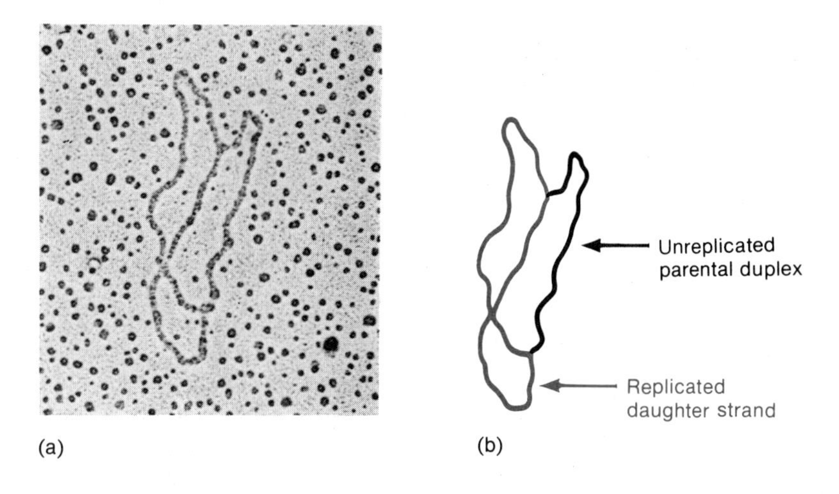

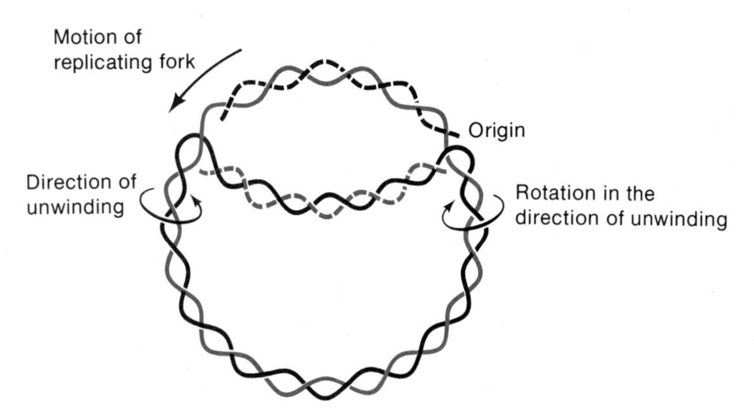

Figure 7-4
Drawing showing that the unwinding motion (curved arrows) of the daughter branches of a replicating circle lacking positions at which free rotation can occur causes overwinding of the unreplicated portion.

had passed the nick; otherwise a daughter strand would be lost, as shown in the figure.

Most DNA molecules replicate as circular structures (Figures 7-3 and 7-4). These circles resemble the Greek letter θ (theta), so the term **θ replication** is used to describe this replication mode. Replication of a circle introduces a topological problem that is more severe than that for a linear molecule. This problem is solved by an enzyme, DNA gyrase.

The unwinding problem in θ replication is formidable because lack of a free end makes rotation of the unreplicated portion impossible. An advancing nick, as in Figure 7-2, or a swivel at the replication origin (the point of initiation of replication) would solve the problem. These suggestions, though not quite correct, anticipate the correct mechanism.

As replication of the two daughter strands proceeds along the helix, in the absence of some kind of swiveling the nongrowing ends of the daughter strands would cause the entire unreplicated portion of the molecule to become overwound (Figure 7-4). This in turn would cause *positive* supercoiling (see Chapter 3) of the unreplicated portion. This supercoiling obviously cannot increase indefinitely because, if it were to do so, the unreplicated portion would become coiled so tightly that no further advance of the replication fork would be possible. This topological constraint ought to be avoidable by the simple nicking-sealing cycle just described but, in general, the twists are removed in another way. As discussed in Chapter 3, most naturally occurring circular DNA molecules are *negatively* supercoiled. Thus, initially the overwinding motion is no problem because it can be taken up by the underwinding already present in the negative supercoil. However, after about 5 percent of the circle is replicated, the negative superhelicity is used up and the topological problem arises.

Most organisms contain one or more enzymes called **topoisomerases.** These enzymes can produce a variety of topological changes in DNA; the most common are production of negative superhelicity and the removal of superhelicity. In *E. coli* there is an enzyme called **DNA gyrase** (Eco topoisomerase II) which is able to produce negative superhelicity, and it is DNA gyrase that is responsible for removing the positive superhelicity generated during replication. The evidence for this point comes from experiments using drugs that inhibit DNA gyrase (e.g., coumermycin, nalidixic acid, oxolinic acid, and novobiocin) and from the study of gyrase mutants. Addition of any of these drugs to a growing bacterial culture inhibits DNA synthesis; in a phage-infected cell they prevent supercoiling of injected phage DNA molecules (that is, the molecules circularize but do not supercoil). Proof that these effects are due to inhibition of DNA gyrase comes from isolating a strain of *E. coli* that can grow normally in the presence of these antibiotics. With this strain the antibiotics have no effect on either DNA replication or supercoiling. If DNA gyrase is isolated from both wild-type *E. coli* and the mutant *E. coli* strain and each form is tested for binding of the antibiotics, it is found that only the wild-type enzyme binds the antibiotic. (This approach of probing for an activity with an inhibitor and comparing *in vitro* and *in vivo* results is used in many problems in molecular biology.)

DNA gyrase does have a nicking-sealing activity but this activity cannot by itself introduce negative superhelical turns. How twisting is accomplished by this enzyme is explained in *MB*, pages 262–266.

ENZYMOLOGY OF DNA REPLICATION

The enzymatic synthesis of DNA is a complex process, primarily because of the need for high fidelity in copying the base sequence and for physical separation of the parental strands. The number of steps that must be completed is far too great to be accomplished by a single enzyme and, in fact, about twenty proteins are known at present to be necessary. Thus, in an effort to provide some understanding with a minimum of confusion, each step in the process will be treated separately. We will consider the basic chemistry of polymerization, the source of the precursors, the problems raised by the chemistry of polymerization, the means of initiating and terminating synthesis, and the mechanisms for eliminating replication errors.

The Polymerization Reaction and the Polymerases

In 1957, Arthur Kornberg showed that in extracts of *E. coli* there exists a DNA polymerase (now called **polymerase I** or **pol I**). This enzyme was able to synthesize DNA from four precursor molecules—namely, the four deoxynucleoside 5′-triphosphates (dNTP), dATP, dGTP, dCTP, and dTTP—as long as a DNA molecule to be copied (a **template** DNA) was provided. Neither 5′-monophosphates nor 5′-diphosphates, nor 3′-(mono-, di-, or tri-) phosphates can be polymerized—only the 5′-triphosphates are substrates for the polymerization reaction; soon we will see why this is the case. Some years later, it was found that pol I, though playing an essential role in the replication process, is not the major polymerase in *E. coli*; instead, the enzyme responsible for advance of the replication fork is polymerase III or pol III.* Pol III also exclusively uses 5′-triphosphates as precursors and requires a DNA template before polymerization can occur. Pol I and pol III have many features in common and, in fact, a few types of DNA molecules replicate by using only pol I. The overall chemical reaction catalyzed by both DNA polymerases is:

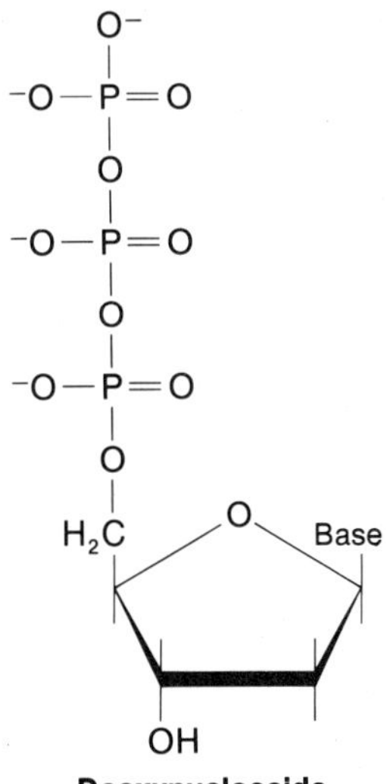

$$\text{Poly(nucleotide)}_n\text{-3'-OH} + \text{dNTP} \rightarrow \text{Poly(nucleotide)}_{n+1}\text{-3'-OH} + \text{PP}$$

in which PP represents pyrophosphate cleaved from the dNTP.

Deoxynucleoside triphosphate

*The numbers I and III refer only to the order in which the enzymes were first isolated, not to their relative importance. Another polymerase, pol II, has also been isolated from *E. coli*. It plays no role in DNA replication. Its biological function is unknown; mutant cells lacking pol II grow normally in the laboratory. For convenience, the abbreviation "pol" will be used in this chapter to refer to specifically named polymerases.

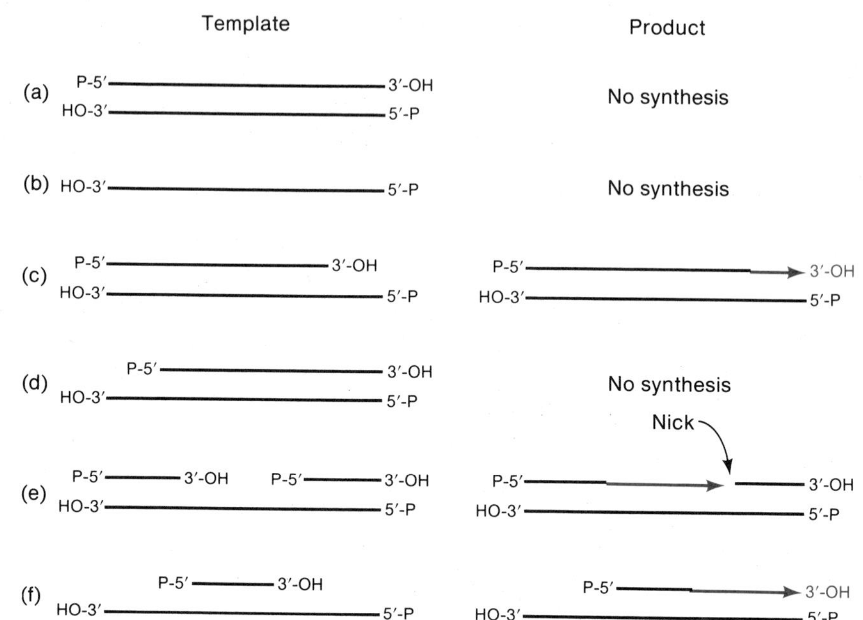

Figure 7-5
The effect of various templates used in DNA polymerization reactions. A free 3′-OH on a hydrogen-bonded nucleotide at the strand terminus and a non-hydrogen-bonded nucleotide at the adjacent position on the template strand is needed for strand growth. Newly synthesized DNA is red.

Pol I and pol III have many features in common. Both enzymes only polymerize deoxynucleoside 5′-triphosphates and can do so only while copying a template DNA. Furthermore, polymerization can only occur by addition to a **primer**—that is, an oligonucleotide hydrogen-bonded to the template strand and whose terminal 3′-OH group is available for reaction (that is, a "free" 3′-OH group). The meaning of a primer is made clear in Figure 7-5, which depicts six potential template molecules; of these, only three can be said to be active—(c), (e), and (f)—each of which has a free 3′-OH group. The lack of activity with (d) and the direction of synthesis with (e) and (f) indicate that nucleotides do not add to a free 5′-P group. The lack of any synthesis with (a) or (b) indicates that addition to a 3′-OH group cannot occur if there is nothing to copy. Thus, we draw two conclusions:

1. Both a primer with a free 3′-OH group and a template are needed.
2. Polymerization consists of a reaction between a 3′-OH group at the end of the growing strand and an incoming nucleoside 5′-triphosphate. When the nucleotide is added, it supplies another free 3′-OH group. Since each DNA strand has a 5′-P terminus and a 3′-OH terminus, strand growth is said to proceed in the 5′→3′ (5′-to-3′) direction.

These two points are summarized in Figure 7-6.

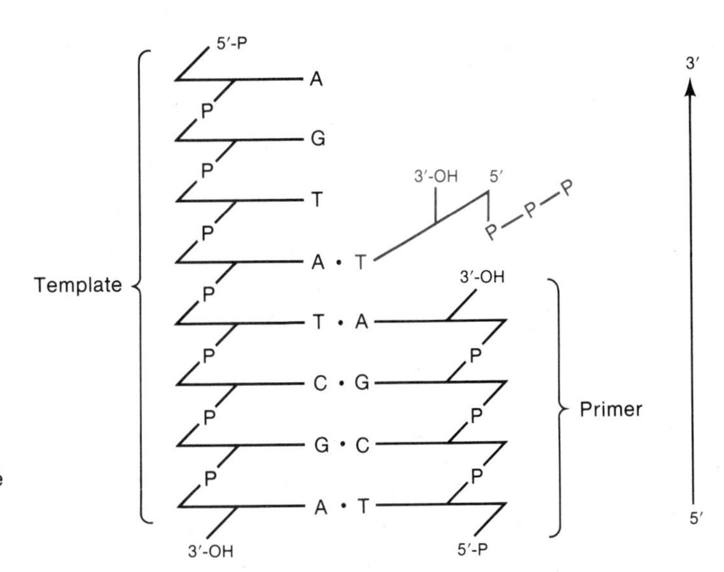

Figure 7-6
Schematic diagram of a replicating DNA molecule showing the distinction between template and primer and the meaning of 5′→3′ synthesis.

Occasionally polymerases add a nucleotide terminus that cannot hydrogen-bond to the corresponding base in the template strand. This may be purely a mistake or may result from the tautomerization of adenine and thymine discussed in Chapter 10. In any case, it is important that the unpaired base be removable while its incorrectness is recognizable—namely, as an unpaired base at the 3′-OH terminus of a growing strand.

Pol I responds to an unpaired terminal base by terminating polymerizing activity, because the enzyme requires a primer that is correctly hydrogen-bonded. When such an impasse is encountered, a 3′→5′ exonuclease activity, which may be thought of simply as pol I running backwards or in the 3′→5′ direction, is stimulated, and the unpaired base is removed (Figure 7-7). After removal of this base, the exonuclease activity stops, polymerizing activity is restored, and chain growth begins again. This exonuclease activity is called the **proofreading** or **editing function** of pol I.

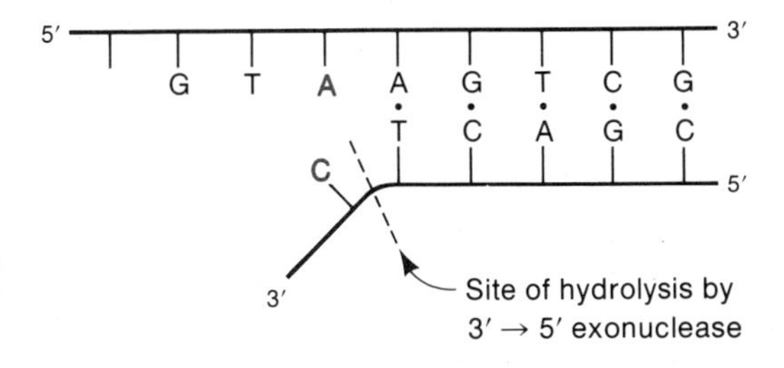

Figure 7-7
The 3′→5′ exonuclease activity of DNA polymerase I showing the site of hydrolysis. The C that is removed (red) does not base-pair with the A being copied (red).

Another function of polymerase I is that of a 5′→3′ exonuclease. This activity has the following features:

1. Nucleotides are removed from the 5′-P terminus only, one by one.
2. More than one nucleotide can be removed by successive cutting.
3. The nucleotide removed must have been base-paired.
4. The nucleotide removed can be either of the deoxy- or the ribo-type.
5. Activity can be at a nick as long as there is a 5′-P group.

We will see shortly that the main function of the 5′→3′ exonuclease activity is to remove ribonucleotide primers.

The 5′→3′ exonuclease activity at a single-strand break (nick) can occur simultaneously with polymerization. That is, as a 5′-P nucleotide is removed, a replacement can be made by the polymerizing activity (Figure 7-8). Since pol I cannot form a bond between a 3′-OH group and a 5′-monophosphate, the nick moves along the DNA molecule in the direction of synthesis. This movement is called **nick translation.**

Experimental conditions can be chosen so that polymerization will occur at a single-strand break without concomitant 5′→3′ exonuclease activity. The growing strand then displaces the parental strand (Figure 7-8). This is thought to be an important step in the mechanism of genetic recombination. Of all *E. coli* polymerases known to date, polymerase I is the only one capable of carrying out an unaided displacement reaction. In other strand displacement reactions, auxiliary proteins are required and ATP is cleaved to fuel the unwinding of the helix; this will be discussed when the events at a replication fork are described.

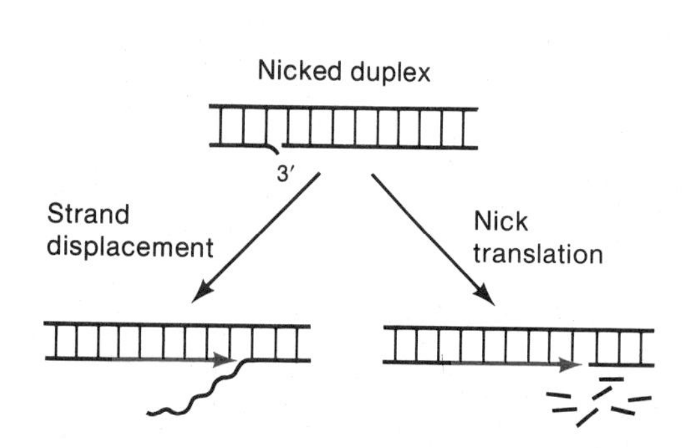

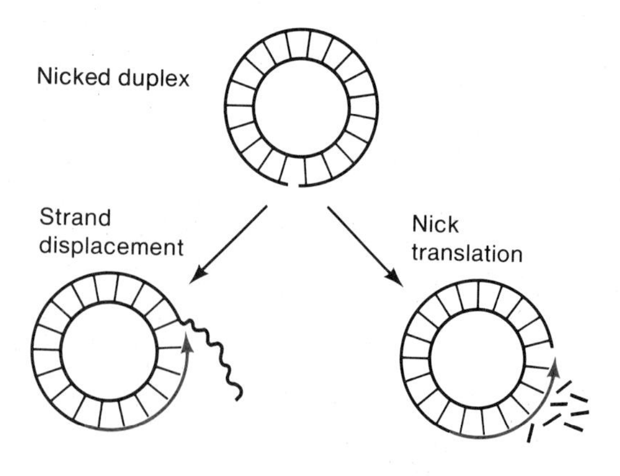

Figure 7-8
Strand displacement and nick translation on linear and circular molecules. In nick translation a nucleotide is exonucleolytically removed for each nucleotide added. The growing strand is shown in red.

Pol III is a very complex enzyme. In its most active form it is associated with eight other proteins to form the **pol III holoenzyme,** occasionally termed pol III. The term holoenzyme refers to an enzyme that contains several different subunits and retains some activity even when one or more subunits is missing. The smallest aggregate having enzymatic activity is called the **core enzyme.** The activities of the core enzyme and the holoenzyme are usually very different. Genes encoding five of the subunits have been identified; these are called *dnaE*, *dnaN*, *dnaQ*, *dnaX*, and *dnaZ*. The *dnaE* protein possesses the major polymerizing activity but each of the subunits, except for the *dnaQ* protein is essential for replication. Pol III shares with pol I a requirement for a template and a primer but its substrate specificity is much more limited. For instance, pol III cannot act at a nick nor is it active with single-stranded DNA primed by either a DNA or RNA nucleotide fragment. The principal activity *in vitro* is on gapped DNA in which the gap is less than 100 nucleotides long. Such a gap is akin to the state of the DNA at a replication fork—that is, the parental strands are separated and bear short single-stranded regions ahead of the growing daughter chain (Figure 7-1). Pol III cannot carry out strand displacement either, and another system is needed to unwind the helix in order that a replication fork will be able to proceed (this will be discussed in detail in a later section). The enzyme, like pol I, possesses a 3′→5′ exonuclease activity which performs the major editing function in DNA replication. This function is carried out by the *dnaE* subunit, which is also the major polymerizing subunit, as we have just mentioned. The *dnaQ* subunit plays an important role in editing, also, but the biochemical basis of the role is not yet known; it probably interacts with the *dnaE* subunit. The principal evidence supporting the view that it is involved in providing fidelity to the replication process is that bacteria containing a mutation in the *dnaQ* gene have a somewhat higher mutation frequency. Pol III also possesses a 5′→3′ exonuclease activity; however, the enzyme acts only on single-stranded DNA so that it cannot carry out nick translation. The biological role of the 5′→3′ exonuclease activity of pol III is unknown at present.

Although pol III holoenzyme is the major replicating enzyme in *E. coli*, much less is known about it than about pol I, because it is a more complex enzyme. Study of pol III is currently an active field of research.

Later in this chapter, when the events at the growing fork catalyzed by the *E. coli* replication system are examined, it will be seen that pol I and pol III holoenzyme are both essential for *E. coli* replication. However, a requirement for two polymerases is not common to all organisms; for instance, *E. coli* phage T4 synthesizes its own polymerase and this enzyme is capable of carrying out all necessary polymerization functions.

All known polymerases (for both DNA and RNA) are capable of chain growth in only the 5′→3′ direction; that is, the growing end of the polymer must have a free 3′-OH group. It is possible for the following reasons that the enzymes evolved in this way to facilitate editing. If 3′→5′ growth were to occur, the growing strand would also be terminated with a 5′-triphosphate and the 3′-OH group of the incoming nucleotide would react with it. Chemically this is certainly acceptable but since the bonds formed contain only a single phosphate, an editing function would leave a free 5′-monophosphate. In order for chain growth to proceed, an enzymatic system would be needed to enter the replication fork and convert the monophosphate to a triphosphate. There is already a great deal going on in the replication fork, so that it would seem more economical for the cell to require 5′→3′ growth exclusively. However, the observation that chain growth proceeds in only one direction introduces what is probably the greatest complication in the entire replication process; this will be described shortly.

DNA LIGASE

Neither replication from a primed circular single strand nor gap filling results in a continuous daughter strand. Discontinuity results because no known polymerase can join a 3′-OH and a 5′-monophosphate group. The joining of these groups is accomplished by the enzyme **DNA ligase,** which functions in replication and other important processes.

E. coli DNA ligase can join a 3′-OH group to a 5′-P group as long as both are termini of adjacent base-paired deoxynucleotides—the enzyme cannot bridge a gap (Figure 7-9).

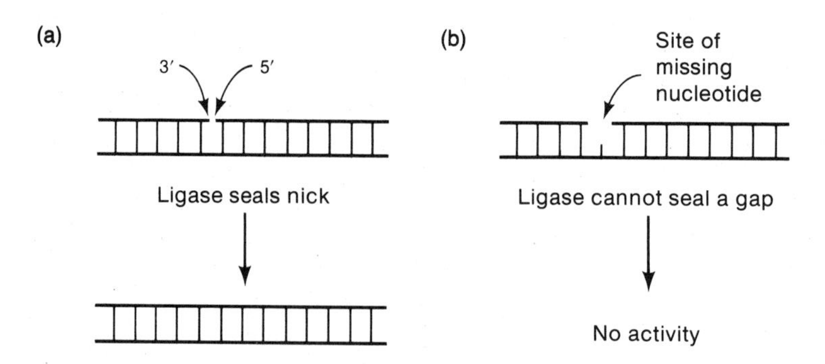

Figure 7-9
The action of DNA ligase. (a) A nick having a 3′-OH and a 5′-P terminus is sealed. (b) If one or more nucleotides are absent, the gap cannot be sealed.

In the usual polymerization reaction, the activation energy for phosphodiester bond formation comes from cleaving the triphosphate. Since DNA ligase has only a monophosphate to work with, it needs another source of energy. It obtains this energy by hydrolyzing either ATP or nicotine adenine dinucleotide (NAD); the energy source depends upon the organism from which the DNA ligase is obtained. The *E. coli* DNA ligase uses NAD.

DISCONTINUOUS REPLICATION

In the model of replication shown in Figure 7-1 both daughter strands are drawn as if replicating continuously. However, no known DNA molecule replicates in this way—instead,

> One of the daughter strands is made in short fragments, which are then joined together.

Polymerase I and polymerase III can add nucleotides only to a 3′-OH group. Examination of the growing fork indicates that if both daughter strands grew in the same overall direction, only one of these strands would have a free 3′-OH group; the other strand would have a free 5′-P group because the two strands of DNA are antiparallel (Figure 7-10). Thus one of the following must be true:

1. There is another polymerase that can add a nucleotide to the 5′ end; that is, it would catalyze strand growth in the 3′→5′ direction.

2. The two strands both grow in the 5′→3′ direction but from opposite ends of the parental molecule, as shown in Figure 7-11. If this

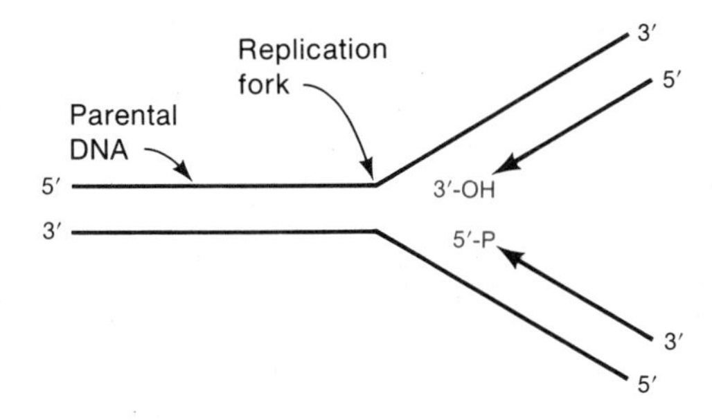

Figure 7-10
The termini (red) that would be present in a replication fork if both strands were to grow in the same overall direction.

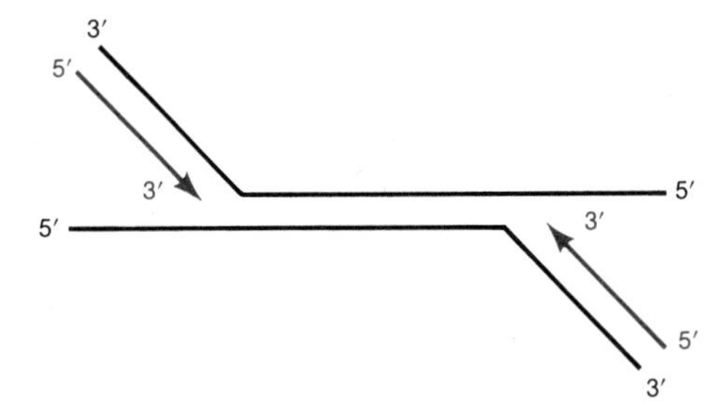

Figure 7-11
One way to replicate an antiparallel DNA molecule by means of 5′→3′ chain growth. Daughter strands are shown in red.

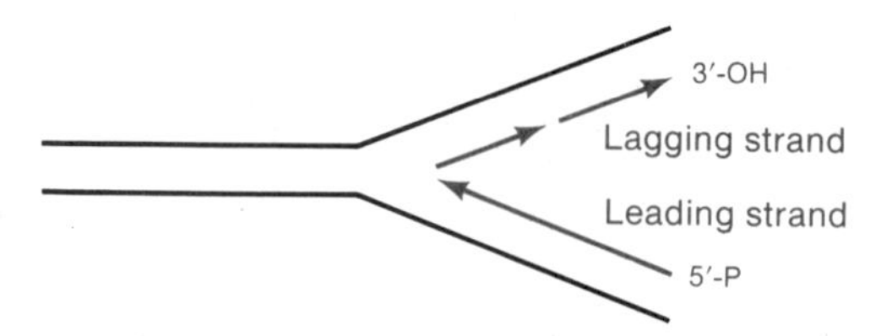

Figure 7-12
A growing fork showing the direction of growth of the leading and lagging strands (both in red).

were correct, a significant fraction of the unreplicated molecule would have to be single-stranded.

3. The two strands both grow in the 5′→3′ direction at a single growing fork and hence do not grow in the same direction along the parental molecule. One way to accomplish this is shown in Figure 7-12. This mode of synthesis is called **discontinuous synthesis.** The **leading strand** is synthesized continuously, and the **lagging strand** is synthesized discontinuously in the form of short fragments, each growing in the correct 5′→3′ direction. In time, the fragments can be joined together to form a continuous strand.

A great deal of evidence eliminates the first two models and supports the idea of discontinuous synthesis. No DNA polymerase has been discovered that adds nucleotides to the 5′-P terminus, and at most about 0.05 percent of intracellular DNA is ever in single-stranded form. The most critical evidence is the following series of experiments.

In 1968 R. Okazaki demonstrated that in *E. coli* newly synthesized DNA consists of fragments that later attach to one another to generate continuous strands. The presence of fragments supported the discontinuous replication model of Figure 7-12. Okazaki did two experiments to demonstrate this. In the first experiment, [^{3}H]dT was added to a growing bacterial culture in order to label the new DNA strands with radioactivity, and 30 seconds later the cells were collected and all of their DNA was isolated. This is called a **pulse-labeling experiment**. The DNA was then sedimented in alkali, which causes strand separation. The type of data obtained (Figure 7-13(a)) showed that the most recently made ("pulse-labeled") DNA sediments very slowly in comparison with single strands obtained from parental DNA (even though these strands usually break in the course of isolation); from the s value* it was estimated that the fragments of pulse-labeled DNA range in size from 1000 to 2000 nucleotides, whereas the isolated parental DNA

* The s value, or sedimentation coefficient, is a measure of the speed with which a particle moves under the influence of centrifugal force. For DNA, it can be related simply to molecular weight.

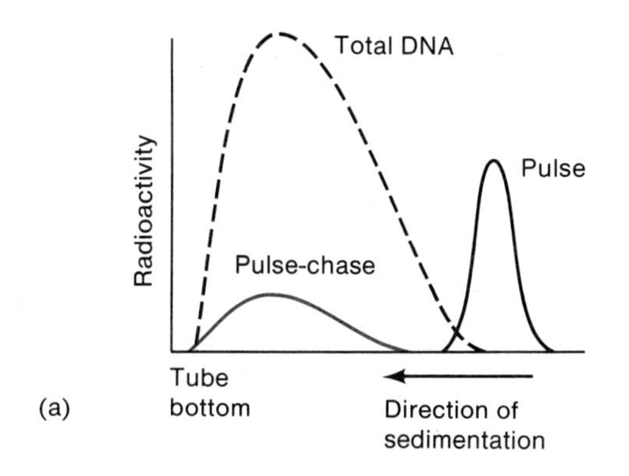

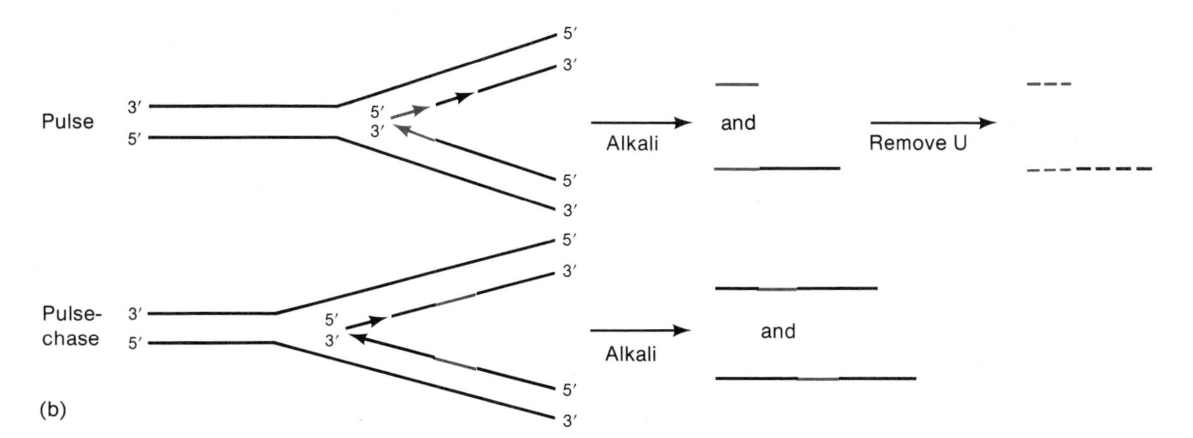

Figure 7-13
(a) The type of data obtained by alkaline sedimentation of pulse-labeled DNA (black) and pulse-chased DNA (red). Total DNA is the sum of the nonradioactive and radioactive DNA, as might be indicated by the optical absorbance. The *s* value of the sedimenting material increases from right to left.
(b) The location of radioactive DNA (red) at the time of pulse-labeling and after a chase. The radioactive molecules present in alkali are shown. The fragmentation resulting from removal of uracil (see later in text) accounts for the fact that all pulse-labeled DNA has a low *s* before the chase.

is usually 20 to 50 times as large. In the second experiment, the bacteria were pulse-labeled for 30 seconds; then, the $[^3H]$dT was replaced with nonradioactive dT, and the bacteria were allowed to grow for several minutes. This is called a **pulse-chase experiment** and it allows one to examine the current state of molecules synthesized at an earlier time. Okazaki observed that the *s* value of the radioactive material increased with time of growth in the nonradioactive medium. These experiments are presented and interpreted in Figure 7-13 in terms of the discontinuous replication model shown in Figure 7-12. Apparently the joining of the pulse-labeled fragments to the growing daughter strands caused the label to sediment with the bulk of the DNA.

These fragments, called **precursor fragments** or **Okazaki fragments,** have the properties predicted by the discontinuous model: they are initially small and then become large as they are attached to previously made DNA. However, the model predicts that only half of the radioactivity should be found in small fragments whereas the data shown in Figure 7-13(a) indicate that all of the newly synthesized DNA consists of fragments. This result should be surprising because there is no reason why the DNA of the 3′-OH-terminated strand should be synthesized discontinuously. In fact, it is not. Occasionally DNA polymerase adds a uracil nucleotide instead of a thymine nucleotide to the end of the growing strand when copying an adenine in the template strand. When this occurs, a repair system excises the uracil, replacing it with thymine. This repair system does not act at the terminus (in contrast with the proofreading 3′–5′ exonuclease) but instead when the uracil is well within the growing strand. The excision process produces a transient single-strand break, which is sealed by DNA ligase after the correct nucleotide is inserted. Thus, the leading strand is made continuously but, after synthesis, is fragmented at the rare uracil sites.

Additional evidence for the discontinuous-synthesis model comes from high-resolution electron micrographs of replicating DNA molecules, showing a short single-stranded region on one side of the replication fork (Figure 7-14). This region results from the fact that synthesis of the discontinuous strand is initiated only periodically; perhaps a particular base sequence or some other signal is required for initiation. In fact, the 3′-OH terminus of the continuously replicating strand is always ahead of the discontinuous strand. This is the origin of the terms leading strand and lagging strand for the continuously and discontinuously replicating strands, respectively.

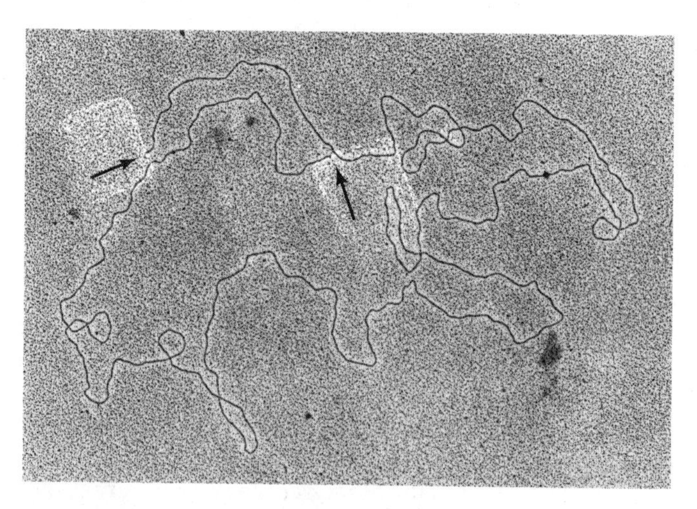

Figure 7-14
A replicating θ molecule of phage λ DNA. The arrows show the two replicating forks. The segment between each pair of thick lines at the arrows is single-stranded DNA; note that it appears thinner and lighter. (Courtesy of Manuel Valenzuela.)

Pol III cannot initiate chain growth, because, like all DNA polymerases, it needs a primer. This problem applies to initiating both the leading strand and the precursor fragments. Either a ribo- or deoxynucleotide, hydrogen-bonded to the template strand, could serve as a primer, because all known DNA polymerases can add nucleotides to the 3′-OH termini of both types. In every case examined, the primer for both leading and lagging strand synthesis is a short RNA oligonucleotide that consists of 1 to 60 bases; the exact number depends on the particular organism. This RNA primer is synthesized by copying a particular base sequence from one DNA strand and differs from a typical RNA molecule in that after its synthesis the *primer remains hydrogen-bonded to the DNA template.* In bacteria two different enzymes are known that synthesize primer RNA molecules—**RNA polymerase,** which is the same enzyme that is used for synthesis of most RNA molecules such as messenger RNA, and **primase** (the product of the *dnaG* gene). Experimentally, these enzymes can be distinguished *in vivo* by their differential sensitivities to the antibiotic rifampicin—RNA polymerase, but not primase, is inhibited by the antibiotic. In *E. coli,* initiation of leading-strand synthesis is rifampicin-sensitive, presumably because RNA polymerase is used. Initiation of precursor-fragment synthesis is resistant to the drug, as it uses primase. A precursor fragment has the following structure, while it is being synthesized:

Precursor fragments are ultimately joined to yield a continuous strand. This strand contains no ribonucleotides so that assembly of the lagging strand must require removal of the primer ribonucleotides, replacement with deoxynucleotides, and then joining. In *E. coli* the first two processes are accomplished by DNA pol I and joining is catalyzed by DNA ligase. How this is done is shown in Figure 7-15. Pol III extends the growing strand until the RNA nucleotide of the primer of the previously synthesized precursor fragment is reached. Pol III can go no further since its 5′→3′ exonuclease is inactive on base-paired DNA; it cannot join a 5′-triphosphate at the terminus of a polymer (i.e., on the primer) to a 3′-OH group on the growing strand and it cannot carry out strand displacement. Thus pol III dissociates from the DNA, leaving a nick. *E. coli* DNA ligase cannot seal the nick because a triphosphate is present; even if an additional enzyme could cleave the triphosphate to a monophosphate, DNA ligase would be inactive when one of the nucleotides is in the ribo form. However, pol I works efficiently at a nick as long as there is a 3′-OH terminus. In this case the enzyme carries out nick translation, probably proceeding into the deoxy section, but there

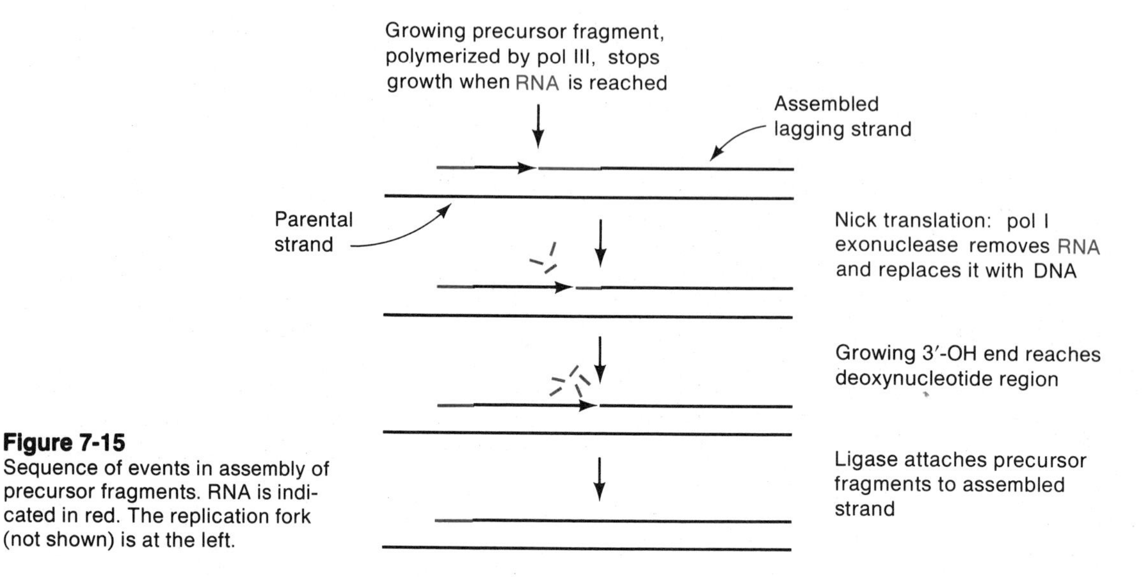

Figure 7-15
Sequence of events in assembly of precursor fragments. RNA is indicated in red. The replication fork (not shown) is at the left.

DNA ligase can compete with pol I and seal the nick. Thus, the precursor fragment is assimilated into the lagging strand. By this time, the next precursor fragment has reached the RNA primer of the fragment just joined and the sequence begins anew.

Synthesis of precursor fragments follows synthesis of the leading strand. In the next section, how the leading strand advances into the parental double helix is described. Once this is understood, we can return to the question of how RNA primer synthesis gets started—a surprisingly complicated process.

EVENTS IN THE REPLICATION FORK

DNA replication requires not only an enzymatic mechanism for adding nucleotides to the growing chains but also a means of unwinding the parental double helix. In this section we will see that these are distinct processes and that the unwinding mechanism is closely related to the initiation of synthesis of precursor fragments.

Polymerase III cannot carry out strand displacement, as has already been discussed, because it is unable to unwind the helix. In order for a helix to be unwound, hydrogen bonds and hydrophobic interactions must be eliminated, and this requires energy. Pol I is able to utilize both the free energy of hydrolysis of the triphosphate group and the binding energy of forming a new hydrogen bond to unwind the parental molecule as it synthesizes the leading strand. No other

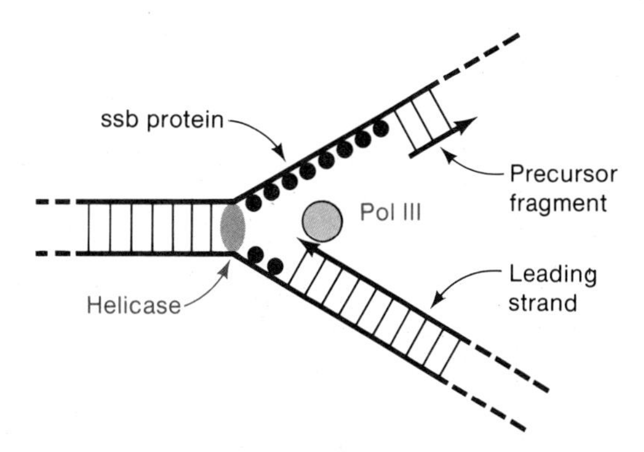

Figure 7-16
The unwinding events in a replication fork.

polymerase can do that; usually a helix must first be unwound in order for a DNA polymerase to advance. Helix-unwinding is accomplished by enzymes called **helicases.** The helicase active in *E. coli* DNA replication hydrolyzes ATP and utilizes the free energy of hydrolysis in an unknown way to unwind the helix, hydrolyzing two ATP molecules per base pair broken.

In *E. coli* the pol III molecule synthesizing the leading strand is not immediately behind the advancing helicase (Figure 7-16). Thus, the helicase leaves in its wake two single-stranded regions, a large one on the strand to be copied by precursor fragments and a smaller one just ahead of the leading strand. In order to prevent the single-stranded regions from annealing or from forming intrastrand hydrogen bonds, the single-stranded DNA is coated with *E. coli* single-strand binding protein (**ssb protein**). This is one of a class of binding proteins that bind tightly to both single-stranded DNA and to one another. As pol III advances, it displaces the ssb protein so that base pairing of the nucleotide being added can occur.

Some replication systems possess a single protein that is both a helicase and a ssb protein. This is true of *E. coli* phage T4, which synthesizes a protein called the gene-*32* protein. The 32-protein binds very tightly to single-stranded DNA and exceedingly tightly to itself. Its binding energy is great enough to unwind the helix.

The mechanism of initiation of replication of precursor fragments is closely connected to both the unwinding of the helicase and the action of the ssb protein. A direct study of initiation of synthesis of precursor fragments has not yet been carried out because experimentally the problem is very complex. Instead, attention has been directed to several *E. coli* phages having single-stranded DNA, in which the first step in replication is the conversion of single-stranded circular DNA to a double-stranded covalent circle. Since some of the phages studied use *E. coli* enzymes exclusively, it is thought that they can serve as a model for understanding initiation of precursor fragment synthesis.

The most complicated system studied so far and the one that is most closely related to *E. coli* DNA synthesis is that of the circular single-stranded DNA of phage ϕX174. In this system primer synthesis requires seven proteins. Primer synthesis begins by the formation of a complex called a **preprimosome,** which contains six proteins: the n, n′, n″, i, DnaB, and DnaC proteins. The protein n′ binds to single-stranded DNA and acquires a bound ATP molecule, and then primase joins the preprimosome, forming a unit called the **primosome.** The protein n′ uses the energy of ATP hydrolysis to move the primosome along the DNA until a priming site, which is chosen at random, is found. Then, the DnaB protein alters the structure of the DNA at the site that has been selected, and this alteration enables primase to initiate synthesis of an RNA primer. Replication then proceeds by pol III, pol I, and ligase, in sequence. The important characteristics of the prepriming reaction are: the lack of a specific sequence at which initiation occurs; the use of a complex of six proteins; the requirement for ATP; and synthesis of RNA by primase.

A SUMMARY OF EVENTS AT THE REPLICATION FORK

Figure 7-17 summarizes the proposed events that occur in or near a replication fork in *E. coli.* A helicase, driven by ATP hydrolysis and probably aided by binding of ssb protein, unwinds the helix. The unpaired bases are coated with ssb protein. The leading strand advances along one parental strand by nucleotide addition catalyzed by polymerase III. The DnaB protein complex moves along the other parental strand, prepriming it so that primase will synthesize a primer RNA. Polymerase III adds nucleotides to the primer, thereby synthesizing a precursor fragment. This synthesis continues up to the primer of the preceding precursor fragment; at this point pol I replaces pol III; by nick-translation the RNA is removed and replaced by DNA. Once the RNA is gone, DNA ligase seals the nick, thereby joining the precursor fragment to the lagging strand. The advance of the replication fork continues until replication is completed.

The reader must surely be impressed with the complexity of this process. However, in some systems replication is somewhat simpler. For instance, with *E. coli* phage T7 (discussed in *MB*, pages 304–306, 628–636), priming is accomplished by a phage RNA polymerase and all replication is performed by a single DNA polymerase encoded in the

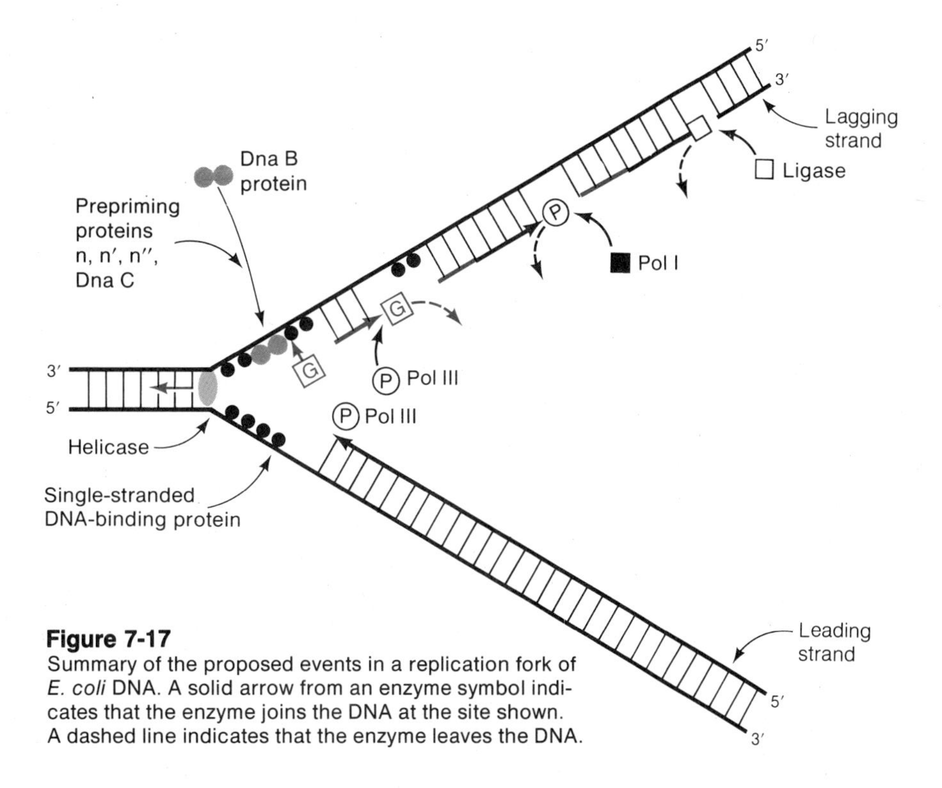

Figure 7-17
Summary of the proposed events in a replication fork of *E. coli* DNA. A solid arrow from an enzyme symbol indicates that the enzyme joins the DNA at the site shown. A dashed line indicates that the enzyme leaves the DNA.

phage DNA; none of the known *E. coli* enzymes are needed. Phages T4 and λ replicate in a somewhat more complicated manner than T7 but still the mechanism is far simpler than *E. coli* replication. Why should the process in *E. coli* be so complex? This may be a consequence of the requirement for fidelity. The larger the DNA molecule, the greater is the probability that a replication error will be made during a round of replication. Furthermore, a bacterial DNA is replicated only once per generation; a phage replicates many times in an infected cell, so that one defective replica would not significantly affect the successful outcome of the infection. Thus, in order to reduce the error frequency to a tolerable level, organisms with a larger DNA molecule and a single replication event need more replication proteins, since each protein can be designed to minimize or correct a particular type of error.

We now leave our analysis of the replication fork and examine how the fork itself is created at the outset; what is the process by which synthesis of double-stranded DNA is initiated? This is one of the least understood subjects of replication especially because it may occur in several ways; it is described in the following section.

INITIATION OF SYNTHESIS OF THE LEADING STRAND

All known double-stranded DNA molecules initiate a round of replication at a unique base sequence, called the **replication origin** or ***ori***. The sequence is specific to each organism, though there is one example in which two organisms have the same sequence.

Initiation can occur in two ways—***de novo* initiation,** in which the leading strand is started afresh, and **covalent extension,** in which the leading strand is covalently attached to a parental strand. We begin with *de novo* initiation.

De Novo Initiation

De novo initiation remains poorly understood. The single feature common to all bacterial and phage systems examined to date is the requirement for an RNA primer synthesized by the host RNA polymerase. Additional proteins are also needed. For instance, initiation of *E. coli* DNA synthesis requires the products of the genes *dnaA*, *dnaH*, *dnaJ*, *dnaK*, and *dnaP*. However, the biochemical activity of these gene products is not known at present.

All known DNA molecules, with only few exceptions, replicate as circles and hence initiate within the helix. Even those molecules that replicate as linear molecules initiate within the helix rather than at one end. In *de novo* initiation, leading-strand synthesis precedes that of the lagging strand. Thus, before synthesis of the first precursor fragment begins, a replication bubble exists that consists of one double-stranded branch, made up of one parental strand paired with the leading strand, and one single-stranded branch, which is the unreplicated, second parental strand (Figure 7-18). Since the leading strand displaces the unreplicated parental strand, the bubble is called a **displacement loop** or **D loop.** Such a configuration is ordinarily a transient one, existing only until synthesis of precursor fragments begins—which may require that the leading strand release a specific sequence (in single-stranded form) that can be used for prepriming. However, in certain circumstances, namely in a replication system that does not employ DNA gyrase to relieve topological constraints, a D loop may be long lived.

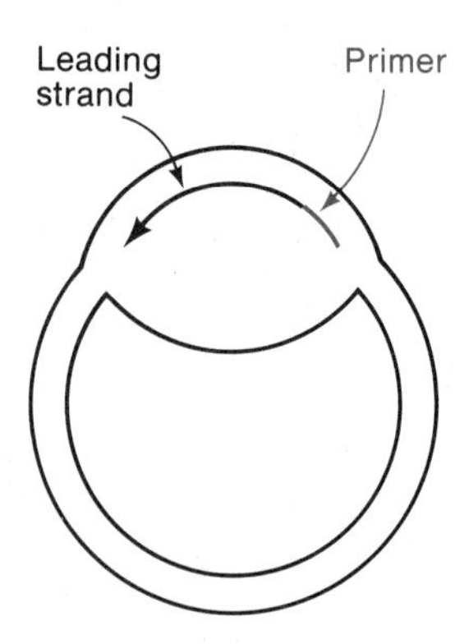

Figure 7-18
A circular DNA molecule with a D loop.

In the initial stages of replication of a naturally occurring circular DNA molecule, advance of the replication fork does not require the presence of DNA gyrase, because such a circular DNA molecule initially is negatively supercoiled and the negative twists compensate for the positive turns introduced by movement of the fork. However,

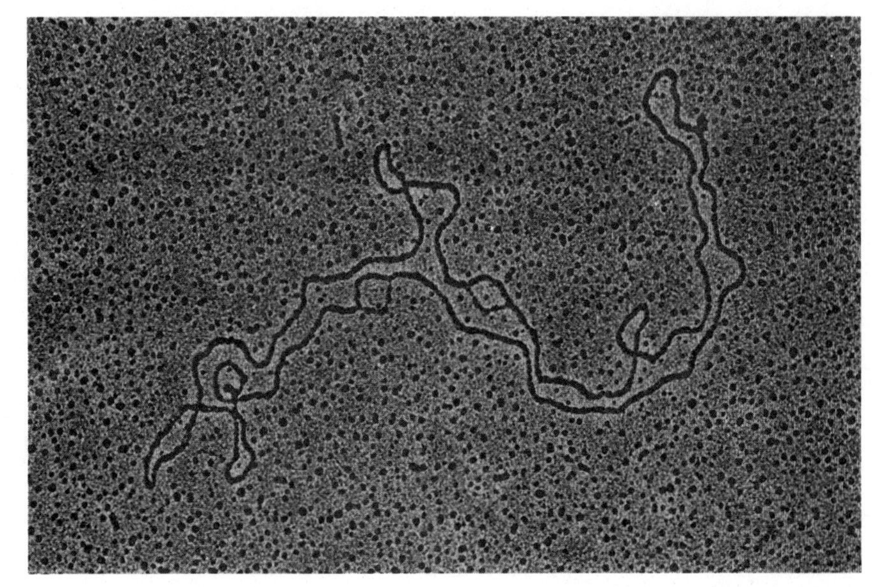

Figure 7-19
An electron micrograph of a dimer of mouse mitochondrial DNA showing diametrically opposing D loops. The single strand in each D loop appears thinner than the double-stranded DNA. The total length of the molecule is 10 μm. (Courtesy of David Clayton.)

once the fork has advanced sufficiently that the negative twists are used up, a topological constraint-relieving system such as gyrase is needed. The superhelix density of all naturally occurring DNA molecules is 0.05 twists per turn of the helix. This means that, if positive superhelicity is forbidden, the leading strand could move roughly 5 percent of the distance along a circular, negatively superhelical molecule before gyrase would be needed. A replication complex cannot in fact move along a double helix if the movement would produce more than one or two turns of overwinding. Thus, if neither gyrase nor any other unwinding system were present, replication would cease when the unreplicated portion had lost its supercoiling. The molecular configuration of the bubble would depend upon whether the first precursor fragment had been initiated; if it had not, the structure in Figure 7-18 would result.

Such a structure is seen in animal cell mitochondrial DNA (Figure 7-19). The molecular weights of these molecules in most species is about 10^7 and the molecular weight of a precursor fragment is typically about 10^6, so the lack of double-stranded DNA in both branches is consistent with the fact that only 4 percent of the DNA (or 2×10^5 molecular weight units of the leading strand) is replicated. The fact that 70 percent of all replicating mitochondrial DNA molecules are in the D-loop configuration indicates that indeed replication has ceased at this point.

Initiation by Covalent Extension: Rolling Circle Replication

There are numerous instances in which, in the course of replication, a circular phage DNA molecule gives rise to linear daughter molecules in which the base sequence of the DNA present in the phage is repeated numerous times, forming a concatemer (Figure 7-20). These concatemers are usually an essential intermediate in phage production. Likewise, in bacterial mating, a linear DNA molecule is transferred by a replicative process from a donor cell to a recipient cell. Both phenomena are consequences of initiation by covalent extension, an event that gives rise to a replication mode known as **rolling circle replication.**

Consider a duplex circle in which, by some initiation event, a nick is made having 3′-OH and 5′-P termini (Figure 7-21). Under the influence of a helicase and ssb protein a replication fork can be generated. Synthesis of a primer is unnecessary because of the 3′-OH group, so leading-strand synthesis can proceed by elongation from this terminus. At the same time, the parental template for lagging-strand synthesis is displaced. The polymerase used for this synthesis is apparently polymerase III. (In this circumstance we can ignore the fact that the enzyme cannot ordinarily carry out a displacement reaction, because in this case, displacement is a result of a coupling of helicase, ssb protein, and polymerase III.) The displaced parental strand is replicated in the usual way by means of precursor fragments. The result of this mode of replication is a circle with a linear branch; it resembles the Greek letter *sigma* and is called σ replication or rolling circle replication.

There are four significant features of rolling circle replication:

1. The leading strand is covalently linked to the parental template for the lagging strand.

Figure 7-20
A concatemer consisting of the repeating unit ABC . . . XYZ. Note that the definition of concatemer does not make any requirements about the terminal sequences.

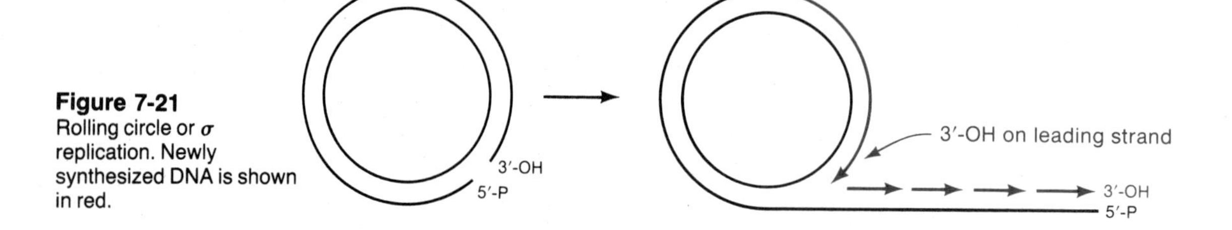

Figure 7-21
Rolling circle or σ replication. Newly synthesized DNA is shown in red.

2. Before precursor fragment synthesis begins, the linear branch has a free 5′-P terminus.
3. Rolling circle replication continues unabated, generating a concatemeric branch.
4. The circular template for leading-strand synthesis never leaves the circular part of the molecule.

BIDIRECTIONAL REPLICATION

In this section we examine the events following a D-looplike initiation step in which chain growth occurs continuously. It will be seen that unless special rules are imposed, replication will be bidirectional—that is, there will be two replication forks.

Somewhat after initiation of synthesis of the leading strand at the origin, the first precursor fragment is synthesized. This is shown in Figure 7-22-I, in which the overall direction in which the replicating fork moves is counterclockwise. In our earlier analysis of lagging-strand replication, it was noted that synthesis of each precursor fragment is terminated when the growing end reaches the primer of the previously synthesized fragment. However, in the case of the first precursor fragment, there can be no previously made fragment. We may consider two possible events: (1) there is a termination signal for synthesis of the first fragment, or (2) there is no such signal and the precursor fragment continues to grow. In the second case, the precursor fragment is equivalent to a leading strand for a second replication fork, moving clockwise, as shown in the figure. Clockwise replication requires the synthesis of precursor fragments in the second replication fork, but this can be achieved by the standard mechanism. The result of these events

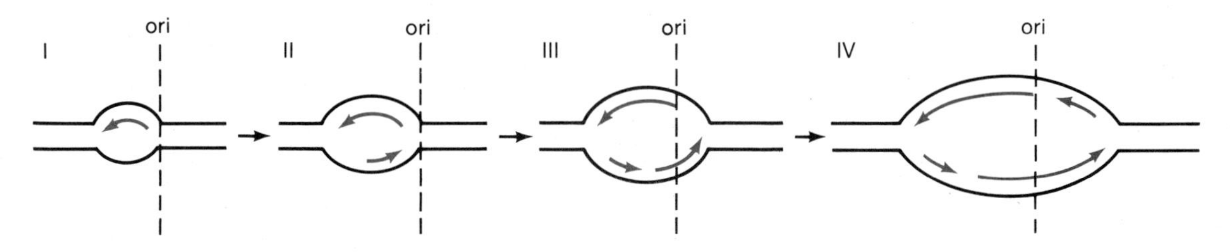

Figure 7-22
The formation of a bidirectionally enlarging replication bubble. I. The leftward-leading strand starts at *ori*. II. The leading strand has progressed far enough that the first rightward precursor fragment begins. III. The leftward-leading strand has progressed far enough that the second rightward precursor fragment has begun. The first rightward precursor fragment has passed *ori* and has become the rightward-leading strand. IV. The rightward-leading strand has moved far enough that the first leftward precursor fragment has begun. There are now two complete replication forks.

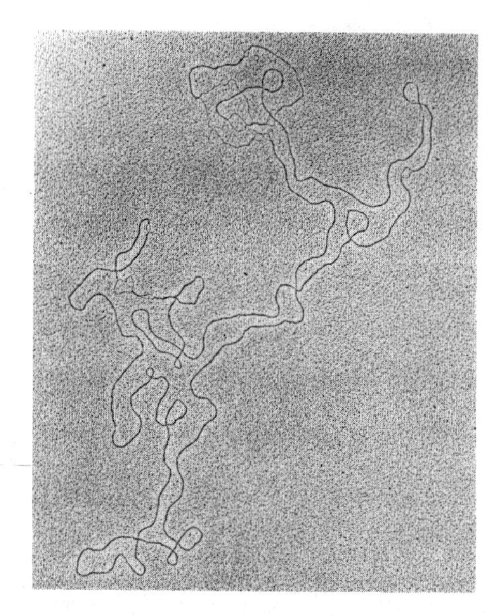

Figure 7-23
An electron micrograph of a partially denatured circular phage λ DNA molecule. With this method of sample preparation the single strands are very thin and faint compared to the double-stranded DNA. (Courtesy of Manuel Valenzuela.)

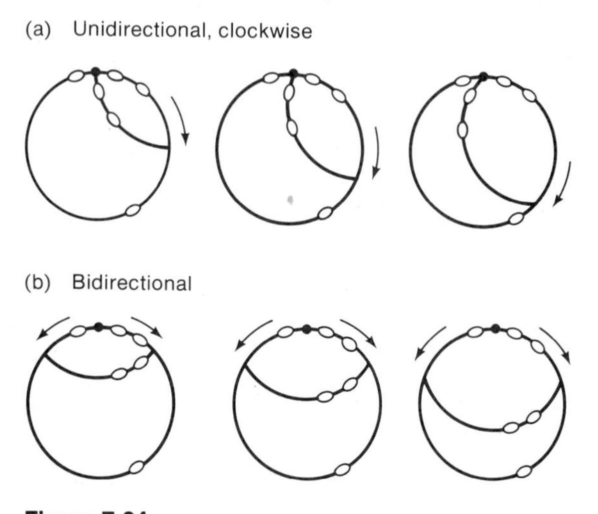

Figure 7-24
A diagram showing the relative positions of branch points and a replication bubble for a DNA molecule replicating (a) unidirectionally and clockwise or (b) bidirectionally. The arrows indicate the direction of movement of the replication fork and the dots indicate the replication origin.

is that the DNA molecule will have two replication forks moving in opposite directions around the circle. This is called **bidirectional replication.** If alternative (1) occurs, there is a single fork and replication is unidirectional. There is no particular reason to have such a stop signal and, in fact, bidirectional replication has the advantage of halving the time required to complete the process.

The method used to study the direction of replication was developed by Ross Inman and Maria Schnös and is called **denaturation mapping.** It is done as follows. If a DNA molecule is heated to a temperature at which melting is just detected, single-stranded bubbles form in the regions having very high A+T content. If formaldehyde is added and the DNA is cooled, the bubbles persist. DNA treated in this way can be observed by electron microscopy and the position of the bubbles can be noted (Figure 7-23). When the technique is applied to a replicating molecule, the bubbles serve as fixed reference points against which the positions of a branch-point can be plotted. Figure 7-24 shows the kinds of molecules that would be expected for unidirectional and bidirectional replication of a hypothetical molecule. Note that (1) the relation between the positions of the bubbles and the branch points differ for the two modes and (2) a replication fork is identified by its changing distance from a bubble.

In unidirectional replication, one branch point remains at a fixed position with respect to the bubbles; this position defines the replication origin. In bidirectional replication both branch points move with respect to the bubbles so that each branch point is a replication fork. If both replication forks move at the same rate, the origin is always at the midpoint of each branch of the replication loop.

Bidirectional replication has been widely observed with phage, bacterial, and plasmid DNA. A small number of phages and plasmids use the unidirectional mode exclusively, so that a stop signal must exist that prevents chain growth of the first precursor fragment past the origin. One plasmid uses a surprising variant of the bidirectional mode in that initiation of movement of one fork occurs at a much later time than the primary initiation event. Bidirectional replication is the major mode of chain growth with eukaryotes.

REPLICATION OF EUKARYOTIC CHROMOSOMES

The replication of eukaryotic chromosomes presents many problems not found in the prokaryotes because of the enormous size of eukaryotic chromosomes and the geometric complexity imposed by the organization of the DNA into nucleosomes. How these problems are handled is described in this section.

The rate of movement of a replication fork in *E. coli* is $\approx 10^5$ base pairs per minute. In eukaryotes the polymerases are much less active and the rate ranges from 500 to 5000 base pairs per minute. Since a typical animal cell contains about fifty times as much DNA as a bacterium, the replication time of an animal cell should be about 1000 times as great as that of *E. coli* or about 30 days. However, the duration of the replication cycle is usually several hours and this is accomplished by having multiple initiation sites. For instance, the DNA of the fruit fly *Drosophila* has about 5000 initiation sites, each separated by about 30,000 bases, and each site replicates bidirectionally. The number of sites is regulated in a way that is not understood. For example, in the round of replication following fertilization of *Drosophila* eggs, the number of initiation points reaches 50,000, and it takes only 3 minutes to replicate all of the DNA. An example of a fragment of this rapidly replicating *Drosophila* DNA is shown in Figure 7-25.

The enormous number of growing forks in eukaryotic cells is reflected in the number of polymerase molecules. In *E. coli* there are between 10 and 20 molecules of pol III holoenzyme. However, a typical

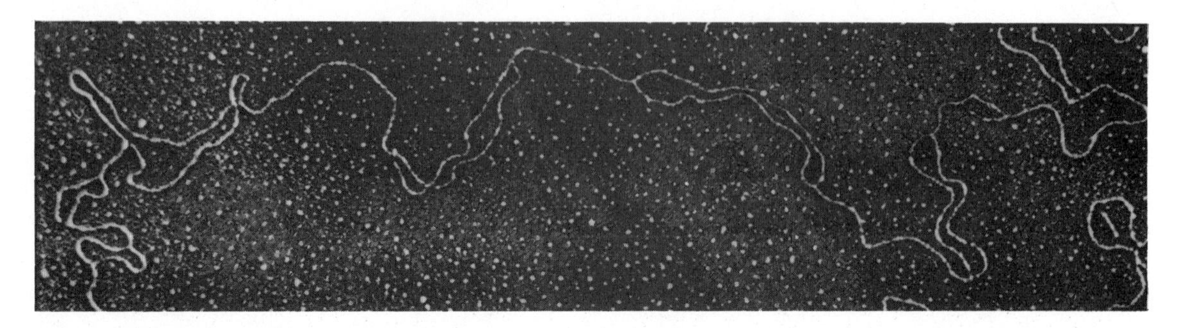

Figure 7-25
Replicating DNA of *Drosophila melanogaster* showing many replicating eyes. The molecular weight of the segment shown is roughly 20×10^6. (Courtesy of David Hogness.)

Figure 7-26
A replicating fork showing nucleosomes on both branches. The diameter of each particle is about 110 Å. (Courtesy of Harold Weintraub.)

animal cell has 20,000 to 60,000 molecules of polymerase α, which is believed to be the major polymerase.

Replication of double-stranded DNA proceeds through both a polymerization step (nucleotide addition) and a dissociation step (strand separation). The replication of chromatin, which is the form that DNA has in eukaryotes, proceeds through an additional dissociation step—namely, dissociation of DNA and histone octamers—and a histone-DNA reassociation step (see Chapter 5 for a discussion of the structure of the octamers contained in nucleosomes). In chromatin, DNA is wrapped around a histone octamer to form a nucleosome and if the DNA were never unwrapped from the histone spool, severe geometric problems would arise at the growing fork. Moreover, after DNA dissociates from the histones, newly formed DNA must rejoin with the nucleosomal octamers so that each daughter molecule will be organized into nucleosomes, just as the parent was.

Examination of replication forks in DNA that has not been deproteinized during isolation indicates that nucleosomes form very rapidly after replication. For example, Figure 7-26 shows that all portions of a replication eye have the beadlike appearance characteristic of nucleosomes.

The synthesis of histones occurs simultaneously with DNA replication—that is, histones are made in the cell as they are needed, so that the cell does not contain an appreciable amount of unassociated histone molecules. In light of this, we would like to know whether newly synthesized histones mix with parental histones in the octamers associated with daughter DNA molecules. The answer is clear: the octamers do not dissociate. Thus, whereas DNA replication is semiconservative, production of histone octamers is conservative.

The result just stated does not give any information about how the conserved histone octamers are arranged on the daughter strands. Two

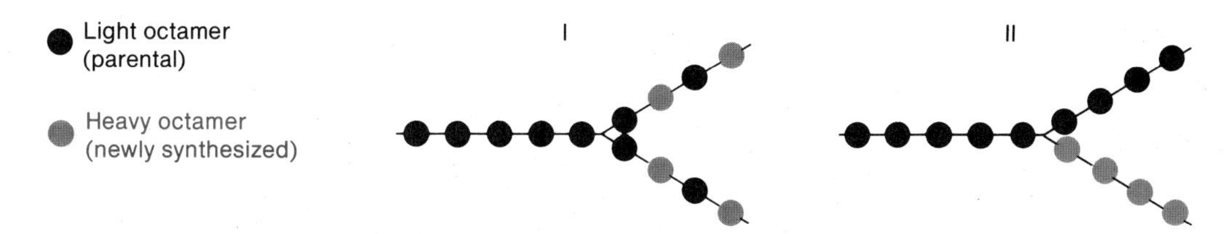

Figure 7-27
Schematic diagram showing two arrangements of parental and newly synthesized octamers in the replicating region. Arrangement II is the correct one.

possibilities are shown in Figure 7-27. In the first panel the parental octamers are distributed between both daughter strands, which would imply that the octamers dissociate completely from the DNA during replication. In the second arrangement all parental octamers are located on one daughter strand; this arrangement is the correct one. A simple experiment, based on the fact that histones are made as needed, shows that octamers do not dissociate from the strand with the parental units. If cycloheximide, an inhibitor of protein synthesis, is added to growing cells, DNA replication continues in those cells in which replication had begun prior to addition of the inhibitor; however, no histones, which are proteins, are made. Replicating chromatin obtained from cycloheximide-treated cells has been observed by electron microscopy. Molecules of the type shown in Figure 7-28 are seen. Each fork in an eye has naked DNA in one branch; the other branch has the parental conserved octamers.

Figure 7-28
Replication of chromatin in the presence of cycloheximide, an inhibitor of protein synthesis. (a) A schematic diagram of an eye. **(b) Replicating chromatin from cycloheximide-treated cells. Left panel: a replication fork. Right panel: interpretive drawing. Particles have a cross-section of 110 Å.** (Courtesy of Harold Weintraub.)

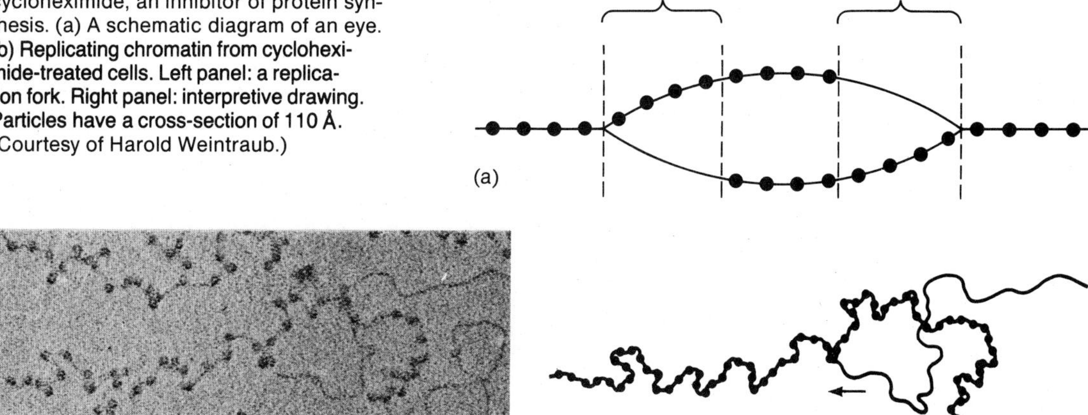

The final question is whether the parental octamers are associated with a particular strand (leading or lagging strand). So far, it has been possible to answer this question for only one system—an animal virus, SV40, whose DNA is also organized in nucleosomes (which may not be true of all viral DNA). In this system the parental octamers are associated exclusively with the leading strand.

Since DNA replication in each eye is bidirectional and the parental octamers are always associated with leading strand synthesis, then each branch of an eye is covered with a long tract of parental octamers and a long tract of new octamers. The dividing point between these tracts is the replication origin of each eye, as shown in Figure 7-29.

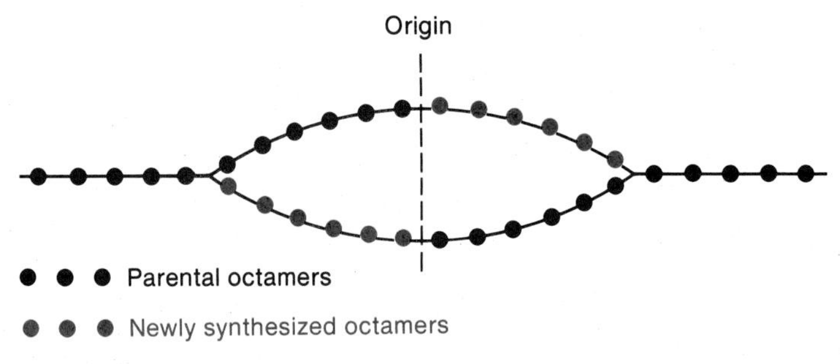

Figure 7-29
The arrangement of parental and newly synthesized octamers in an eye, when parental octamers are associated with the leading strand and newly synthesized octamers are associated with the lagging strand.

8 Repair

There is no single molecule whose integrity is as vital to the cell as DNA. Thus, in the course of hundreds of millions of years there have evolved efficient systems for correcting occasional replication errors (for example, the editing functions of DNA polymerases) and for eliminating damage by environmental agents and by intracellular chemicals. A few of these systems are examined in this chapter. We begin by describing the principal kinds of damage that can occur in a DNA molecule.

ALTERATIONS OF DNA MOLECULES

There are three distinct mechanisms for altering the structure of DNA: (1) base substitutions during replication, (2) base changes resulting from the inherent chemical instability of the bases or of the *N*-glycosylic bond, and (3) alterations resulting from the action of other chemicals and environmental agents. These mechanisms are responsible for the occurrence of the following defects:

1. *An incorrect base in one strand that cannot form hydrogen bonds with the corresponding base in the other strand.* This defect can result from a replication error that by chance is not corrected by the editing

function, or by spontaneous loss of an amino group, converting cytosine to uracil or adenine to hypoxanthine.

2. *Missing bases.* The N-glycosylic bond of a purine nucleotide is spontaneously broken at physiological temperatures, though at a very low rate. This process is called **depurination** because the purine is lost from the DNA. The rate of spontaneous depurination is about one purine removed per 300 purines per day at pH 7 and 37°C, which amounts to about 10^4 purines per day in a mammalian cell and 0.25 purines per day per generation time for a bacterium. The rate is increased as the pH is lowered or as the temperature is elevated.

3. *Altered bases.* Bases can be changed into strikingly different compounds by a variety of chemical and physical agents. For instance, ionizing radiation (such as the β particles emitted by naturally occurring radioisotopes or laboratory x rays) can break purine and pyrimidine rings and can cause several types of chemical substitutions. The best-studied altered base is the dimer formed by two pyrimidines as a result of ultraviolet radiation. The most prominent of these dimers is the thymine dimer shown in Figure 8-1. The significant effects of the presence of thymine dimers are the following: (1) the DNA helix becomes distorted as the thymines, which are in the same strand, are pulled toward one another (Figure 8-2); and (2) as a result of the distortion, hydrogen-bonding to adenines in the opposing strand, though possible (because the hydrogen-bonding groups are still pres-

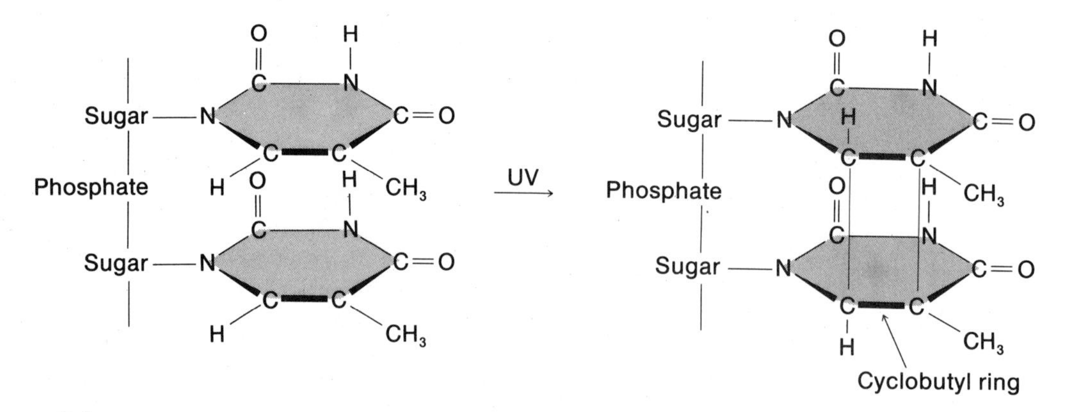

Figure 8-1
Structure of a cyclobutylthymine dimer. Following ultraviolet (UV) irradiation, adjacent thymine residues in a DNA strand are joined by formation of the bond shown in red. Although not drawn to scale, these bonds are considerably shorter than the spacing between the planes of adjacent thymines, so that the double-stranded structure becomes distorted. The shape of the thymine ring also changes as the C=C double bond of each thymine is converted to a C—C single bond in each cyclobutyl ring.

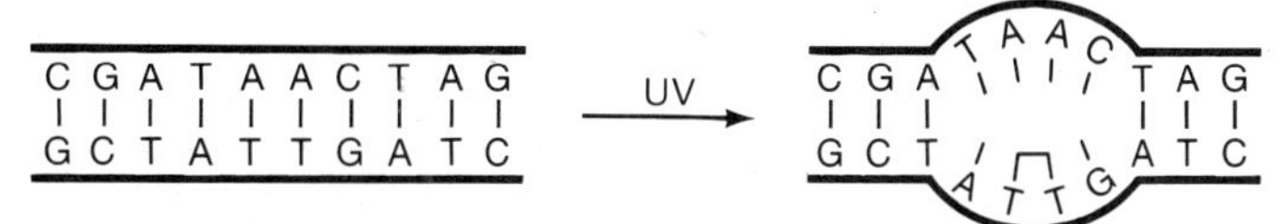

Figure 8-2
Distortion of the DNA helix caused by two thymines moving closer together when joined in a dimer. The dimer is shown as two joined lines.

ent), is significantly weakened; this weakening causes inhibition of advance of the replication fork, as will be discussed in a later section.

4. *Single-strand breaks.* A variety of agents can break phosphodiester bonds. Among the more common chemicals are peroxides, sulfhydryl-containing compounds (for example, cysteine) and metal ions such as Fe^{2+} and Cu^{2+}. Ionizing radiation produces strand breaks both by the action of secondary electrons produced by a β particle or an x-ray photon passing nearby and by production of free radicals in water (e.g., OH) that can attack the bond. Single-strand breaks are usually repaired by the action of DNA ligase.

5. *Double-strand breaks.* If a DNA molecule receives a sufficiently large number of randomly located single-strand breaks, two breaks may be situated opposite one another, resulting in breakage of the double helix. This is an important type of damage in cancer radiotherapy. Double-strand breaks are usually not repaired.

6. *Cross-linking.* Some antibiotics (for example, mitomycin C) and some reagents (the nitrite ion) can form covalent linkages between a base in one strand and an opposite base in the complementary DNA strand. This prevents strand separation during DNA replication and also causes a local distortion of the helix. Repair of cross links will not be discussed (for information, see *MB*, pages 335–336).

Repair of incorrect, missing, and altered bases will be discussed in the following sections.

REPAIR OF INCORRECT BASES

As has already been described, polymerases I and III occasionally catalyze incorporation of an "incorrect" base which cannot form a hydrogen bond with the template base in the parental strand; such errors are usually corrected by the editing function of these enzymes. However, the integrity of the base sequence of DNA is so important that a second system exists for correcting the occasional error missed by the editing function. This correction system is called **mismatch repair.** In mismatch repair a pair of non-hydrogen-bonded bases is recognized as incorrect and a polynucleotide segment is excised from one strand, thereby removing one member of the unmatched pair. The resulting gap is filled in by pol I, which presumably uses this "second chance to get it right" to form only correct base pairs; then the final seal is made by DNA ligase.

If it is to correct but not create errors, the mismatch repair system must be able to distinguish the correct base in the parental strand from the incorrect base in the daughter strand. The critical information is the location of particular adenines in the sequence G-A-T-C. These bases carry methyl groups not found elsewhere in the DNA. Methylation of the bases in this sequence is associated with replication, occurring in the *daughter* DNA strands. However, it does not take place in the replication fork but is delayed somewhat. The effect is that parent strands are fully methylated, whereas there is a gradient of methylation along a newly synthesized daughter strand, with the least methylation near the replication fork. The mismatch repair system recognizes the degree of methylation of each strand, and when a mismatch is found, it preferentially excises nucleotides from the undermethylated strand—that is, from the daughter strand. Thus, the parental strand is always the template, enabling the repair system to correct misincorporation errors.

The mismatch repair system is very important for another reason—namely, because certain unusual molecules called base analogues (see Chapter 11) can be incorporated into DNA without being recognized by the editing function. An example of this is a base having both an enol and a keto form in equilibrium and in which only one of these forms can hydrogen-bond with a standard DNA base. We can imagine a situation in which the enol-keto equilibrium lies far toward the keto form and only the enol form can base-pair with (let us say) a cytosine. Substitution of this base for a guanine would be very rare but nonetheless this base might be incorporated into the DNA during a brief period in which the base is in the enol form. Since in that form the odd base would nevertheless be able to pair, the editing function would not recognize it as incorrect. Afterwards, however, the base would tend to be predominately in the keto form so there would be a mismatched base pair most of the time.

Cytosine sometimes loses an amino group (*deamination*), forming uracil. After one round of replication this would lead to replacement of a G · C pair by an A · U pair, which would become an A · T pair after another round of replication. Since this would be mutagenic, cells have evolved a mechanism for replacing the unwanted U by a C. The

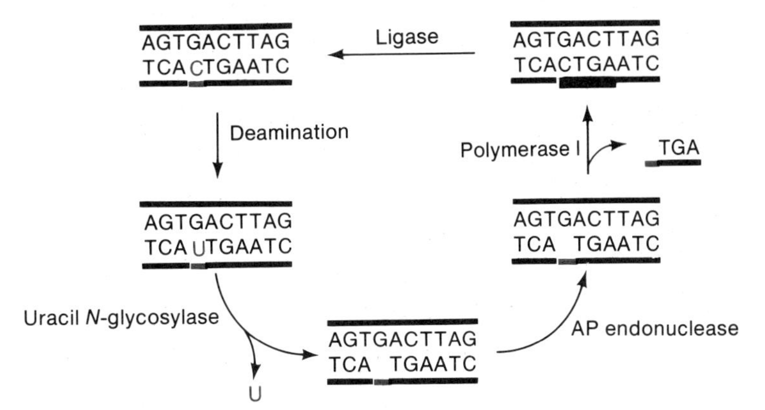

Figure 8-3
Scheme for repair of cytosine deamination. The same mechanism could remove a uracil that is accidentally incorporated.

first step of this repair cycle (Figure 8-3) is removal of the uracil by the enzyme **uracil *N*-glycosylase**. This enzyme cleaves the N-glycosylic bond and leaves the deoxyribose in the backbone. A second enzyme, **AP endonuclease**,* makes a single cut, freeing one end of the deoxyribose. This is followed by removal of the deoxyribose and several adjacent nucleotides (possibly by the endonuclease activity of pol I), after which pol I fills the gap with the correct nucleotides. (This sequence, endonuclease–exonuclease–polymerase, is an example of a general repair mechanism called excision-repair, which will be described shortly). Adenine is also occasionally deaminated, yielding hypoxanthine; a specific glycosylase is also available to remove the hypoxanthine. The remainder of the repair pathway is that used for uracil.

The latter part of the repair sequence shown in Figure 8-3, namely, that beginning with AP endonuclease, is also used in the repair of missing bases arising by depurination.

REPAIR OF THYMINE DIMERS

The best-understood repair of altered bases is that of thymine dimers, for which there are four pathways, which can be subdivided into two classes—light-induced repair (photoreactivation) and light-independent repair (dark repair). The latter can be accomplished by three distinct mechanisms: (1) excision of the damaged bases (**excision repair**), (2) reconstruction of a functional DNA molecule from undamaged fragments (**recombinational repair**), and (3) disregard of the

*AP stands for apurinic acid, a polynucleotide from which purines have been removed by hydrolysis of the *N*-glycosylic bonds.

damage (**SOS repair**). The chemical mechanisms for each of these repair processes are described in this section.

Photoreactivation

Photoreactivation is an enzymatic cleavage of thymine dimers activated by visible light (300–600 nm). An enzyme called the photoreactivating or **PR enzyme** has been isolated from almost all cells, from bacteria to animals. The PR enzyme itself does not absorb light nor does it bind any light-absorbing compound. In a way that is not known, the enzyme-DNA complex nevertheless absorbs light and uses the light energy to cleave the C—C bonds of the cyclobutyl rings shown in Figure 8-1.

Excision Repair

Excision repair is a multistep enzymatic process. The four steps for *E. coli*, summarized as **cut—patch—cut—seal**, are shown in Figure 8-4. In the first step, an **incision** step, a repair endonuclease recognizes the distortion produced by a thymine dimer and makes a single cut in the sugar-phosphate backbone ahead of the dimer. At the incision site there

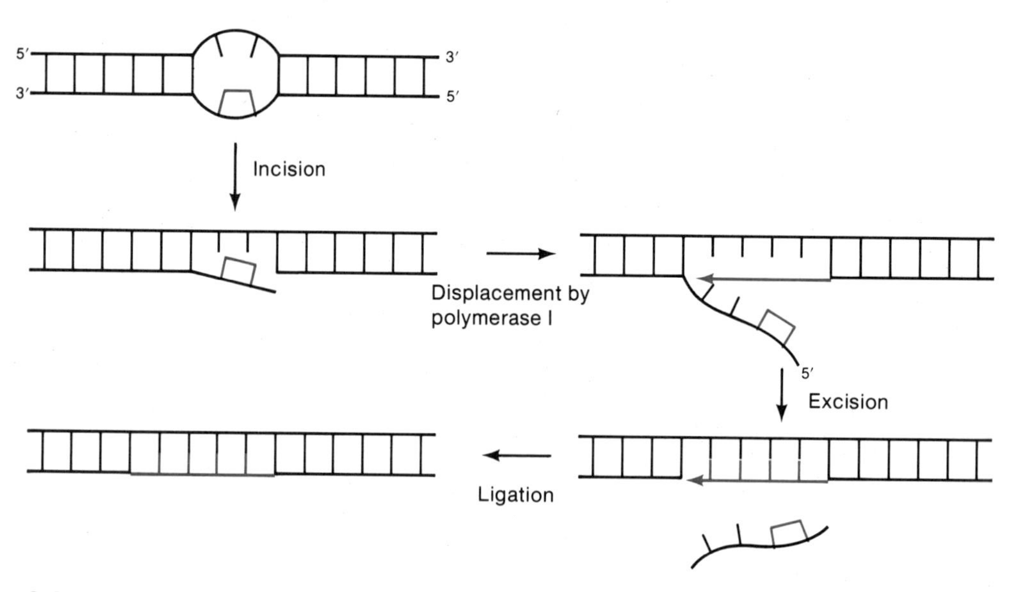

Figure 8-4
Scheme for excision repair of a thymine dimer by the cut—patch—cut—seal mechanism. The thymine dimer and the displacing segment synthesized by polymerase I are both shown in red.

is a 5′-P group on the side of the cut containing the dimer and a 3′-OH group on the other side. The 3′-OH group is recognized by polymerase I, which then synthesizes a new strand while displacing a DNA segment consisting of about twenty nucleotides and carrying the thymine dimer. This segment is excised in part by the 5′→3′ exonuclease activity of pol I and in part by a variety of other nucleases. The final step is joining the newly synthesized segment to the original strand by DNA ligase. This scheme, shown for *E. coli*, varies slightly in other organisms.

The incision activity of *E. coli* is determined by three genes called *uvrA*, *uvrB*, and *uvrC*. The products of the *uvrA* and *uvrB* genes are two subunits of a protein complex (called ultraviolet-endo I) that has endonuclease activity. The UvrC product is necessary for maximum endonuclease activity *in vivo*, but its precise role is not yet known.

Incision enzymes have been isolated from many sources. Two particularly well-studied ultraviolet-repair enzymes are those of *E. coli* phage T4 and of the bacterium *Micrococcus luteus*. The incision enzymes of yeast and of many mammalian cells have also been characterized in detail. The mammalian systems seem to be more complicated and require a larger collection of enzymes.

Many human diseases may result from inability to carry out excision repair. The best studied, xeroderma pigmentosum, is a result of mutations in genes that encode the ultraviolet-light incision system. Cells cultured from patients with the disease are killed by much smaller ultraviolet doses than are cells from a normal person. Furthermore, their ability to remove thymine dimers from their DNA is very much reduced. Patients with this disease develop skin lesions when exposed to sunlight and commonly develop one of several kinds of skin cancer.

Recombinational Repair

The excision repair systems just described are responsible for the removal of many of the thymine dimers. However, before sufficient time has elapsed for their repair, many dimers have interfered with various cellular processes. The deleterious effects of these remaining dimers are eliminated by recombinational repair, which is carried out by the system responsible for genetic exchange.

In order to discuss the mechanism of recombinational repair, it is necessary to know the effect of a thymine dimer on DNA replication. When polymerase III reaches a thymine dimer, the replication fork fails to advance. A thymine dimer is still capable of forming hydrogen bonds with two adenines because the chemical change in dimerization does not alter the groups that engage in hydrogen bonding. However, the dimer introduces a distortion into the helix and when an adenine is

Figure 8-5
Blockage of replication by thymine dimers (represented by joined lines) followed by re-starts several bases beyond the dimer. The black region is a segment of ultraviolet-light-irradiated parental DNA. The red region represents synthesis of a daughter molecule from right to left. The daughter strand contains gaps.

added to the growing chain, polymerase III reacts to the distorted region as if a mispaired base had been added; the editing function then removes the adenine. The cycle begins again—an adenine is added and then it is removed; the net result is that the polymerase is stalled at the site of the dimer. (The same effect would occur if, instead of a dimer, radiation or chemical damage resulted in formation of a base to which no nucleoside triphosphate could base pair.) Evidence that such a phenomenon occurs after ultraviolet irradiation is the existence of an ultraviolet-light-induced idling process—that is, rapid cleavage of deoxynucleoside triphosphates to monophosphates without any net DNA synthesis (i.e., without advance of the replication fork). A cell in which DNA synthesis is permanently stalled cannot complete a round of replication and does not divide.

There are two different ways in which DNA synthesis can get going again—**postdimer initiation** and **transdimer synthesis.** These are responsible for **recombination repair** and **SOS repair,** respectively. In this section recombination repair is described; SOS repair is discussed in the succeeding section.

One way to deal with a thymine dimer block is to pass it by and initiate chain growth beyond the block (Figure 8-5). Such postdimer initiation does occur after a pause of about five seconds per thymine dimer but the mechanism, which appears to involve unprimed reinitiation, is unknown. The result of this process is that the daughter strands have large gaps, one for each unexcised thymine dimer. There is no way to produce viable daughter cells by continued replication alone, because the strands having the thymine dimer will continue to turn out gapped daughter strands, and the first set of gapped daughter strands would be fragmented when the growing fork enters a gap. However, by a recombination mechanism called **sister-strand exchange** proper double-stranded molecules can be made.

The essential idea in sister-strand exchange is that a single-stranded segment free of any defects is excised from a "good" strand on the homologous DNA segment at the replication fork and somehow inserted into the gap created by excision of a thymine dimer (Figure 8-6). The combined action of polymerase I and DNA ligase joins this inserted piece to adjacent regions, thus filling in the gap. The gap

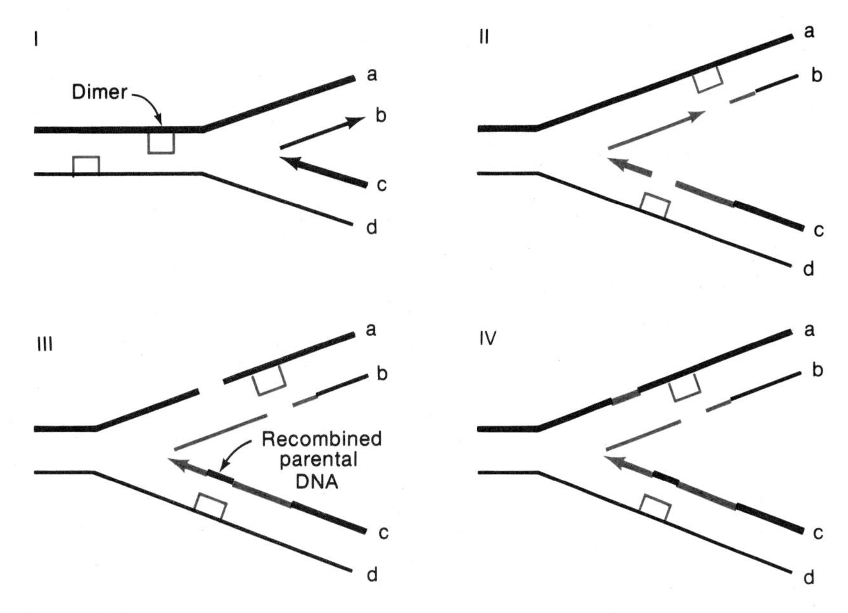

Figure 8-6
Recombinational repair. I. A molecule containing two thymine dimers (red boxes) in strands a and d is being replicated. II. By postdimer initiation, a molecule is formed whose daughter strands b and c have gaps. If repair does not occur, in the next round of replication, strands a and d would yield gapped daughter strands, and strands b and c would again be fragmented. III. By sister-strand exchange a continuous segment of a is excised and inserted into strand c. (In a second round of replication, strand c would be a template for synthesis of functional DNA.) IV. The gap in a is next filled. Such a DNA molecule would probably engage in a second sister-strand exchange in which a segment of c would fill the gap in b. The gap in c would then be filled. In this way strand c also becomes a functional template. DNA synthesized after irradiation is shown in red. Heavy and thin lines are used for purposes of identification only.

formed in the donor molecule by excision is also filled in completely by polymerase I and ligase. If this exchange and gap filling are done for each thymine dimer, two complete daughter single strands can be formed and each can serve in the *next* round of replication as a template for synthesis of normal DNA molecules. Note that the system fails if two dimers in opposite strands are very near one another because then no undamaged sister strand segments are available to be excised. The molecular details of recombinational repair are not known with precision, so that the model shown in Figure 8-6 must be considered to be a working hypothesis that is at present consistent with the facts.

Recombinational repair is an important mechanism because it eliminates the necessity for delaying replication for the many hours that would be needed for excision repair to remove all thymine dimers. It may also be the case that some kinds of damage cannot be eliminated by excision repair—for example, alterations that do not cause helix distortion but do stop DNA synthesis. Recombination repair has been demonstrated in several bacteria, but it is not known whether it occurs in animal cells.

Since recombinational repair occurs after DNA replication, in contrast with excision repair, it is often called **postreplicational repair.**

SOS Repair

SOS repair is a **bypass system** that allows DNA chain growth across damaged segments at the cost of fidelity of replication. It is an error-prone process; even though intact DNA strands are formed, the strands are often defective. (The principle involved is that survival with some loss of information is better than no survival at all.) SOS repair is not understood particularly well at present but is thought to invoke a relaxation of the editing system in order to allow polymerization to proceed across a dimer (transdimer synthesis) despite the distortion of the helix. In the previous section the idling caused by the response of the editing system to a dimer was described. When the SOS system is activated, apparently the editing system does not react to the distortion, the growing fork advances, and two adenines are placed in the growing chain at the sites specified by the thymines in the dimer. An early hypothesis about the mechanism of SOS repair suggested that the polymerase placed any bases in the growing chain at the damaged site and that this random replacement was the cause of mutagenesis. However, recently an important observation has been made about the base sequence in the daughter strand—namely, that incorrect nucleotides are indeed occasionally added to the growing chain but the mistakes are not necessarily at the sites of the adenines that would normally be added if the DNA were not damaged. The current hypothesis is that damaged DNA induces an error-prone replication system that has less proofreading activity (if any) than the normal replication system. This lower-fidelity system causes *all* newly made DNA strands to have a higher-than-normal number of mispaired bases. It seems likely that polymerase III is modified in some way (possibly by loss or alteration of a subunit needed for high-fidelity replication) so it can continue chain growth without being stalled at damaged sites (Figure 8-7). Because of the presence of mispaired bases, most progeny will be mutants (although the base-pair mismatch repair system presumably can correct many of these errors). How long this error-prone replication continues is not known. The SOS repair system is thought to be a major cause of ultraviolet-induced mutagenesis.

SOS repair (and probably recombinational repair as well) differs from excision repair in that SOS repair is *induced* as a result of the

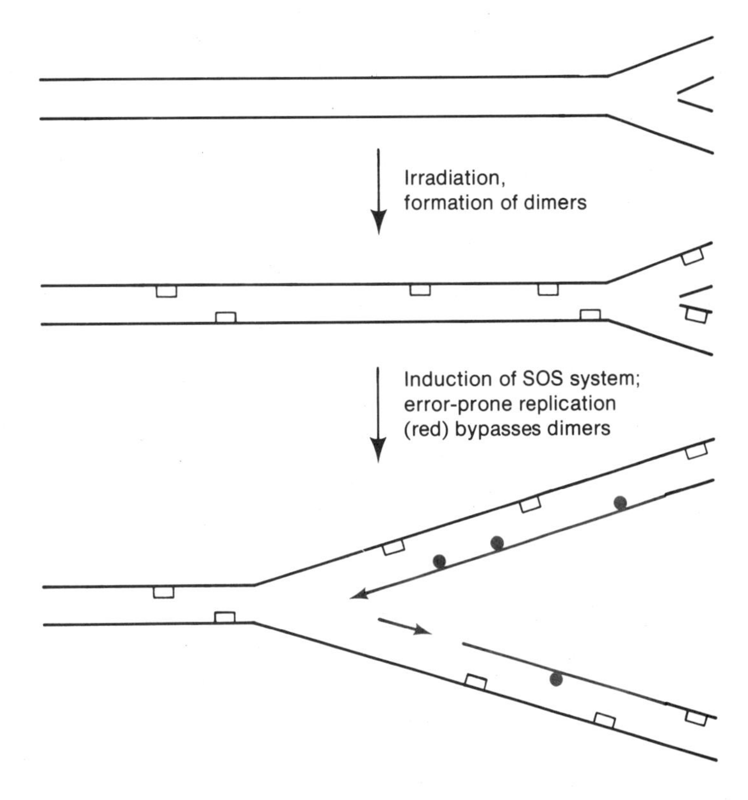

Figure 8-7
SOS repair. A DNA molecule in an early stage of replication is irradiated with ultraviolet light and thymine dimers are formed. The SOS system is induced and all subsequent replication (shown in red) has a higher-than-usual number of misincorporated bases (red dots). Some of these bases can be removed by the mismatch repair system.

damage to the DNA. That is, the responsible enzymes are not present until after a cell has been damaged. The best evidence for this point comes from an analysis of the repair of ultraviolet-light-inactivated phage by irradiated bacterial host cells.

Most phages do not need a fully functional (or even viable) host to produce progeny phage particles. For instance, a study of the result of infecting ultraviolet-irradiated *E. coli* with phage λ showed that, if the host is irradiated with a dose of ultraviolet light yielding 10 percent survival of the ability to form colonies, there is no loss of the ability to support phage infection. However, if the λ particles are also irradiated, the irradiated *E. coli* cells are better able to support growth of the irradiated λ than are unirradiated cells. That is, the survival of an ultraviolet-irradiated λ is higher on an irradiated host than on an unirradiated host (Figure 8-8). This phenomenon is called **UV-reactivation.** A notable feature of this phenomenon is that although more phage survive, the surviving population contains a higher frequency of mutations than is observed with unirradiated bacteria; this is to be expected if the repair mechanism utilizes an error-prone replication system.

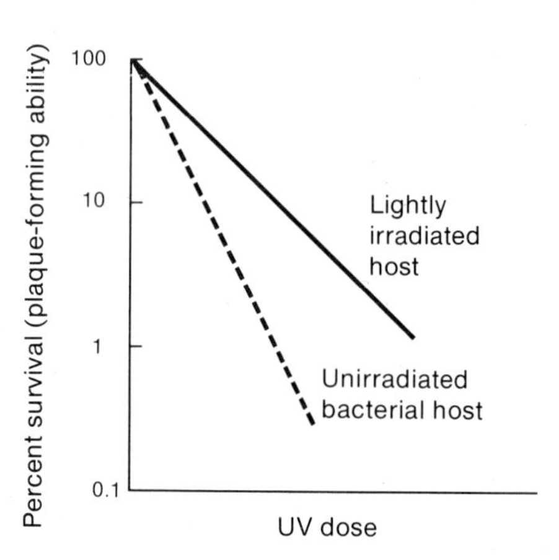

Figure 8-8
UV reactivation of ultraviolet-light-irradiated phage λ. The dashed line shows the survival curve (for plaque-forming ability) obtained when λ phage irradiated with various doses of ultraviolet light are plated on unirradiated bacteria. The solid line represents survival of plaque-forming ability, when ultraviolet-light-irradiated λ are plated on lightly irradiated bacteria.

It is reasonable that the SOS repair should normally be shut down. Clearly, if cells have an elegant editing function for maintaining high fidelity in replication, they should also have a means of making all error-prone replicative processes inactive when relaxation of fidelity is not essential to survival. The mechanism by which the SOS system is turned on is a multistep process that is not yet fully understood; some information can be found in *MB*, pages 332–333.

9 Transcription

Gene expression is accomplished by the transfer of genetic information from DNA to RNA molecules and then from RNA to protein molecules. RNA molecules are synthesized by using the base sequence of one strand of DNA as a template in a polymerization reaction that is catalyzed by enzymes called DNA-dependent RNA polymerases or simply **RNA polymerases.** The process by which RNA molecules are initiated, elongated, and terminated is called **transcription.**

Three aspects of transcription will be considered in this chapter: (1) the enzymology of RNA synthesis, (2) the signals that determine at what points on a DNA molecule transcription starts and stops, and (3) the types of transcription products and how they are converted to the RNA molecules needed by the cell. We begin with the first item.

ENZYMATIC SYNTHESIS OF RNA

In this section we describe the basic features of the polymerization of RNA, the identity of the precursors, the nature of the template, the properties of the polymerizing enzyme, and the mechanisms of initiation, elongation, and termination of synthesis of an RNA chain.

The essential chemical characteristics of the synthesis of RNA are the following:

1. The precursors in the synthesis of RNA are the four ribonucleoside 5′-triphosphates (rNTP) ATP, GTP, CTP, and UTP. On the ribose portion of each NTP there are two OH groups—one each on the 2′- and 3′-carbon atoms (Figure 2-5).
2. In the polymerization reaction a 3′-OH group of one nucleotide reacts with the 5′-triphosphate of a second nucleotide; a pyrophosphate is removed and a phosphodiester bond results (Figure 9-1). This is the same reaction that occurs in the synthesis of DNA.
3. The sequence of bases in an RNA molecule is determined by the base sequence of the DNA. Each base added to the growing end of the RNA chain is chosen by its ability to base-pair with the DNA strand used as a template; thus, the bases C, T, G, and A in a

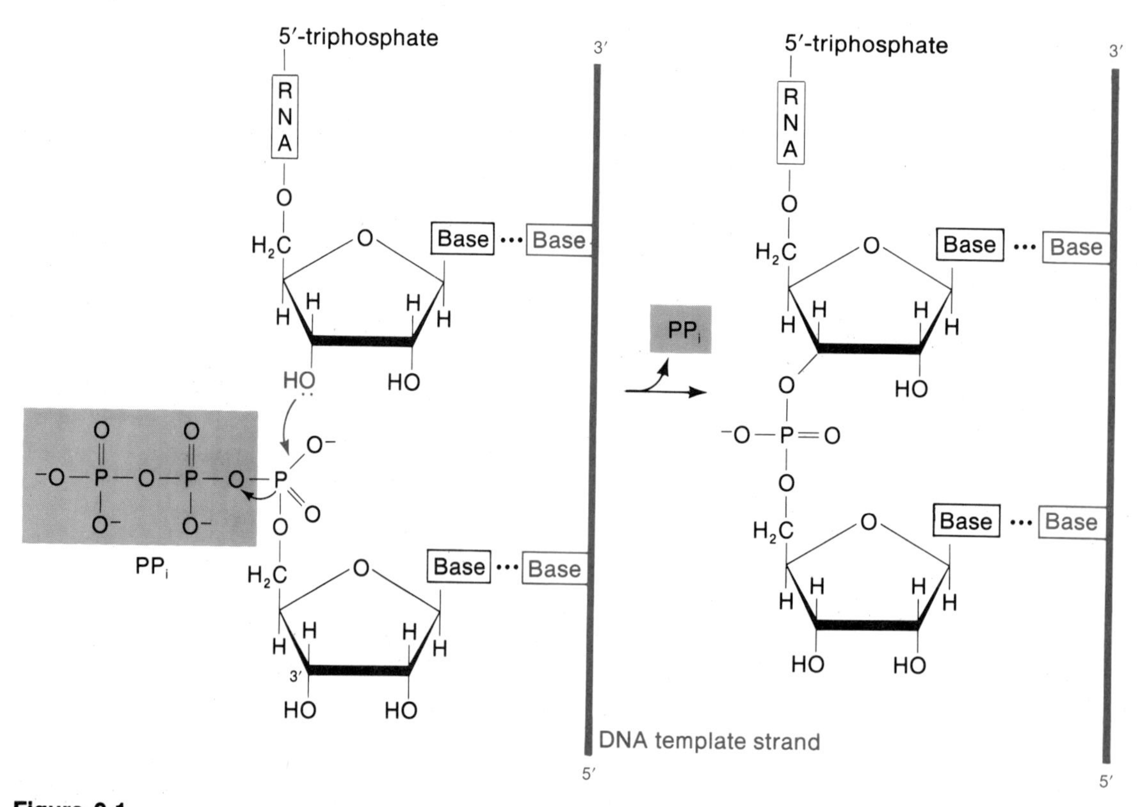

Figure 9-1
Mechanism of the chain-elongation reaction catalyzed by RNA polymerase. The red arrow joins the reacting groups. The pyrophosphate group (shaded in red) and the red hydrogen atom do not appear in the RNA strand. The DNA template and the RNA strands are antiparallel, as in double-stranded DNA.

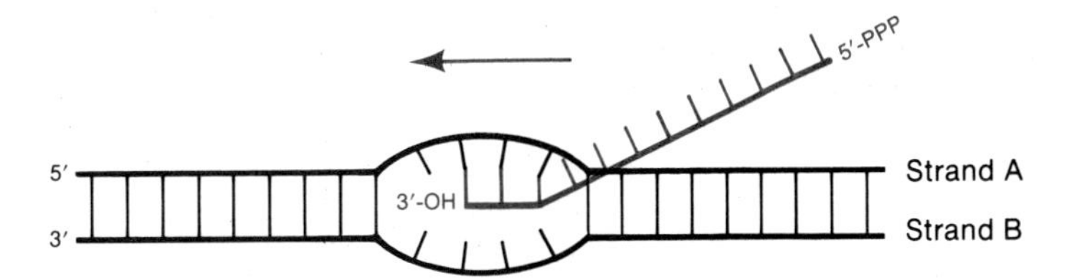

Figure 9-2
An RNA strand (shown in red) is copied only from strand A of a segment of a DNA molecule. No RNA is copied from strand B in that region of the DNA molecule. However, elsewhere, for example, in a different gene, strand B might be copied; in that case, strand A would not be copied in that region of the DNA. The RNA molecule is antiparallel to the DNA strand being copied and is terminated by a 5′-*tri*phosphate at the nongrowing end. The red arrow shows the direction of RNA chain growth.

DNA strand cause G, A, C, and U, respectively, to appear in the newly synthesized RNA molecule.

4. The DNA molecule being transcribed is double-stranded, yet in any particular region only one strand serves as a template. The meaning of this statement is shown in Figure 9-2.
5. The RNA chain grows in the 5′→3′ direction: that is, nucleotides are added only to the 3′-OH end of the growing chain—this is the same as the direction of chain growth in DNA synthesis. Furthermore, the RNA strand and the DNA template strand are antiparallel to one another.
6. RNA polymerases, in contrast with DNA polymerases, are able to initiate chain growth; that is, no primer is needed.
7. Only ribonucleoside 5′-triphosphates participate in RNA synthesis and the first base to be laid down in the initiation event is a triphosphate. Its 3′-OH group is the point of attachment of the subsequent nucleotide. Thus, the 5′ end of a growing RNA molecule terminates with a triphosphate (Figure 9-2).

The overall polymerization reaction may be written as

$$n\text{NTP} + \text{XTP} \xrightarrow[\text{Mg}^{2+}]{\text{DNA, RNA-P}} (\text{NMP})_n\text{—XTP} + n\text{PP}_i$$

in which XTP represents the first nucleotide at the 5′ terminus of the RNA chain, NMP is a mononucleotide in the RNA chain, RNA-P is RNA polymerase, and PP_i is the pyrophosphate released each time a nucleotide is added to the growing chain. The Mg^{2+} ion is required for all nucleic acid polymerization reactions.

E. coli RNA polymerase consists of five subunits—two identical α subunits and one each of types β, β', and σ. The σ subunit dissociates from the enzyme easily and in fact does so during one stage of polymerization. The term **holoenzyme** is used to describe the complete enzyme and **core enzyme** for the σ-free unit. We use the name RNA polymerase when the holoenzyme is meant. RNA polymerase is one of the largest enzymes known and can easily be seen by electron microscopy (Figure 9-3).

The synthesis of RNA consists of four discrete stages: (1) binding of RNA polymerase to a template at a specific site, (2) initiation, (3) chain elongation, and (4) chain termination and release. A discussion of these stages follows.

TRANSCRIPTION SIGNALS

The first step in transcription is binding RNA polymerase to a DNA molecule. Binding occurs at particular sites called **promoters,** which are specific sequences of 20–200 bases at which several interactions occur. Figure 9-4 shows portions of several promoter sequences in *E. coli* and *E. coli* phages (each promoter sequence is recognized by *E. coli* RNA polymerase) and their important features. In a region from five to ten bases to the left of the first base copied into mRNA is the right end of a sequence called the **Pribnow box.** All sequences found in Pribnow boxes are considered to be variants of a basic sequence TATAA$\underline{\text{T}}$G. The underscored T, at base 6 in the Pribnow box, from six to nine bases to the left of the first base transcribed (the distance depending on the distance from the Pribnow box to the transcription start point), is present in all promoters sequenced to date. It is called the "conserved T" and different sequences are usually compared by aligning conserved T's vertically. The Pribnow box is thought to be the sequence that orients RNA polymerase, so that synthesis proceeds from left to right (as the sequence is drawn), and the region at which the double helix opens to form the open-promoter complex (see below).

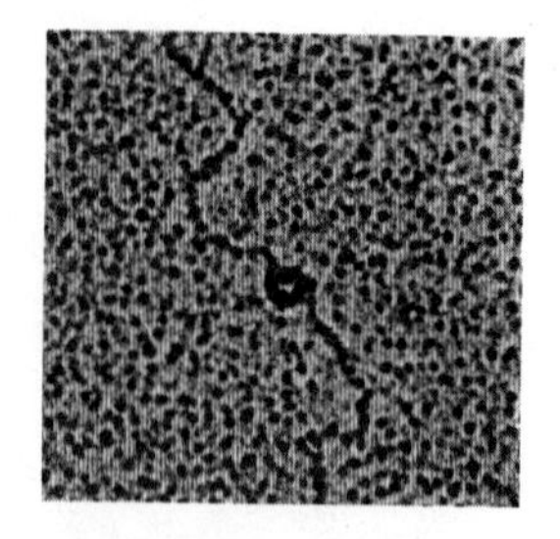

Figure 9-3
E. coli RNA polymerase molecule bound to DNA. (× 160,000). (Courtesy of Robley Williams.)

The Pribnow sequence TATAATG is an example of a **consensus sequence**—that is, a pattern of bases from which actual sequences observed in many different systems differ by usually no more than one or two bases. In eukaryotes the corresponding consensus sequence is TATAAAA (the **TATA** or **Hogness box**).

A second important sequence present in most prokaryotic promoters is called the **−35 sequence** because of its location with respect to the transcription start site (Figure 9-4). It contains nine bases and is thought to be the initial site of binding of RNA polymerase. Details of the initial binding process are unclear but it seems likely that the σ subunit first binds to the −35 sequence in a highly specific interaction and then, owing to the great size of the enzyme, the appropriate region of the polymerase can come in contact with the Pribnow box region

```
CCAGGCTTTACACTTTATGCTTCCGGCTCGTATGTTGTGTGGAATTG
CTTTTTGATGCAATTCGCTTTGCTTCTGACTATAATAGACAGGGTAA
GGCGGTGTT|GACATAAATACCACTGGCGGTGATACTGAGCACATCAG
GTGCGTGTT|GACTATTTTACCTCTGGCGGTGATAATGGTTGCATGTA
ATTGTTGTTGTT|AACTTGTTTATTGCAGCTTATAATGGTTACAAATA
CGTAACACTTTACAGCGGCGCGTCATTTGATATGATGCGCCCCGCTT
```

−35 Sequence   mRNA start

Figure 9-4
Base sequences in the noncoding strand of six different *E. coli* promoters, showing the three important regions.

(a) Template binding

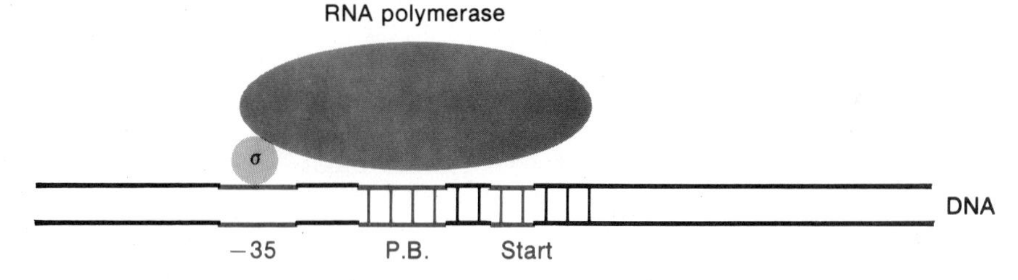

(b) Dissociation of σ subunit from −35 sequence; movement to Pribnow box.

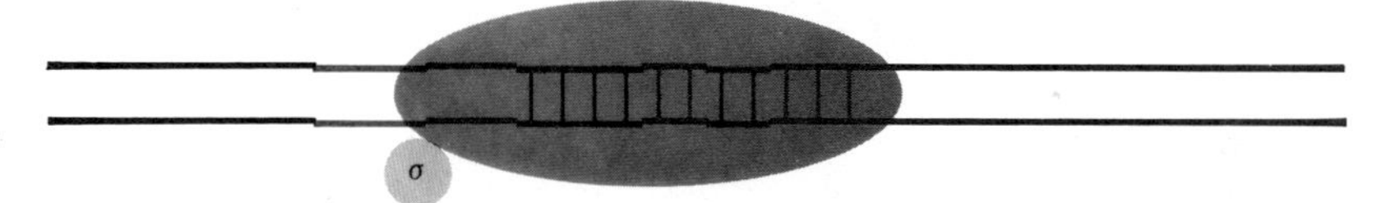

(c) Establishment of open-promoter complex

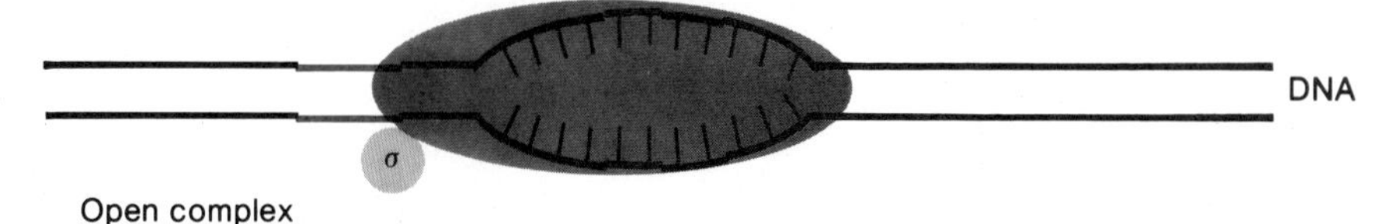

Figure 9-5
A proposed scheme for the binding of RNA polymerase to a promoter to form an open-promoter complex. Regions of the DNA molecule important for binding are shown in red. The shape of RNA polymerase is idealized for schematic purposes. The enzyme covers the region from bases −45 to +15, and the unpaired region in (c) extends from (roughly) −12 to +2. The enzyme is shown in contact with both strands because the strands are actually wrapped around one another in a helical array; however, true binding occurs only to bases in the coding strand. "PB" indicates the Pribnow box.

(Figure 9-5). Once bound to the Pribnow box, the polymerase then dissociates from the leftmost recognition site.

The **open-promoter complex** is a highly stable complex and is the active intermediate in chain initiation. In this complex a local unwinding ("melting") of the DNA helix occurs starting about ten base pairs from the left end of the Pribnow box and extending to the position of the first transcribed base. This melting is necessary for pairing of the incoming ribonucleotides. The base composition of the Pribnow box sequence (A+T-rich) renders the DNA strand susceptible to denaturation. Presumably RNA polymerase itself induces this conformational change.

When the promoter lacks a −35 sequence, a positive effector molecule is usually needed for initiating transcription. For example, the λ *pre* promoter is active only when the λ cII protein is present. Although the mechanism of action of these effectors is not well understood, it is clear that the effector must first bind to a specific sequence in the promoter and that this binding facilitates the interaction between RNA polymerase and the Pribnow box. When the effector is absent, RNA polymerase still binds to the Pribnow box, but the open-promoter complex does not form and transcription does not begin. An important effector protein, which will be discussed further in Chapter 14, is the CAP protein; this protein is needed for the activation of many promoters for genes required for sugar metabolism, and regulation of the binding ability of CAP is a major means of regulating the expression of these genes.

Once an open-promoter complex has formed, RNA polymerase is ready to initiate synthesis. RNA polymerase contains two nucleotide binding sites called the **initiation site** and the **elongation site.** The initiation site binds only purine triphosphates, namely ATP and GTP, and one of these (usually ATP) is the first nucleotide in the chain. (UTP has been observed but it is rare.) Thus, the first DNA base that is transcribed is usually thymine. The initiating nucleoside triphosphate binds to the enzyme in the open-promoter complex and forms a hydrogen bond with the complementary DNA base (Figure 9-6). The

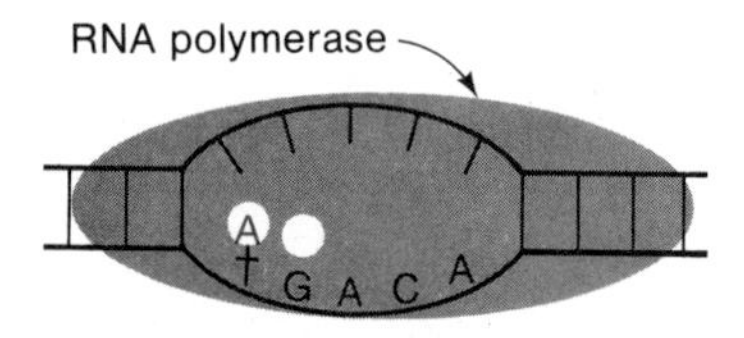

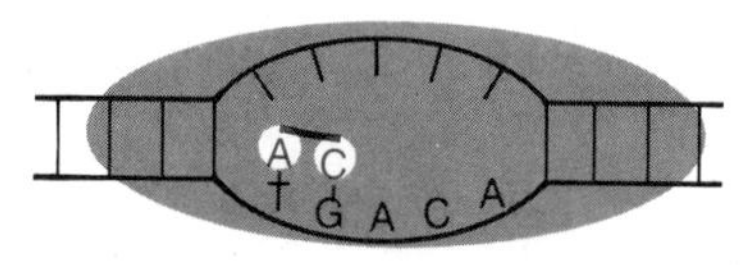

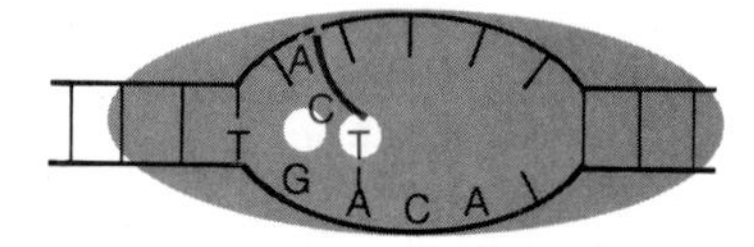

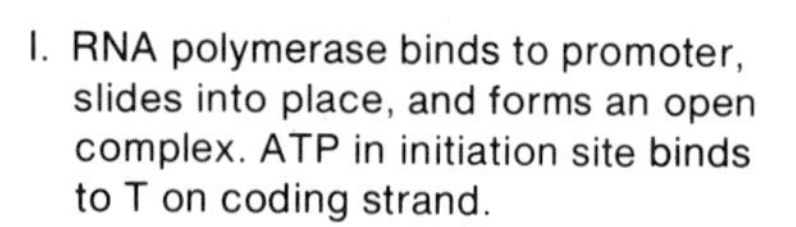
I. RNA polymerase binds to promoter, slides into place, and forms an open complex. ATP in initiation site binds to T on coding strand.

II. A NTP is added to the elongation site and is covalently linked to the A.

III. RNA polymerase moves over to the next DNA base. The initiating dinucleotide is released. A NTP enters the elongation site and is covalently linked to the dinucleotide. Then movement of RNA polymerase continues.

Figure 9-6
A scheme for initiation of RNA synthesis. The enzyme is drawn smaller than in Figure 11-9 and without the σ factor. Only the bases in the lower strand are recognized.

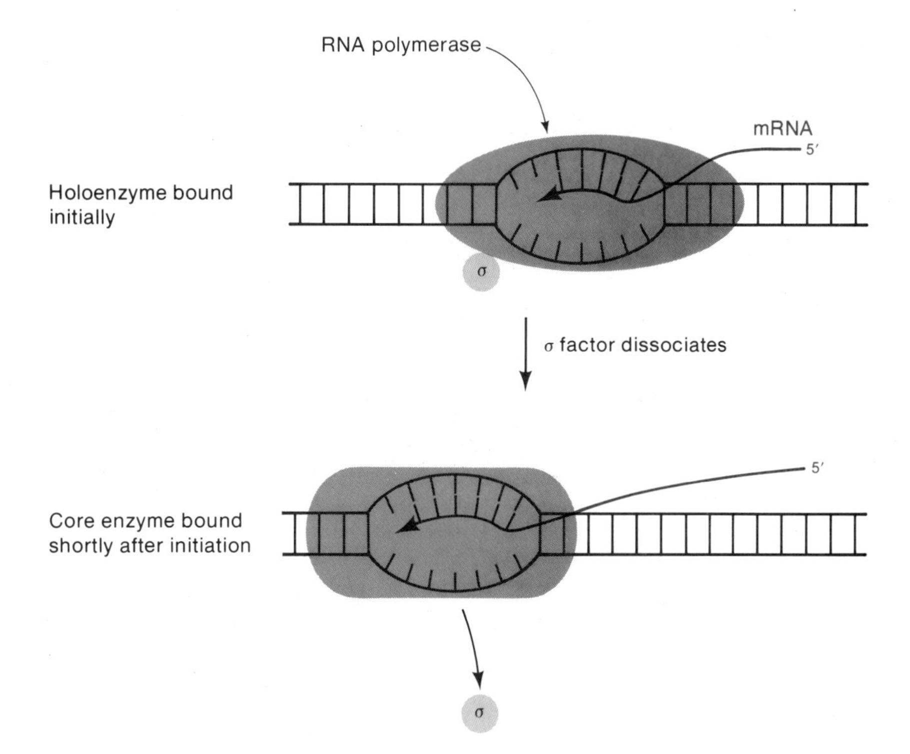

Figure 9-7
Diagram of a transcription bubble shortly after transcription begins and, at a slightly later stage, when the core enzyme has changed shape and the σ subunit has dissociated from the core enzyme.

elongation site is then filled with a nucleoside triphosphate that is selected strictly by its ability to form a hydrogen bond with the next base in the DNA strand. The two nucleotides are then joined together, the first base is released from the initiation site, and initiation is complete. The dinucleotide remains hydrogen-bonded to the DNA. The elongation phase then begins when the polymerase releases the base and then moves along the DNA chain.

After several nucleotides (approximately eight) are added to the growing chain, RNA polymerase changes its structure and loses the σ subunit. Thus, most elongation is carried out by the core enzyme (Figure 9-7). The core enzyme moves along the DNA, binding a nucleoside triphosphate that can pair with the next DNA base and opening the DNA helix as it moves. The open region extends only over a few base pairs; that is, the DNA helix recloses just behind the enzyme.

The newly synthesized RNA is released from its hydrogen bonds with the DNA as the helix re-forms; however, a few RNA bases remain paired with the DNA template during RNA synthesis.

Termination of RNA synthesis occurs at specific base sequences in the DNA molecule. Twenty termination sequences have so far been

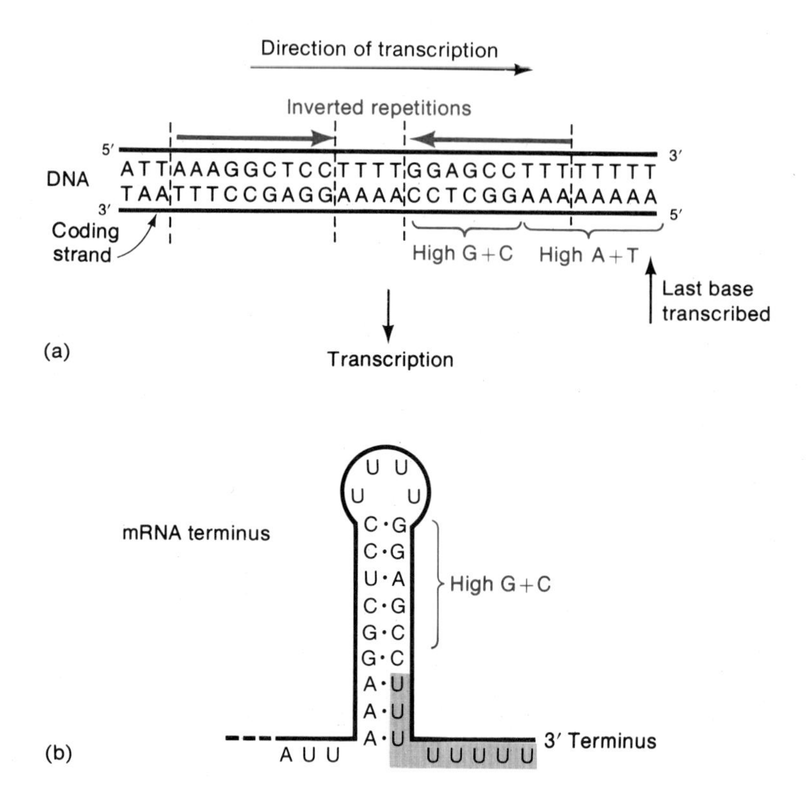

Figure 9-8
Base sequence of (a) the DNA of the *E. coli trp* operon at which transcription termination occurs and of (b) the 3′ terminus of the mRNA molecule. The inverted repeat sequence is indicated by reversed red arrows. The mRNA molecule is folded to form a stem-and-loop structure thought to exist. The relevant regions are labeled in red; the terminal sequence of U's in the mRNA is shaded in red.

determined and each has the characteristics shown in Figure 9-8. There are three important regions:

1. First, there is an inverted-repeat base sequence containing a central nonrepeating segment; that is, the sequence in one DNA strand would read like ABCDEF-XYZ-F′E′D′C′B′A′ in which A and A′, B and B′, and so on, are complementary bases. Thus, this sequence is capable of intrastrand base pairing, forming a "stem-and-loop" configuration in the RNA transcript and possibly in the DNA strands.
2. The second region is near the loop end of the putative stem (sometimes totally within the stem) and is a sequence having a high G+C content.
3. A third region (sometimes absent) is a sequence of A • T pairs (which may begin in the putative stem) that yields in the RNA a sequence of six to eight uracils often followed by an adenine.

There are two types of termination events—those that depend only on the DNA base sequence and those that require the presence of a termination protein called **Rho.** Both types of events occur at specific but distinct base sequences. Rho-dependent termination is important in certain regulatory mechanisms, for the termination activity of Rho can be prevented by inhibitors called antiterminators. An example of such an event will be seen in Chapter 15.

The final step in the termination process is dissociation of the core enzyme from the DNA. Following this event, the core enzyme interacts with a free σ subunit to re-form the holoenzyme. Thus there is a σ cycle—σ falls off the holoenzyme after elongation begins and rejoins a core enzyme (not necessarily the same one) that has dissociated from the DNA. The holoenzyme is then available for initiating synthesis again. In general, a particular holoenzyme does not bind to the promoter of the gene that it has just transcribed.

CLASSES OF RNA MOLECULES

There are three major classes of RNA molecules—messenger RNA (mRNA), ribosomal RNA (rRNA), and transfer RNA (tRNA). All are synthesized from DNA base sequences, which are termed RNA genes, but they have very different functions in protein synthesis, as will be seen in this and the next chapter. There are also significant differences between the structures and modes of synthesis of the RNA molecules of prokaryotes and eukaryotes, though the basic mechanisms of their functions are nearly the same. The greatest amount of information has been obtained from studies with bacteria and bacterial cell extracts, so that it is here that we begin. Transcription in eukaryotes and the structure and synthesis of eukaryotic RNA molecules are discussed in a later section.

Messenger RNA

The base sequence of a DNA molecule determines the amino acid sequence of every polypeptide chain in a cell, though amino acids have no affinity for DNA. Thus, instead of a direct pairing between amino acids and DNA, a multistep process is used in which the information contained in the DNA is converted to a form in which amino acids can be arranged in an order determined by the DNA base sequence. This process begins with the transcription of the base sequence of one of the DNA strands (the **coding strand** or **sense strand**) into the base sequence of an RNA molecule and it is from this molecule—messenger RNA—that the amino acid sequence is obtained by the protein-synthesizing machinery of the cell. (The DNA strand that is not transcribed is called the **antisense strand.**) As we will see in Chapter 10, the base sequence of the mRNA is then read in groups of three bases (a group of three is called a **codon**) from a start codon to a stop point, with each codon corresponding either to one amino acid or a stop signal.

A DNA segment corresponding to one polypeptide chain plus the start and stop signals is called a **cistron** and a mRNA encoding a single polypeptide is called monocistronic mRNA. It is very common for a mRNA molecule to encode several different polypeptide chains; in this case it is called a **polycistronic mRNA** molecule.

In addition to cistrons and start and stop sequences for translation, other regions in mRNA are significant. For example, translation of a mRNA molecule (that is, protein synthesis) seldom starts exactly at one end of the RNA and proceeds to the other end; instead, initiation of synthesis of the first polypeptide chain of a polycistronic mRNA may begin hundreds of nucleotides from the 5′-P terminus of the RNA. The section of nontranslated RNA before the coding regions is called a **leader;** in some cases, the leader contains a regulatory region (called an **attenuator**) that determines the rate of protein synthesis. Untranslated sequences are found at both the 5′-P and the 3′-OH termini and a polycistronic mRNA molecule may contain intercistronic sequences (**spacers**) hundreds of bases long.

An important characteristic of prokaryotic mRNA is its short lifetime; usually within a few minutes after an mRNA molecule is synthesized, nuclease degradation occurs. Although this means that continuous synthesis of a particular protein requires ongoing synthesis of the corresponding mRNA molecule, rapid degradation of mRNA is nonetheless advantageous to bacteria, whose environment and needs often fluctuate widely. Synthesis of a necessary protein can be regulated as needed, simply by controlling transcription. When a particular protein is needed by a cell, the appropriate mRNA molecule will be made. However, when the protein is no longer needed, inhibition of synthesis of the mRNA is sufficient to prevent wasted synthesis, because the previously made mRNA molecules will gradually be degraded.

Stable RNA: Ribosomal RNA and Transfer RNA

During the synthesis of proteins genetic information is supplied by messenger RNA. RNA also plays other roles in protein synthesis. For example, proteins are synthesized on the surface of an RNA-containing particle called a **ribosome;** these particles consist of three classes of ribosomal RNA (**rRNA**), which are stable molecules. Also, amino acids do not line up against the mRNA template independently during protein synthesis but are aligned by means of a set of about fifty adaptor RNA molecules called transfer RNA (**tRNA**), also a stable species. Each tRNA molecule is capable of "reading" three adjacent mRNA bases (a codon) and placing the corresponding amino acid at a site on the ribosome at which a peptide bond is formed with an adjacent amino acid. Neither rRNA nor tRNA is used as a template. The roles of ribosomes and of tRNA molecules will be explained in the following chapter; here we are concerned only with their synthesis, which involves posttranscriptional modification of a remarkable sort.

The synthesis of both rRNA and tRNA molecules is initiated at a promoter and completed at a terminator sequence and, in this respect, their synthesis is no different from that of mRNA. However, the following three properties of these molecules indicate that neither rRNA nor tRNA molecules are the **primary transcripts** (immediate products of transcription):

1. The molecules are terminated by a 5′-monophosphate rather than the expected triphosphate found at the ends of all primary transcripts.
2. Both rRNA and tRNA molecules are much smaller than the primary transcripts (the transcription units).
3. All tRNA molecules contain bases other than A, G, C, and U and these "unusual" bases (as they are called) are not present in the original transcript.

All of these molecular changes are made after transcription by a process called **posttranscriptional modification** or, more commonly, **processing.**

Both rRNA and tRNA molecules are excised from large primary transcripts. Often a single transcript contains the sequences for several different molecules—for example, different tRNA molecules, or both tRNA and rRNA molecules. Formation of rRNA molecules is a result of fairly straightforward excision of a single continuous sequence. However, tRNA molecules—which contain bases other than A, U, G, and C—are not only cut by several enzymes acting in a particular order, but also chemical modification of various bases in the primary transcript occurs. An example of the production of a particular tRNA molecule is given in Figure 9-9, to illustrate the complexity of the process. A detailed discussion of processing is beyond the scope of this book but can be found in *MB*, pages 394–399.

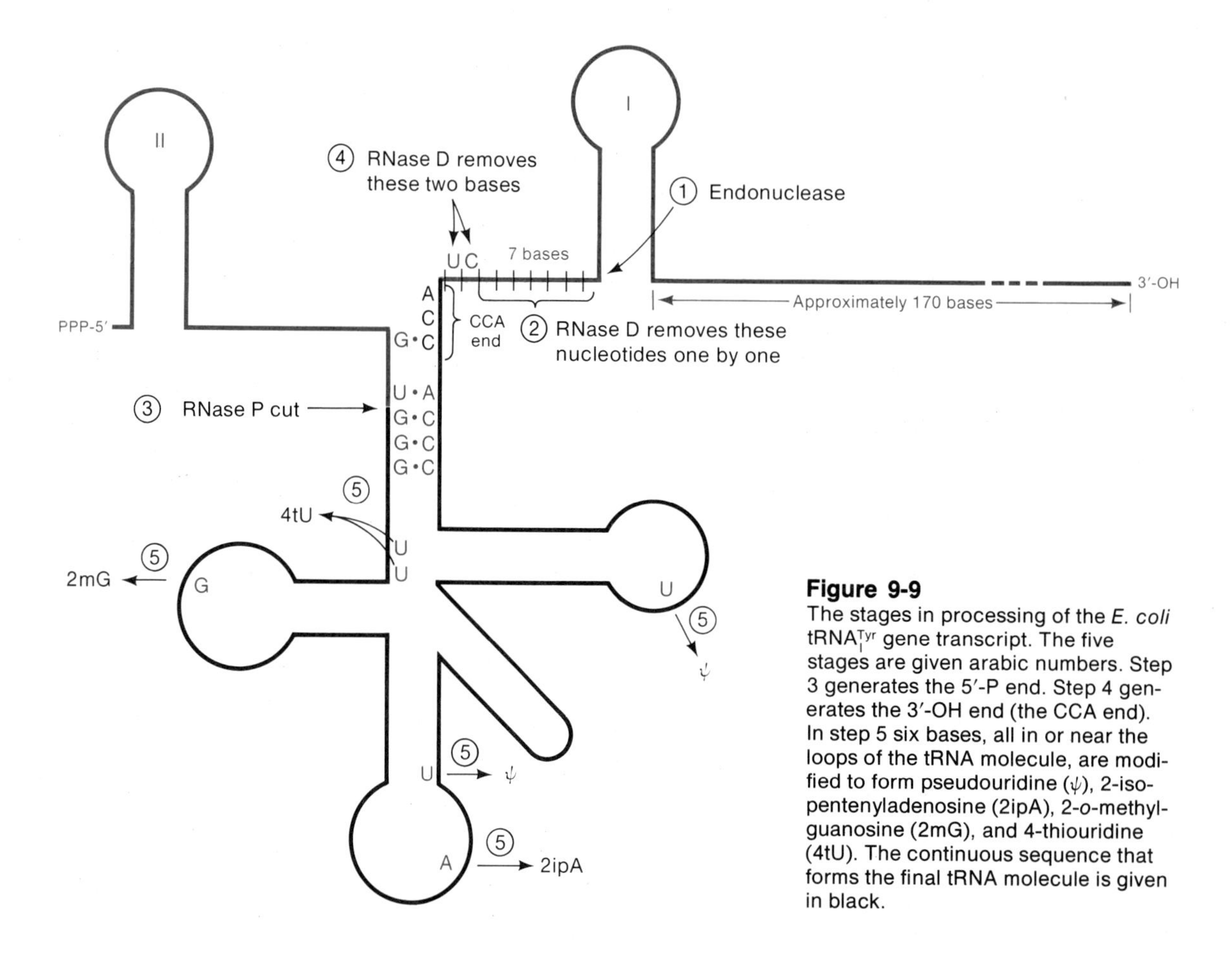

Figure 9-9
The stages in processing of the *E. coli* $tRNA_I^{Tyr}$ gene transcript. The five stages are given arabic numbers. Step 3 generates the 5′-P end. Step 4 generates the 3′-OH end (the CCA end). In step 5 six bases, all in or near the loops of the tRNA molecule, are modified to form pseudouridine (ψ), 2-isopentenyladenosine (2ipA), 2-*o*-methylguanosine (2mG), and 4-thiouridine (4tU). The continuous sequence that forms the final tRNA molecule is given in black.

TRANSCRIPTION IN EUKARYOTES

The basic features of the transcription and the structure of mRNA in eukaryotes are similar to those in bacteria. However, there are five notable differences, which are the following:

1. Eukaryotic cells contain three classes of nuclear RNA polymerase and these are responsible for the synthesis of different classes of RNA.
2. Many mRNA molecules are very long lived.
3. Both the 5′ and 3′ termini are modified; a complex structure called the **cap** is found at the 5′ end and a long (up to 200 nucleotides) sequence of polyadenylic acid—poly(A)—is found at the 3′ end.

4. The mRNA molecule that is used as a template for protein synthesis is usually about one-tenth the size of the primary transcript. During mRNA processing intervening sequences called **introns** are excised and the fragments are rejoined.
5. All eukaryotic mRNA molecules are monocistronic.

These points are illustrated in Figure 9-10, which shows a schematic diagram of a typical eukaryotic mRNA molecule and how it is produced. Clearly the production of mRNA in eukaryotes is not a simple matter of transcribing the DNA.

Initiation of transcription in eukaryotes is not well understood. A sequence—TATAAAAA—analogous to the Pribnow box is required for initiation. Other sequences called upstream activation sites and enhancers, which will be discussed in Chapter 16, affect the efficiency of initiation, but their mechanism of action is not known. Poorly understood proteins, called transcription factors, are also involved in the initiation of transcription of certain classes of genes. The fact that DNA is in the form of chromatin certainly must make transcription more complicated in eukaryotes than in prokaryotes, and evidence exists suggesting that chromatin structure is altered in regions being transcribed; how it is altered is far from clear though. There is no information available about termination nor is there any evidence for Rho-like termination factors.

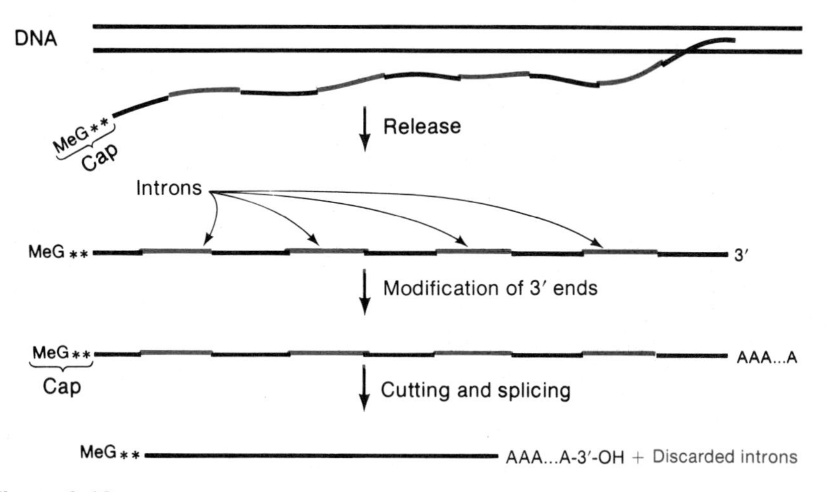

Figure 9-10
Schematic drawing showing production of eukaryotic mRNA. The primary transcript is capped before it is released. Then, its 3′-OH end is modified, and finally the intervening regions are excised. MeG denotes 7-methylguanosine and the two asterisks indicate the two nucleotides whose riboses are methylated.

The three major classes of eukaryotic RNA polymerases are denoted I, II, and III; they can be distinguished by the ions required for their activity, the optimal ionic strength, and their sensitivity to inhibition by various antibiotics. All are found in the eukaryotic nucleus. Minor RNA polymerases, which have not yet been studied in detail, are found in mitochondria and chloroplasts. The locations and products of the nuclear RNA polymerases are listed below.

	Class I	*Class II*	*Class III*
Location:	Nucleolus	Nucleoplasm	Nucleoplasm
Product:	rRNA	mRNA	tRNA, 5S RNA

Note that RNA polymerase II is the enzyme responsible for all mRNA synthesis.

The biochemical reaction catalyzed by the eukaryotic RNA polymerases is the same as that catalyzed by *E. coli* RNA polymerase.

The 5′ terminus of a eukaryotic mRNA molecule carries a methylated guanosine derivative, 7-methylguanosine (7-MeG) in an unusual 5′-5′ linkage to the 5′-terminal nucleotide of the primary transcript. Occasionally the sugars of the adjacent nucleotides are also methylated. The unit

(7-MeG)-5′-PPP-5′-(G or A, with possibly methylated ribose)-3′-P-

in which P and PPP refer to mono- and triphosphate groups respectively, is called a **cap.**

Capping occurs shortly after initiation of synthesis of the mRNA, possibly before RNA polymerase II leaves the initiation site, and precedes all excision and splicing events. The biological significance of capping has not yet been unambiguously established but it is believed that it is required for efficient protein synthesis. Capping may function to protect the mRNA from degradation by nucleases and to provide a feature for recognition by the protein-synthesizing machinery.

Most, but not all, animal mRNA molecules are terminated at the 3′ end with a poly(A) tract, which is added to the primary mRNA by a nuclear enzyme, poly(A) polymerase. The adenylate residues are not added to the 3′ terminus of the primary transcript. Transcription normally passes the site of addition of poly(A), so that an endonucleolytic cleavage must occur before the poly(A) is added. A base sequence–AAUAAA–10 to 25 bases upstream from the poly(A) site is a component of the system for recognizing the site of poly(A) addition. Interestingly, some primary transcripts contain two or more sites at which

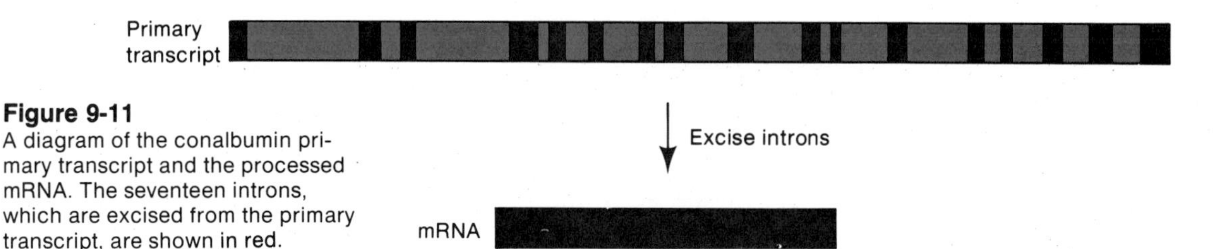

Figure 9-11
A diagram of the conalbumin primary transcript and the processed mRNA. The seventeen introns, which are excised from the primary transcript, are shown in red.

poly(A) can be added. The differentially terminated mRNA molecules usually play different roles in the life cycle of the particular organism. The length of the poly(A) segment can be from 20 to 200 nucleotides. The significance of the poly(A) terminus is unknown at present, but it is believed that it increases the stability of mRNA; a possible role in attachment of mRNA to intracellular membranes has also been proposed. Some cellular mRNA molecules lack poly(A), so its presence cannot be obligatory for successful translation.

Most of the primary transcripts of higher eukaryotes contain untranslated intervening sequences (**introns**) that interrupt the coding sequence and are excised in the conversion of the primary transcript to mRNA (Figures 9-10 and 9-11). The amount of discarded RNA ranges from 50 percent to nearly 90 percent of the primary transcript. The remaining segments (**exons**) are joined together to form the finished mRNA molecules. The excision of the introns and the formation of the final mRNA molecule by joining of the exons is called **RNA splicing.**

The number of introns per gene varies considerably (Table 9-1) and for a given protein is not the same in all organisms. Furthermore, within

Table 9-1
Translated eukaryotic genes in which introns have been demonstrated

Gene	*Number of introns*
α-Globin	2
Immunoglobulin L chain	2
Immunoglobulin H chain	4
Yeast mitochondria cytochrome *b*	6
Ovomucoid	6
Ovalbumin	7
Ovotransferrin	16
Conalbumin	17
α-Collagen	52

Note: At present the genes for histones and for interferon are the only known translated genes in the higher organisms that do not contain introns.

a particular gene the introns have many different sizes and are usually larger than exons.

Splicing occurs in the nucleus after capping and addition of poly(A). The existence of splicing explains the fact the nucleus contains an enormous number of different RNA molecules, whose size distribution is very great. All animal mRNA molecules are monocistronic, so that one might expect the mRNA to have a small range of sizes. However, the amount of RNA that is discarded as a result of excision and splicing ranges widely, so the sizes of the primary transcripts cover a range of molecular weights from about 6×10^4 to 3×10^6. Furthermore, introns are excised one by one, and apparently ligation occurs before the next intron is excised, so the number of different nuclear RNA molecules present at any instant is huge. Translation does not occur until processing is complete, so that before processing was recognized, the nucleus appeared to contain a significant amount of possibly useless RNA; these molecules were given the general name **heterogeneous nuclear RNA** (HnRNA), a term that is still used. After processing is completed, the mature mRNA is transported to the cytoplasm to be translated; it is not known why the primary transcript and partially processed molecules are not transported.

The necessity to remove an intron without altering the coding sequence of the resulting mRNA molecule demands great fidelity in the cleavage process. For example, a cutting error in which the cutting site is displaced by a single base would completely destroy the reading frame of the bases in the mRNA. This fidelity is provided by the base sequence itself. In all genes observed, in which splicing occurs, the splice sites of the primary transcript are marked by a sequence resembling

$$5'-{}^{\mathrm{A}}_{\mathrm{C}}\mathrm{AG}\overset{\downarrow}{}\underline{\mathrm{GT}}{}^{\mathrm{A}}_{\mathrm{G}}\mathrm{AGT}\ \overset{\text{intron}}{\cdots\cdots}\ (\mathrm{Py})_6\mathrm{XC}\underline{\mathrm{AG}}\overset{\downarrow}{}\mathrm{G}^{\mathrm{G}}_{\mathrm{T}}-3'$$

in which Py is any pyrimidine, X is any base, and the arrows are the splice points. The underlined bases are the same for almost observed introns (that is, they are conserved), whereas the remainder is a consensus sequence in which there is some variation from one splice site to the next.

In the nucleus many RNA molecules are complexed with specific proteins, forming a **ribonucleoprotein particle** called an **RNP.** In the course of the splicing reaction primary transcripts are assembled into particles called HnRNP. A second small nuclear RNP, called **U1 snRNP,** is essential to splicing. This particle contains seven proteins and a single small RNA molecule with 165 nucleotides; part of the RNA base sequence is complementary to the sequence at the 5′ side of the intron and forms a double-stranded RNA-RNA hybrid in an early stage of the splicing process. It is unlikely that the U1 RNA also binds to the 3′ splice site; what recognizes this site and what brings the two adjacent exons together is not known. The chemical events in the splicing process have recently been elucidated, but will not be presented.

The existence of introns seems to be fairly wasteful, especially in view of the relative amounts of intron and exon RNA. In a few cases, the presence of introns increases the coding capacity of a particular stretch of genes in that alternative splicing patterns yield different mRNA molecules. This is common in DNA viruses, but not particularly frequent in cellular DNA, though about ten examples are known at present.

Many hypotheses have been presented for the origin of interrupted genes. One prominent theory is that each exon represents the coding sequence of an ancestral small protein and that in the course of evolution rearrangements occurred that brought the sequences together from various parts of the genome, forming new proteins. If no mechanism for splicing RNA had existed, production of the coding sequence of a new, functional protein would have required a precise recombination of sequences; most rearrangements would not have yielded a functional protein and in time they would have been lost. An alternative is that splicing became possible and that the sequences were instead brought initially into a single transcription unit. With such a rearrangement the ancestral proteins would have still have been synthesized, and the cell would have grown and reproduced normally. If splicing had occurred occasionally, different splicing patterns would have been tried in the course of time, and any splicing event that produced an improved protein and thereby a superior organism would have been advantageous and therefore maintained by natural selection. Examination of the structure of many eukaryotic proteins suggests that such a scenario may have occurred repeatedly. Many proteins are known whose polypeptide chains fold to form **domains.** That is, the three-dimensional structure of these proteins in such that the molecules appear to consist of several independently folded regions, each separated by a short polypeptide segment (Figure 9-12). Furthermore, the binding sites of proteins with multiple binding sites and of enzymes having several catalytic activities are often located in separate domains. The significant point is that in many cases the amino acid sequence of each domain is encoded in one exon, suggesting that each domain is derived from an ancestral protein, as proposed.

Earlier, it was pointed out that prokaryotic tRNA is excised from a larger primary transcript, and, as indicated in Figure 9-9, that the tRNA sequence does not contain intervening sequences. However, in eukary-

Figure 9-12
A diagram of a hypothetical polypeptide chain folded into two domains, one black and one red. In multifunctional proteins each domain may contain a binding site. In proteins having a single function a binding site frequently occurs at the contact point between the domains (arrow).

otes the tRNA primary transcript contains a single intron (in the anticodon segment), which is excised. The tRNA primary transcript differs from most primary transcripts in that the intrastrand base-pairing and folding are identical to that of the finished tRNA molecule. Furthermore, the folding is such that the splice sites are next to one another, which facilitates the splicing process. With these molecules splicing is a two-stage reaction. In the first step, an RNA endonuclease cleaves the primary transcript and the intron is released; in the second step, the two ends of the nearly finished molecule are joined by an RNA ligase. There is mounting evidence that tRNA splicing occurs when the molecule is bound to the nuclear membrane. This is an important observation because it may indicate that splicing is linked to transport of RNA from the nucleus to the cytoplasm.

MEANS OF STUDYING INTRACELLULAR RNA

Several procedures are used to study RNA metabolism *in vivo*. In most of the techniques, radioactive RNA is prepared by adding ^{3}H-labeled or ^{14}C-labeled uridine to growth medium. However, this also produces radioactive DNA, because uridine can be metabolized to cytosine and thymine; this is significant when studying RNA metabolism because, when RNA is isolated from cells, it is usually contaminated with DNA. The contaminating DNA can be removed, though, by treatment of the sample with pancreatic DNase, for this enzyme degrades the DNA to mononucleotides and small oligonucleotides, which can be easily separated from the RNA.

Most investigations of RNA *in vivo* are concerned with either the presence, synthesis, or degradation of specific mRNA species—for example, the mRNA from a particular gene. In order to carry out such analyses, it is necessary to distinguish the particular RNA molecule from all other RNA molecules. This is usually done by **DNA-RNA hybridization.** With this technique, DNA with a sequence complementary to the RNA molecule of interest is permanently fixed to a nitrocellulose filter. The filter is subsequently incubated with an extract containing radioactive RNA, using conditions leading to renaturation, and then washed repeatedly. RNA that is not present in DNA-RNA hybrids is removed; the amount of radioactivity remaining on the filter is a measure of the amount of the particular RNA present in the extract.

A variety of procedures are used to obtain the DNA used in a hybridization experiment. In the most useful experiments, a DNA segment containing a sequence complementary to the RNA of interest is inserted (by the genetic engineering techniques described in Chapter 13) into a piece of foreign DNA that has no sequences in common with the RNA and that is easily purified. In this case, the DNA sequence of interest is said to be **cloned.** Typical foreign DNA molecules are plasmids, and phage and viral DNA. A DNA molecule carrying a cloned DNA sequence to be used in a hybridization experiment is called a **probe.**

The technique used most commonly is the **Southern transfer,** or **Southern blotting,** procedure (named after E. Southern, its developer), a method in which hybridization is performed with a large number of distinct DNA segments simultaneously. In this procedure the total DNA of an organism is broken into discrete fragments by restriction enzymes (Chapter 13), which make cuts in unique base sequences. Fragments are separated by gel electrophoresis (as in the Maxam-Gilbert base-sequencing technique, Chapter 3) and hybridization is carried out to all of the fragments. Various techniques, which will not be described, enable the positions of particular fragments to be identified and the location of hybridized radioactive RNA to be matched with these fragments. The technique is performed as follows (Figure 9-13).

DNA is enzymatically fragmented and then electrophoresed through an agarose gel. Following electrophoresis the gel is soaked in a denaturing solution (usually NaOH), so that all DNA in the gel is converted to single-stranded DNA, which is needed for hybridization. A large sheet of nitrocellulose paper is placed on top of several sheets of ordinary filter paper; the gel, which is typically in the form of a broad flat slab, is then placed on the nitrocellulose filter and covered with a glass plate to prevent drying. A weight is then placed on the top of the stack and the liquid is squeezed out of the gel. The liquid passes downward through the nitrocellulose filter and the denatured DNA binds tightly to the nitrocellulose; the remaining liquid passes through the nitrocellulose and is absorbed by the filter paper. DNA molecules do not diffuse very much, so that if the gel and the nitrocellulose are in firm

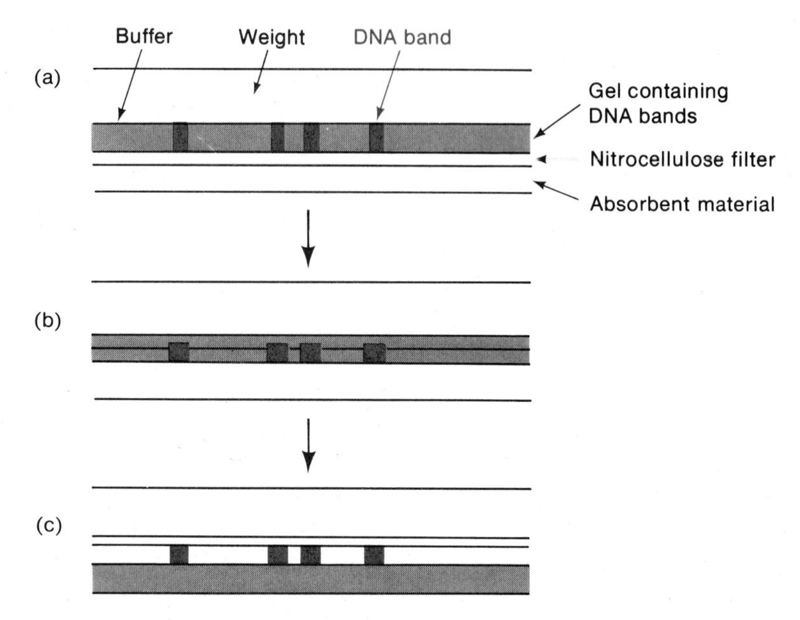

Figure 9-13
The Southern transfer technique. (a) A stack—consisting of a weight, a gel, a filter, and absorbent material—at the time the weight is applied. (b) A later time—the weight has forced the buffer (shaded area), which carries the DNA, into the nitrocellulose. (c) The lowest layer has absorbed the buffer but not the DNA, which remains bound to the nitrocellulose.

contact, the positions of the DNA molecules on the filter are identical to their positions in the gel. The nitrocellulose filter is then dried in vacuum, which insures that the DNA remains on the filter during the hybridization step. The dried filter is then moistened with a very small volume of a solution of ^{32}P-labeled RNA, placed in a tight-fitting plastic bag to prevent drying, and held at a temperature suitable for renaturation (usually for 16–24 hours). The filter is then removed, washed to remove unbound radioactive molecules, dried, and autoradiographed with x-ray film. The blackened positions of the film indicate the locations of the DNA molecules whose DNA base sequences are complementary to the sequences of the added radioactive molecules. Since usually the genes contained in each fragment are known, specific mRNA molecules can be identified. The degree of blackening of the film is easily measured quantitatively and is proportional to the amount of RNA that has hybridized. Thus, the amount of mRNA transcribed from each region of a DNA molecule can be measured. Note that by this technique many different mRNA molecules can be studied simultaneously.

The Southern transfer method has many other uses, particularly for DNA-DNA hybridization analysis.

10 Translation

The synthesis of every protein molecule in a cell is directed by intracellular DNA. There are two aspects to understanding how this is accomplished—the **information** or **coding problem** and the **chemical problem.** By the information problem is meant the mechanism by which a base sequence in a DNA molecule is translated into an amino acid sequence of a polypeptide chain. The chemical problem refers to the actual process of synthesis of the protein: the means of initiating synthesis; linking together the amino acids in the correct order; terminating the chain; releasing the finished chain from the synthetic apparatus; folding the chain; and, often, postsynthetic modification of the newly synthesized chain. The overall process is called **translation.**

The approach in this chapter is to present an outline of the process, in order to introduce the terminology, and then to present the major features of the coding and decoding systems, and the mechanism for polypeptide synthesis.

OUTLINE OF TRANSLATION

Protein synthesis occurs on intracellular particles called **ribosomes.** These particles, which in prokaryotes consist of three RNA molecules and about 55 different protein molecules, contain the enzymes needed to form a peptide bond between amino acids, a site for binding the

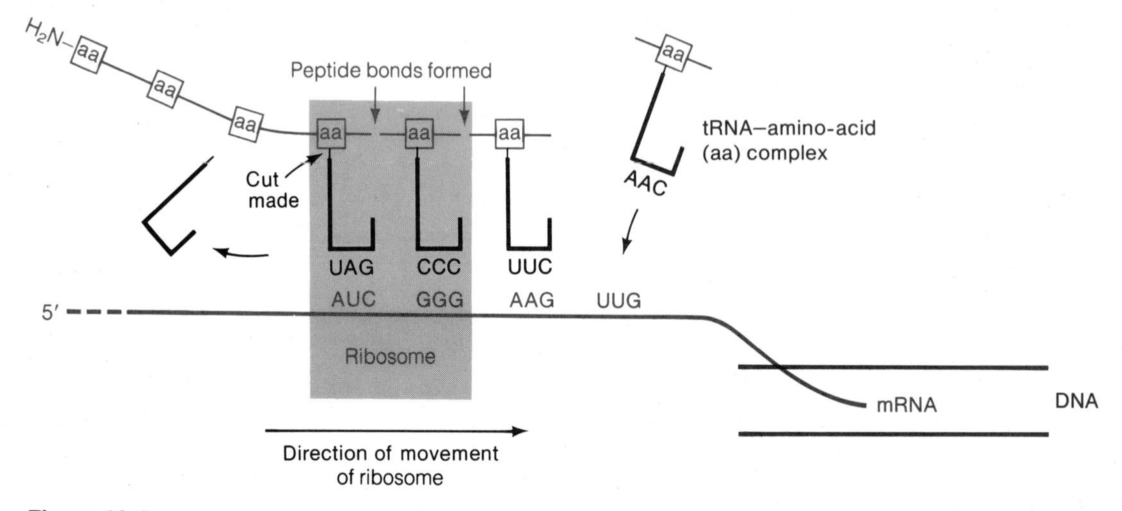

Figure 10-1
A diagram showing how a protein molecule is synthesized. Note that the relative directions of polypeptide and RNA synthesis are such that polypeptide synthesis can occur before the mRNA is completed. This is, in fact, the case in prokaryotes.

mRNA, and sites for bringing in and aligning the amino acids in preparation for assembly into the finished polypeptide chain. Amino acids themselves are unable to interact with the ribosome and cannot recognize bases in the mRNA molecule. Thus there exists the collection of carrier molecules mentioned in the previous chapter, **transfer RNA (tRNA).** These molecules contain a site for amino acid attachment and a region called the **anticodon** that recognizes the appropriate base sequence (the **codon**) in the mRNA. Thus, proper selection of the amino acids for assembly is determined by the positioning of the tRNA molecules, which in turn is determined by hydrogen-bonding between the anticodon of each tRNA molecule and the corresponding codon of the mRNA.

A schematic diagram showing the events that occur on the ribosome is given in Figure 10-1. The scheme shown applies equally to prokaryotes and eukaryotes, with a single exception. In eukaryotes, since transcription occurs in the nucleus and protein synthesis occurs in the cytoplasm, the mRNA is not attached to the DNA during protein synthesis, in contrast with what is shown in the figure.

Clearly there must be many different tRNA molecules, because each amino acid must be brought in to the ribosome in a way that ensures that it corresponds to the base sequence of the mRNA. Thus,

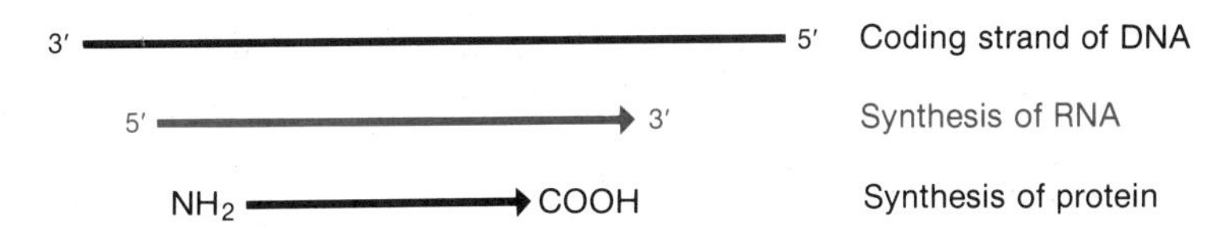

Figure 10-2
Directions of synthesis of RNA and protein with respect to the coding strand of DNA.

there are specific tRNA molecules that correspond to each amino acid. Furthermore, the linkage of each amino acid to its tRNA molecule is catalyzed by a specific enzyme, which ensures that the appropriate amino acid will be attached to the correct tRNA molecule.

In earlier chapters several examples of directional synthesis of macromolecules were seen, for example, DNA and RNA. This is also the case for polypeptide synthesis, which begins at the amino terminus. Furthermore, translation of an mRNA molecule occurs in only one direction, namely, 5′ to 3′. Figure 10-2 summarizes the polarity of synthesis of mRNA and protein with respect to the DNA coding strand. These directions were also shown in Figure 10-1.

THE GENETIC CODE

By the **genetic code** one means the collection of base sequences (codons) that correspond to each amino acid and to translation signals.

Since there are 20 amino acids, there must be more than 20 codons to include signals for starting and stopping the synthesis of particular protein molecules. If one assumes that all codons have the same number of bases, then each codon must contain at least three bases. The argument for this conclusion is the following. A single base cannot be a codon because there are 20 amino acids and only four bases. Pairs of bases also cannot serve as codons because there are only $4^2 = 16$ possible pairs of four bases. Triplets of bases are possible because there are $4^3 = 64$ triplets, which is more than adequate. In fact, the genetic code is a triplet code and all 64 possible codons carry information of some sort. Furthermore, in translating mRNA molecules the codons do not overlap but are "read" sequentially (Figure 10-3).

The general properties of the code—for example, that each codon contains three bases and that codons do not overlap—were deduced from genetic experiments (see *MB*, pages 429–433). The sequence of

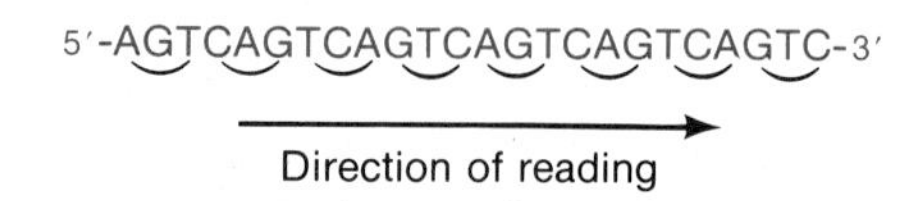

Figure 10-3
Bases in mRNA are read sequentially in the 5′-to-3′ direction, in groups of three.

Table 10-1
The "universal" genetic code

First position (5′ end)	Second position				*Third position (3′ end)*
	U	C	A	G	
U	Phe Phe Leu Leu	Ser Ser Ser Ser	Tyr Tyr Stop Stop	Cys Cys Stop Trp	U C A G
C	Leu Leu Leu Leu	Pro Pro Pro Pro	His His Gln Gln	Arg Arg Arg Arg	U C A G
A	Ile Ile Ile [Met]	Thr Thr Thr Thr	Asn Asn Lys Lys	Ser Ser Arg Arg	U C A G
G	Val Val Val [Val]	Ala Ala Ala Ala	Asp Asp Glu Glu	Gly Gly Gly Gly	U C A G

Note: The boxed codons are used for initiation. GUG is very rare.

each codon was determined from *in vitro* protein-synthesizing experiments in which synthetic mRNA molecules of known sequence were used; for example, when polyuridylic acid was used as an mRNA, the amino acid polyphenylalanine was made, indicating that UUU is the codon for phenylalanine. Use of poly(U) terminated with a 3′ guanine yielded polyphenylalanine with a carboxyl-terminal leucine, indicating that UUG is a leucine codon. These and other experiments (which are described in detail in *MB*, pages 433–438) identified all of the codons. The code is shown in Table 10-1.

The following features of the code should be observed:

1. Most amino acids have more than one codon. In fact, only methionine and tryptophan have a single codon. Furthermore, multiple codons corresponding to a single amino acid usually differ only by the third base. For example, GGU, GGC, GGA, and GGG all code for glycine. Thus, the code is said to be redundant (the term degenerate is also used).

2. Three codons signal termination of polypeptide synthesis–the stop codons UAA, UAG, and UGA.

3. One codon signals initiation of polypeptide synthesis—the start codon AUG, which codes for methionine. An important question is how a particular AUG sequence is designated as a start codon, whereas others located in a coding sequence act only as an internal codon for methione. In prokaryotes, special base sequences, which will be discussed shortly, serve this function; a different mechanism is used for eukaryotes. In some organisms GUG is also used as a start codon for some proteins.

To date, the same codon-amino acid relations exist for all organisms—viruses, prokaryotes, and eukaryotes—and the code is said to be universal. An exception are the codes of mitochondria, the energy-generating organelles of eukaryotes, and chloroplasts, the photosynthetic organelles of plants, in which there are several deviations (see *MB*, pages 486–490). Furthermore, the particular deviations are species-specific. The evolutionary significance of these differences is a widely discussed topic.

Transfer RNA and the Aminoacyl Synthetases

The decoding operation by which the base sequence within an mRNA molecule becomes translated to an amino acid sequence of a protein is accomplished by the tRNA molecules and a set of enzymes, the **aminoacyl tRNA synthetases.**

The tRNA molecules are small, single-stranded nucleic acids ranging in size from 73 to 93 nucleotides. Like all RNA molecules, they have

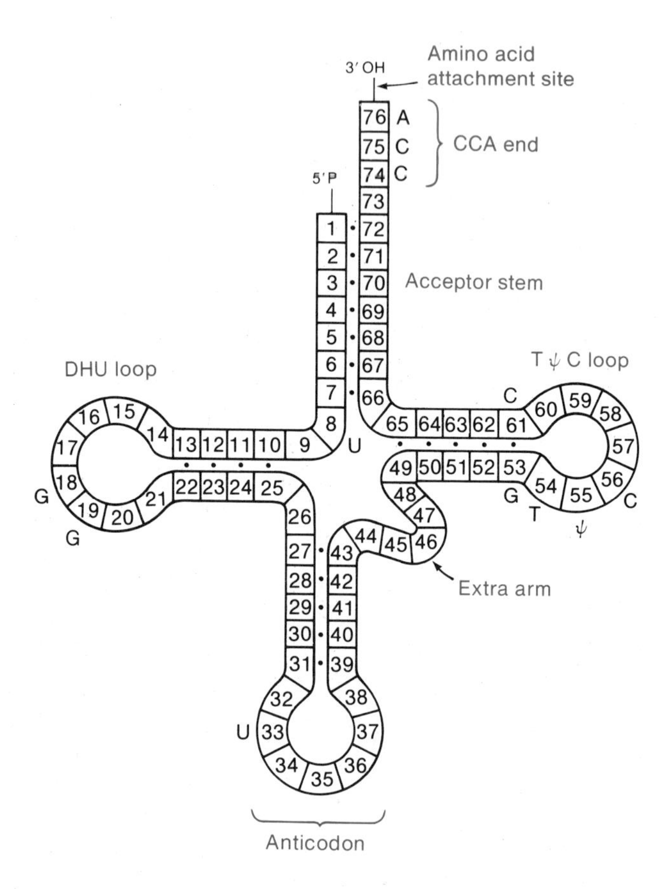

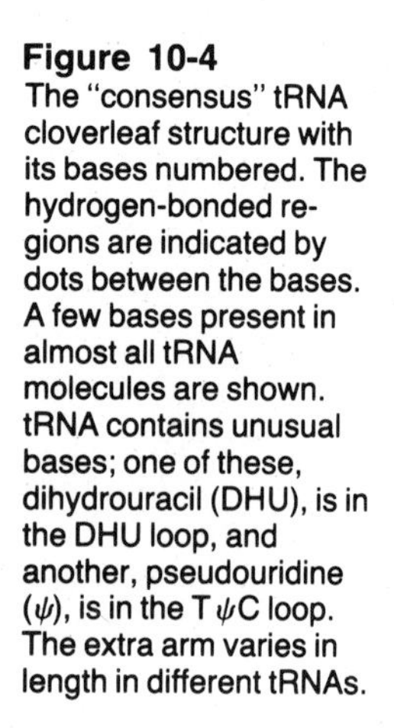

Figure 10-4
The "consensus" tRNA cloverleaf structure with its bases numbered. The hydrogen-bonded regions are indicated by dots between the bases. A few bases present in almost all tRNA molecules are shown. tRNA contains unusual bases; one of these, dihydrouracil (DHU), is in the DHU loop, and another, pseudouridine (ψ), is in the TψC loop. The extra arm varies in length in different tRNAs.

a 3′-OH terminus, but the opposite end terminates with a 5′-monophosphate rather than a 5′-triphosphate, because tRNA molecules are cut from a large primary transcript. Internal complementary base sequences form short double-stranded regions, causing the molecule to fold into a structure in which open loops are connected to one another by double-stranded stems (Figure 10-4). In two dimensions a tRNA molecule is drawn as a planar cloverleaf. Its three-dimensional structure is more complex, as is shown in Figure 10-5. Panel (a) shows the skeletal model of a yeast tRNA molecule that carries phenylalanine, and panel (b) shows an interpretive drawing.

Three regions of each tRNA molecule are used in the decoding operation. One of these regions is the **anticodon,** a sequence of three bases that can form base pairs with a codon sequence in the mRNA. No normal tRNA molecule has an anticodon complementary to any of the stop codons UAG, UAA, or UGA, which is why these codons are stop signals. A second site is the **amino acid attachment site,** the 3′ terminus of the tRNA molecule; the amino acid corresponding to the particular mRNA codon that base-pairs with the tRNA anticodon is covalently

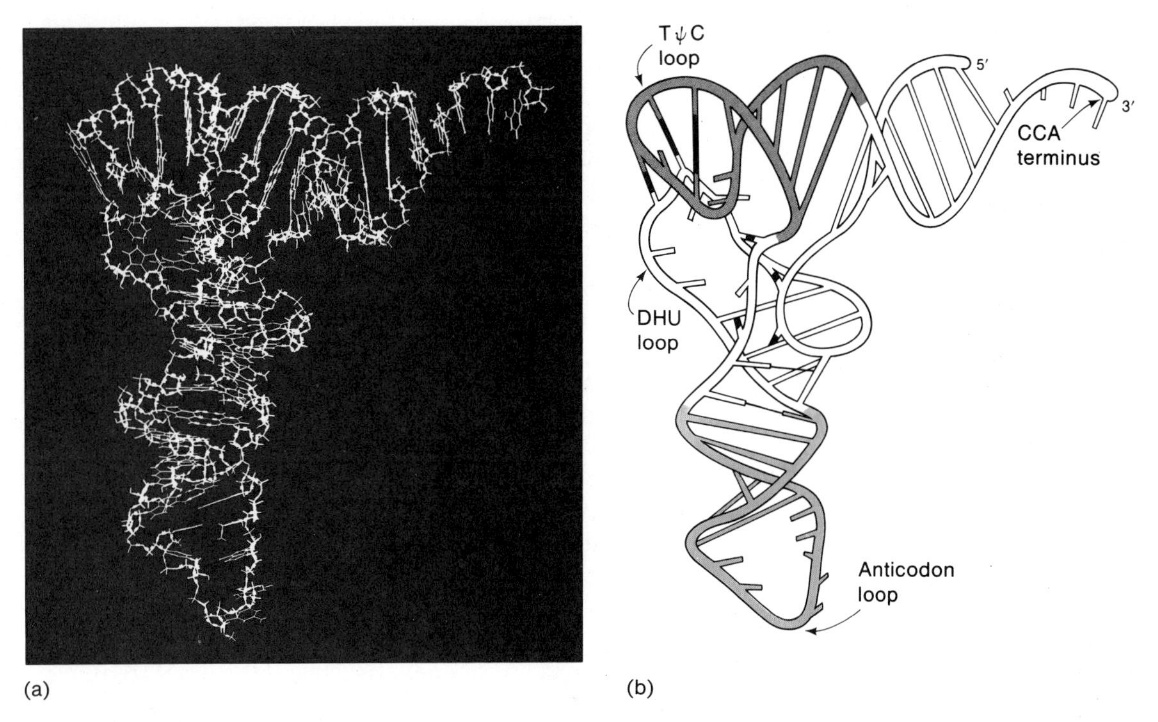

Figure 10-5
(a) Photograph of a skeletal model of yeast $tRNA^{Phe}$. (b) Schematic diagram of the three-dimensional structure of yeast $tRNA^{Phe}$. (Courtesy of Dr. Sung-Hou Kim.)

linked to this terminus. These bound amino acids are joined together during polypeptide synthesis. A specific aminoacyl tRNA synthetase matches the amino acid with the anticodon; to do so, the enzyme must be able to distinguish one tRNA molecule from another. The necessary distinction is provided by an ill-defined region encompassing many parts of the tRNA molecule and called the **recognition region.**

The different tRNA molecules and synthetases are designated by stating the name of the amino acid that can be linked to a particular tRNA molecule by a specific synthetase; for example, leucyl-tRNA synthetase attaches leucine to $tRNA^{Leu}$. When an amino acid has become attached to a tRNA molecule, the tRNA is said to be **acylated** or **charged.** An acylated tRNA molecule is designated in several ways. For example, if the amino acid is glycine, the acylated tRNA would be written glycyl-tRNA or Gly-tRNA. The term **uncharged tRNA** refers to a tRNA molecule lacking an amino acid, and **mischarged tRNA** to one acylated with an incorrect amino acid.

At least one, and usually only one, aminoacyl synthetase exists for each amino acid. For a few of the amino acids specified by more than one codon, more than one synthetase exists.

Accurate protein synthesis, the placement of the "correct" amino acid at the appropriate position in a polypeptide chain, requires (1) attachment of the correct amino acid to a tRNA molecule by the synthetase and (2) fidelity in codon-anticodon binding. An important experiment showed that codon-anticodon binding is based only on base-pair recognition and that the identity of the amino acid attached to the tRNA molecule does not influence this recognition. In this experiment $tRNA^{Cys}$ was charged with radioactive cysteine. The cysteine was then chemically converted to radioactive alanine, yielding alanyl-$tRNA^{Cys}$. This mischarged tRNA molecule was then used in an *in vitro* system containing hemoglobin mRNA and capable of synthesizing hemoglobin, the complete amino acid sequence of which was known. Hemoglobin containing radioactive alanine was made, but the radioactive alanine was present at the sites normally occupied by cysteine rather than at normal alanine positions. This experiment confirmed the hypothesis that tRNA is an adaptor molecule, that the amino acid recognition region and the anticodon are distinct regions, and that the anticodon determines the position of insertion of whatever amino acid is linked to tRNA.

Several amino acids are structurally similar and it is to be expected that synthetases might make occasional mistakes. If the error frequency is high, it seems reasonable to expect that an editing mechanism, as we saw with DNA synthesis, has evolved. Valine and isoleucine constitute such a possibly ambiguous pair of amino acids and, in fact, isoleucyl-tRNA synthetase forms valyl-AMP, which remains bound to the synthetase, at a frequency of about one per 225 activation events. This would mean that 1/225 of all isoleucine positions in proteins could contain a valine. For a typical protein containing 500 amino acids of which 25 were isoleucines, about one copy in nine could be altered. Since there are at least ten known examples of misacylation, there could be an error in almost every molecule. This does not occur, though.

The editing mechanism that corrects the valine-isoleucine error is a hydrolytic step in which valyl-AMP is cleaved and removed from the enzyme. The hydrolysis is carried out by the isoleucyl-tRNA synthetase itself. Interestingly, the signal that activates the hydrolytic function is the attempted binding of valine to $tRNA^{Ile}$. The number of times this editing system fails and valyl-$tRNA^{Ile}$ is formed is about 1 in 800. Thus, the overall error frequency—that is, the fraction of isoleucine sites occupied by valine—is $(1/225)(1/800) = 1/180{,}000$. If all possible amino acid misacylations occur at this frequency, only about 0.17 percent of the proteins would be defective. A similar case, in which methionyl-tRNA synthetase forms threonyl-AMP and homocysteyl-AMP, has also been analyzed. The mechanism by which incorrect misacylation is corrected is explained in *MB*, pages 450–451.

THE WOBBLE HYPOTHESIS

The pattern of the redundancy of the code suggests that something is missing in the explanation of codon-anticodon binding; the most striking aspect of the redundancy is that with only a few exceptions, the identity of the third codon base appears to be unimportant. That is, XYA, XYB, XYC, and XYD usually correspond to the same amino acid.

In 1965, Francis Crick made a proposal, known as the **wobble hypothesis,** that explains the fact that some tRNA molecules respond to several codons and also provides insight into the pattern of redundancy of the code. Up to that time it was generally assumed that no base pair other than G • C, A • T, or A • U would be found in a nucleic acid. This is true of DNA because the regular helical structure of double-stranded DNA imposes two steric constraints: (1) two purines cannot pair with one another because there is not enough space for a planar purine-purine pair, and (2) two pyrimidines cannot pair because they cannot reach one another. Crick proposed that since the anticodon is located within a single-stranded RNA loop, the codon-anticodon interaction might not require formation of a structure with the usual dimensions of a double helix. By model-building he showed that the steric requirements were less stringent at the third position of the codon; by allowing a little play in the structure (this play is called wobble), Crick demonstrated that other base pairs can exist between codon and anticodon. He required, first, that the first two base pairs be of the standard type in order to maximize stability and, second, that the third base pair not produce as much distortion as a purine-purine pair might cause. He included inosine in his model because it was known to be in the anticodons of several tRNA molecules, and he proposed that the base pairs listed in Table 10-2 are possible in the third position of the codon.

The possibility of forming the four base pairs shown in the table, namely A • I, U • I, C • I, and G • U, explains how a single tRNA molecule can respond to several codons. An example is the following.

Table 10-2
Allowed pairings according to the wobble hypothesis

Third position codon base	*First position anticodon base**
A	U,I
G	C,U
U	G,I
C	G,I

*A is not allowed; see 1(c) below.

There are two major species of the alanine tRNA of yeast. One of these responds to the codons GCU, GCC, and GCA. Its anticodon is IGC, which is consistent with the entries in Table 10-2 and shows that only inosine can pair with U, C, and A. (Remember the convention for naming the codon and the anticodon—*always with the 5′ end at the left.* Thus the codon 5′-GCU-3′ is matched by the anticodon 5′-IGC-3′). Similarly, yeast $tRNA^{Ala}_{II}$ responds only to GCG; there are two possible anticodons, CGC and UGC, because both C and U can bond to G. If the anticodon were UGC, $tRNA^{Ala}_{II}$ would respond to both GCG and GCA, which is not the case; thus, the anticodon cannot be UGC. If it were CGC, the only codon recognized would be GCG, which is the case. Thus, the anticodon must be CGC, as indeed it is.

The most striking achievement of the wobble hypothesis is that it explains the arrangement of all synonyms in the code.

POLYCISTRONIC mRNA

Many prokaryotic mRNA molecules are polycistronic—that is, they contain sequences specifying the synthesis of several proteins. Thus, a polycistronic mRNA molecule must possess a series of start and stop codons. If a mRNA molecule encodes three proteins the minimal requirement would be the sequence

Start, protein 1, stop–start, protein 2, stop–start, protein 3, stop.

Actually, such an mRNA molecule is probably never so simple in that the leader sequence preceding the first start signal may be several hundred bases long and there is usually a sequence called a **spacer** of from 5 to 20 bases between one stop codon and the next start codon. Thus the structure of a tricistronic mRNA is more typically that shown in Figure 10-6.

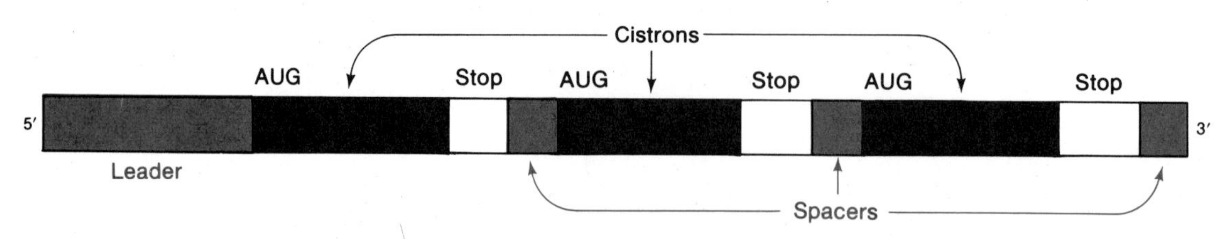

Figure 10-6
Arrangement of cistrons and untranslated regions (red) in a typical polycistronic mRNA molecule.

OVERLAPPING GENES

In all that has been said so far about coding and signal recognition, an implicit assumption has been that the mRNA molecule is scanned for start signals to establish the reading frame and that reading then proceeds in a single direction within the reading frame. The idea that several reading frames might exist in a single segment was not considered for many years, primarily because a mutation in a gene that overlapped another gene would often produce a mutation in the second gene, and mutations affecting two genes had not been observed. The notion of overlapping reading frames was also rejected on the grounds that severe constraints would be placed on the amino acid sequences of two proteins translated from the same portion of mRNA. However, because the code is highly redundant, the constraints are actually not so rigid.

If multiple reading frames were present in an organism, a single DNA segment would be utilized with maximal efficiency. However, a disadvantage is that evolution might be slowed in that single-base-change mutations would be deleterious more often than if there were a unique reading frame. Nonetheless, some organisms—namely, small viruses and the smallest phages, have evolved having overlapping reading frames.

The *E. coli* phage ϕX174 contains a single strand of DNA consisting of 5386 nucleotides whose base sequence is known. If a single reading frame were used, at most 1795 amino acids could be encoded in the sequence and, if we take 110 as the molecular weight of an "average" amino acid, at most 197,000 molecular weight units of protein could be

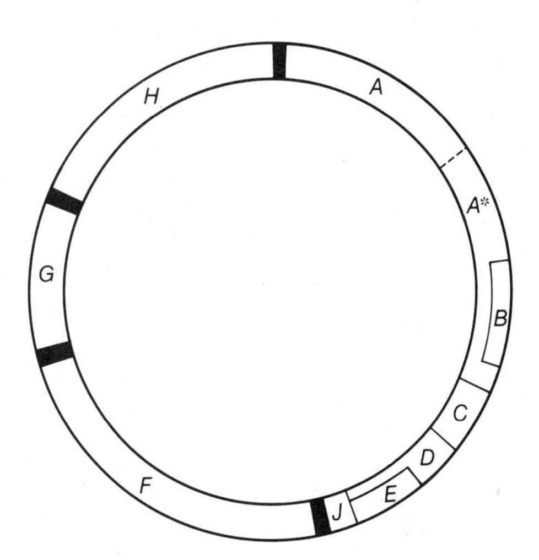

Figure 10-7
Genetic map of ϕX174 showing the overlapping genes. Spacers are blackened.

made. However, the phage makes eleven proteins and the total molecular weight of these proteins is 262,000. This paradox was resolved when it was shown that translation occurs in several reading frames from three mRNA molecules (Figure 10-7). For example, the sequence for protein B is contained totally in the sequence for protein A but translated in a different reading frame. Similarly, the protein E sequence is totally within the sequence for protein D. Protein K is initiated near the end of gene *A*, includes the base sequence of B, and terminates in gene *C*; synthesis is not in phase with either gene *A* or gene *C*. Of note is protein A′ (also called A*), which is formed by reinitiation within gene *A* and in the same reading frame, so that it terminates at the stop codon of gene *A*. Thus, the amino acid sequence of A′ is identical to a segment of protein A. In total, five different proteins obtain some or all of their primary structure from shared base sequences in ϕX174. This phenomenon, known as **overlapping genes,** has been observed in the related phage G4 and in the small animal virus SV40.

It should be realized that the single structural feature responsible for gene overlap is the location of each AUG initiation sequence.

POLYPEPTIDE SYNTHESIS

In the preceding sections how the information in an mRNA molecule is converted to an amino acid sequence having specific start and stop points has been discussed. In this section the chemical problem of attachment of the amino acids to one another is examined.

Polypeptide synthesis can be divided into three stages—(1) **initiation,** (2) **elongation,** and (3) **termination.** The main features of the initiation step are the binding of mRNA to the ribosome, selection of the initiation codon, and the binding of acylated tRNA bearing the first amino acid. In the elongation stage there are two processes: joining together two amino acids by peptide bond formation, and moving the mRNA and the ribosome with respect to one another so that the codons can be translated successively. In the termination stage the completed protein is dissociated from the synthetic machinery and the ribosomes are released to begin another cycle of synthesis.

We begin our discussion with a description of the structure of the ribosome.

Ribosomes

A ribosome is a multicomponent particle that contains several enzymes needed for protein synthesis and that brings together a single mRNA molecule and charged tRNA molecules in the proper position and orientation so that the base sequence of the mRNA molecule is translated into an amino acid sequence (Figure 10-8(a)). The properties of the *E. coli*

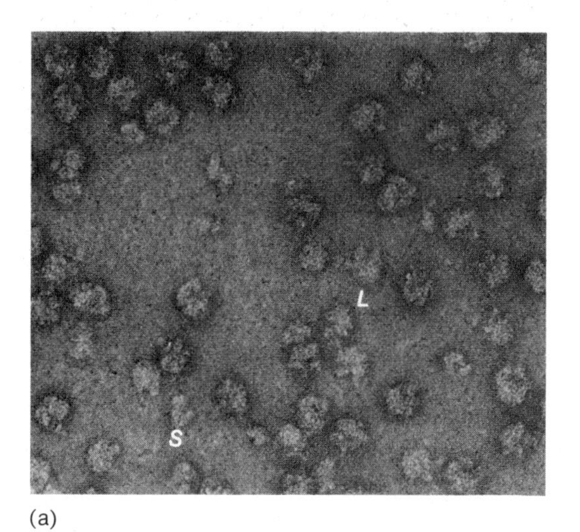

(a)

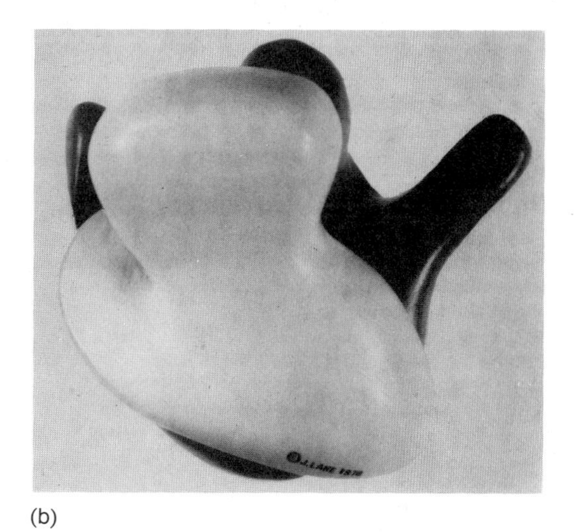

(b)

Figure 10-8
Ribosomes. (a) An electron micrograph of *E. coli* 70S ribosomes. A few ribosomal subunits are also in the field; S denotes a 30S particles and L denotes a 50S particle. (b) A three-dimensional model of the 70S ribosome. The small subunit is light and the large subunit is dark. (Courtesy of James Lake.)

ribosome are best understood, and it serves as a useful model for discussion of all ribosomes.

All ribosomes contain two subunits (Figure 10-8(b)). For historical reasons, the intact ribosome and the subunits have been given numbers that describe how fast they sediment when centrifuged. For *E. coli* (and for all prokaryotes) the intact particle is called a **70S ribosome** (S is a measure of the sedimentation rate) and the subunits, which are unequal in size and composition, are termed **30S and 50S.** A 70S ribosome consists of one 30S subunit and one 50S subunit.

Both the 30S and the 50S particles can be dissociated into RNA (called **rRNA** for ribosomal RNA) and protein molecules under appropriate conditions (Figure 10-9). Each 30S subunit contains one 16S rRNA molecule and 21 different proteins; a 50S subunit contains two RNA molecules—one 5S rRNA molecule and one 23S rRNA molecule—and 32 different proteins. In each particle usually only one copy of each protein molecule is present, though a few are duplicated or modified. Like tRNA molecules, rRNA molecules are cut from large primary transcripts. In many organisms some tRNA molecules are cut also from the large transcripts containing rRNA.

The basic features of eukaryotic ribosomes are similar to those of bacterial ribosomes, but all eukaryotic ribosomes are somewhat larger than those of prokaryotes. They contain a greater number of proteins (about 80) and an additional RNA molecule (four in all). The biological significance of the differences between prokaryotic and eukaryotic

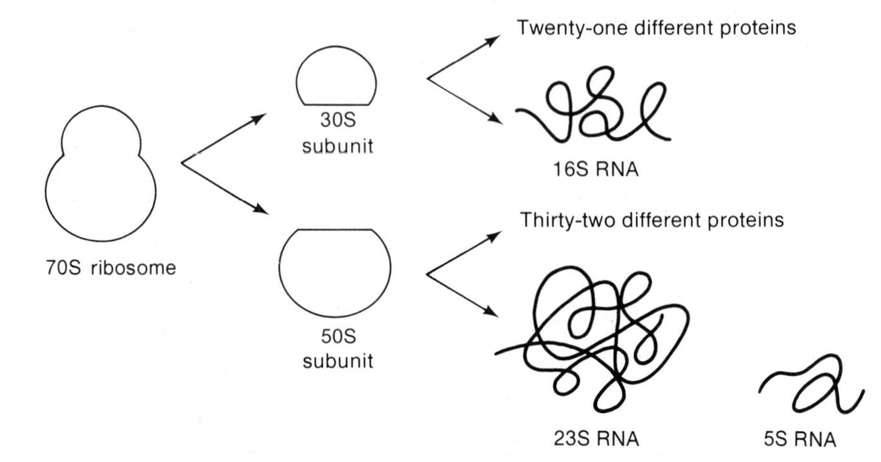

Figure 10-9
Dissociation of a prokaryotic ribosome. The configuration of two overlapping circles will be used throughout this chapter, for the sake of simplicity. The correct configuration is shown in Figure 10-8(b)

ribosomes is unknown. A typical eukaryotic ribosome—an **80S ribosome**—consists of two subunits, **40S** and **60S.** These sizes may vary by as much as $\pm$ 10 percent from one organism to the next, in contrast with bacterial ribosomes, which have sizes that are nearly the same for all bacterial species examined. The best-studied eukaryotic ribosome is that of the rat liver. The 40S and 60S subunits can also be dissociated. A 40S subunit, which is analogous to the 30S subunit of prokaryotes, consists of one 18S rRNA molecule and about 30 proteins, and the 60S subunit contains three rRNA molecules—one 5S, one 5.8S, and one 28S rRNA molecule—and about 50 proteins. The 5.8S, 18S, and 28S rRNA molecules of eukaryotes correspond functionally to the 5S, 16S, and 23S molecules of bacterial ribosomes. The bacterial counterpart to the eukaryotic 5S rRNA is very likely present as part of the 23S rRNA sequence.

Stages of Polypeptide Synthesis in Prokaryotes

Polypeptide synthesis in prokaryotes and eukaryotes follows the same overall mechanism, though there are differences in detail, the most important being the mechanism of initiation. The prokaryotic system is best understood and will serve as a model for our discussion.

An important feature of initiation of polypeptide synthesis is the use of a specific initiating tRNA molecule. In prokaryotes this tRNA molecule is acylated with the modified amino acid *N*-**formylmethionine (fMet); the tRNA is often designated $tRNA^{fMet}$ (Figure 10-10). Both $tRNA^{fMet}$ and $tRNA^{Met}$ recognize the codon AUG, but only** $tRNA^{fMet}$ is used for initiation. The $tRNA^{fMet}$ molecule is first acylated **with** *methionine* and an enzyme (found only in prokaryotes) adds a

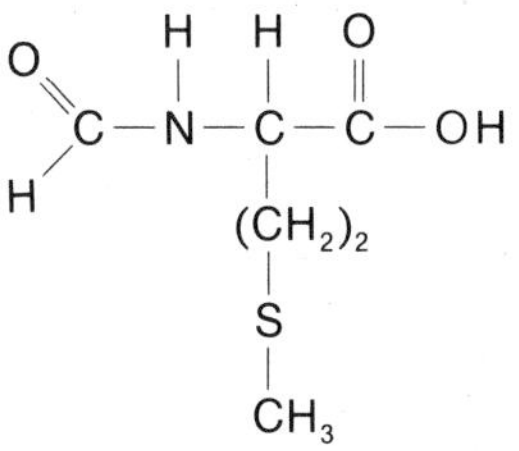

Figure 10-10
Chemical structure of *N*-formylmethionine. If the HC=O group at the left were an H, the molecule would be methionine.

formyl group to the amino group of the methionine. In eukaryotes the initiating tRNA molecule is charged with methionine also, but formylation does not occur. The use of these initiator tRNA molecules means that *while being synthesized,* all prokaryotic proteins have N-formylmethionine at the amino terminus and all eukaryotic proteins have methionine at the amino terminus. However, these amino acids are frequently altered or removed later (this is called **processing**), and all amino acids have been observed at the amino termini of completed protein molecules isolated from cells.

Polypeptide synthesis in bacteria begins by the association of one **30S subunit (not the entire 70S ribosome)**, an mRNA molecule, fMet-tRNA, three proteins known as **initiation factors,** and guanosine 5′-triphosphate (GTP). These molecules constitute the **30S preinitiation complex** (Figure 10-11). Since polypeptide synthesis begins at an AUG start codon and AUG codons are found within coding sequences (that is, methionine occurs within a polypeptide chain), some signal must be present in the base sequence of the mRNA molecule to identify a particular AUG codon as a start signal. The means of selecting the correct AUG sequence differs in prokaryotes and eukaryotes. In prokaryotic mRNA molecules a particular base sequence—AGGAGGU, called the **ribosome binding site** or sometimes the **Shine-Dalgarno sequence**—near the AUG codon used for initiation forms base pairs with a complementary sequence near the 3′ terminus of the 16S rRNA molecule of the ribosome

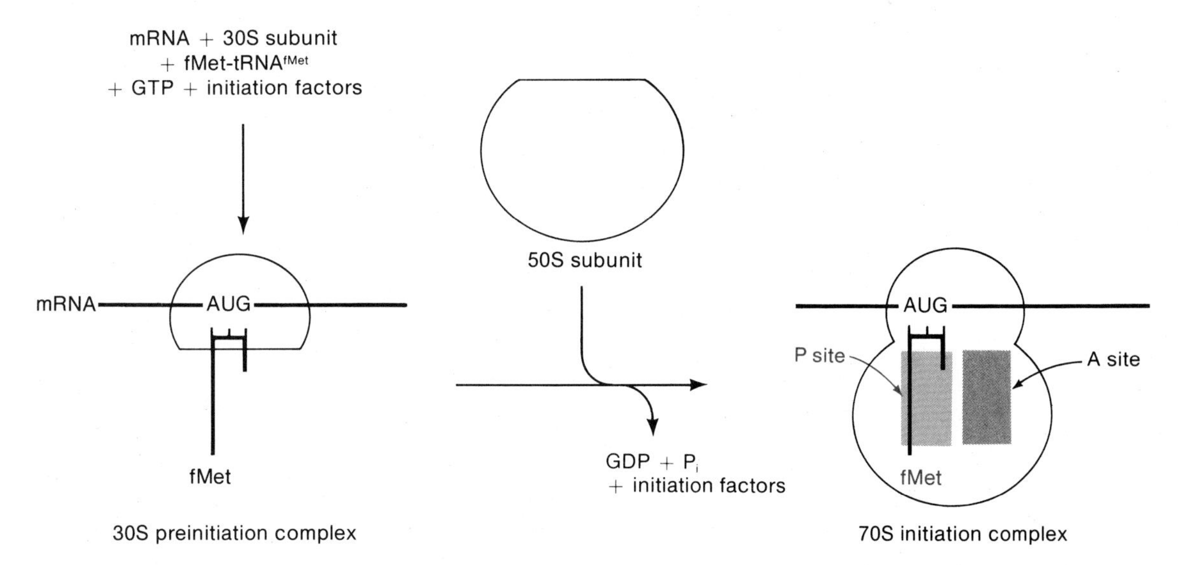

Figure 10-11
Early steps in protein synthesis in prokaryotes: formation of the 30S preinitiation complex and of the 70S initiation complex.

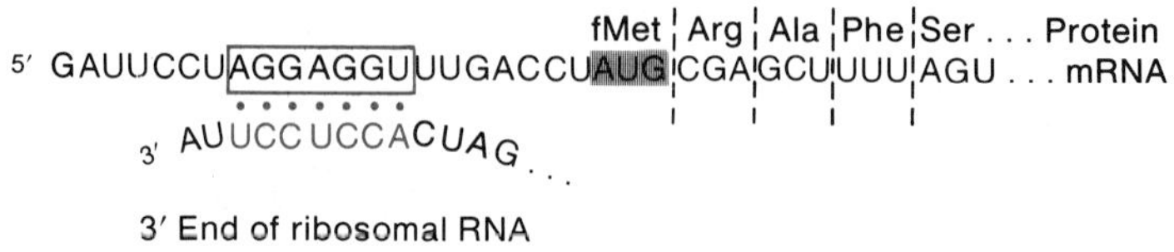

fMet | Arg | Ala | Phe | Ser . . . Protein
5′ GAUUCCUAGGAGGUUUGACCUAUGCGAGCUUUUAGU . . . mRNA
3′ AUUCCUCCACUAG . . .

3′ End of ribosomal RNA

Figure 10-12
Initiation of translation in prokaryotes. Base-pairing between the Shine-Dalgarno sequence (in box) in the mRNA and the complementary region (red) near the 3′ terminus of 16S rRNA. The AUG start codon is shaded.

(Figure 10-12). In eukaryotic mRNA molecules the 5′ terminus binds to the ribosome, after which the mRNA molecule slides along the ribosome until the AUG codon *nearest the 5′ terminus* is in contact with the ribosome; the consequence of this mechanism for initiation in eukaryotes will be explained at the end of this section.

Following formation of the 30S preinitiation complex, a 50S subunit joins with this complex to form a **70S initiation complex** (Figure 10-11).

The 50S subunit contains two tRNA binding sites. These sites are called the **A (aminoacyl) site** and the **P (peptidyl) site.** When joined with the 30S preinitiation complex, the position of the 50S subunit in the 70S initiation complex is such that the fMet-$tRNA^{fMet}$, which was previously bound to the 30S preinitiation complex, occupies the P site of the 50S subunit. Placement of fMet-$tRNA^{fMet}$ in the P site fixes the position of the fMet-tRNA anticodon so that it pairs with the AUG initiator codon in the mRNA. Thus, *the reading frame is unambiguously defined upon completion of the 70S initiation complex.*

Once the P site is filled, the A site of the 70S initiation complex becomes available to any tRNA molecule whose anticodon can pair with the codon adjacent to the initiation codon. After occupation of the A site a peptide bond between N-formylmethionine and the adjacent amino acid is formed by an enzyme complex called **peptidyl transferase.** As the bond is formed, the N-formyl methionine is cleaved from the fMet-tRNA in the P site.

After the peptide bond forms, an uncharged tRNA molecule occupies the P site and a dipeptidyl-tRNA occupies the A site. At this point three movements occur: (1) the $tRNA^{fMet}$ in the P site, now no longer linked to an amino acid, leaves this site, (2) the peptidyl-tRNA moves from the A site to the P site, and (3) the mRNA moves a distance of three bases in order to position the next codon at the A site (Figure 10-13). This step requires the presence of an elongation protein **EF-G** and GTP, and it is likely that mRNA movement is a consequence of the tRNA motion. After mRNA movement has occurred, the A site is again available to accept a charged tRNA molecule having a correct anticodon.

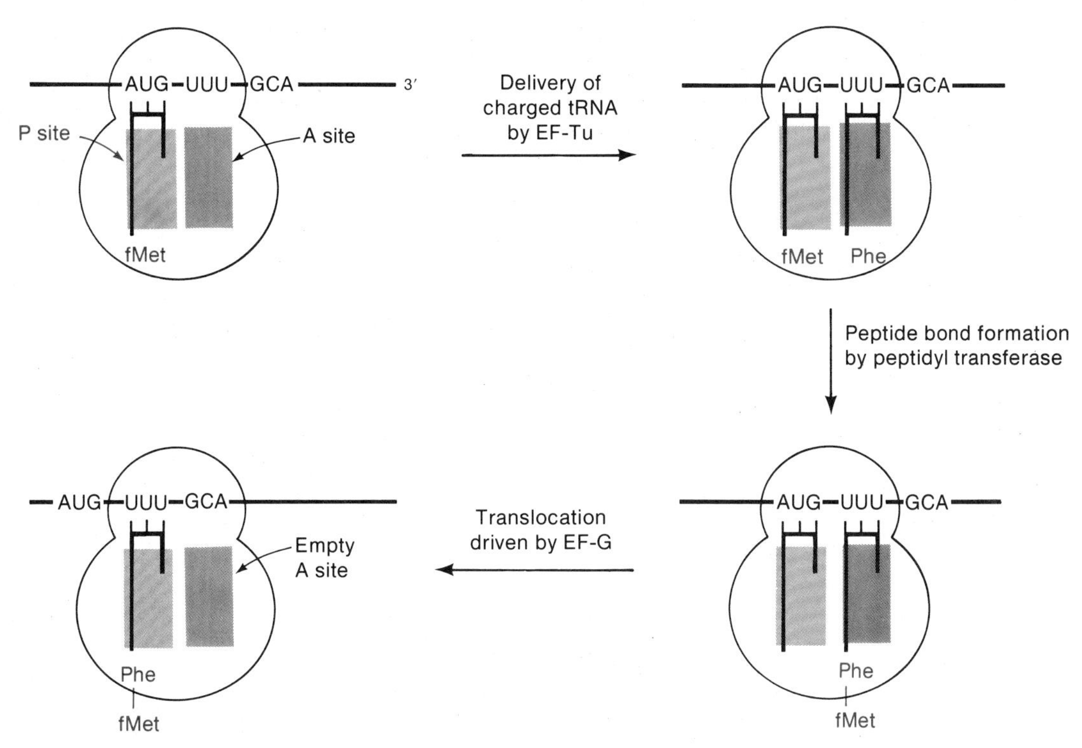

Figure 10-13
Elongation phase of protein synthesis: binding of charged tRNA, peptide bond formation, and translocation.

When a chain termination codon (either UAA, UAG, or UGA) is reached, no acylated tRNA exists that can fill the A site, so chain elongation stops. However, the polypeptide chain is still attached to the tRNA occupying the P site. Release of the protein is accomplished by proteins called **release factors.** In the presence of release factors peptidyl transferase separates the polypeptide from the tRNA; then, the polypeptide chain, which has been held on the ribosome solely by the interaction with the tRNA in the P site, is released from the ribosome. The 70S ribosome then dissociates into its 30S and 50S subunits, completing the cycle.

If the mRNA molecule is polycistronic and the AUG codon initiating the second polypeptide is not too far from the stop codon of the first, the 70S ribosome will not always dissociate but will re-form an initiation complex with the second AUG codon. The probability of such an event decreases with increasing separation of the stop codon and the next AUG codon. In some genetic systems the separation is sufficiently great that more protein molecules are always translated from the first gene than from subsequent genes, and this is a mechanism for maintaining particular ratios of gene products (Chapter 14). Mutations sometimes arise that convert a sense (amino-acid specifying) codon to a stop codon—for example, the mRNA molecule has a UAG codon at the site of a UAC codon. Such mutations, which will be discussed further in Chapter 11, cause premature termination of a polypeptide. If the mutation is sufficiently far upstream from the normal stop codon of the gene containing the mutation, the distance to the next AUG codon may be so great that separation of the translating 70S ribosome and the mRNA molecule is almost inevitable. In this case, genes downstream from the mutation will rarely be translated. Such mutations are the most common type of polar mutation.

In eukaryotes reinitiation of polypeptide synthesis following an encounter of a ribosome with a stop codon does *not* occur. Also, as pointed out earlier in this section, polypeptide synthesis in eukaryotes is initiated when a ribosome binds to the 5′ terminus of an mRNA molecule and slides along to the first AUG codon. There is no mechanism for initiating polypeptide synthesis at any AUG other than the first one encountered; *eukaryotic mRNA is always monocistronic* (Figure 10-14). However, a primary transcript can contain coding sequences for more than one polypeptide chain and in fact this is a frequent arrangement in animal viruses. In these cases, differential splicing generates several different mRNA molecules from one transcript (Chapter 9). For example, if the AUG nearest the 5′ terminus is excised, the second AUG will become available. Another mechanism for producing several proteins from a single transcript is protein processing. In this case, a single giant polypeptide chain, called a **polyprotein,** is made and then cleaved into several component polypeptide chains, each constituting a distinct protein (Chapter 14).

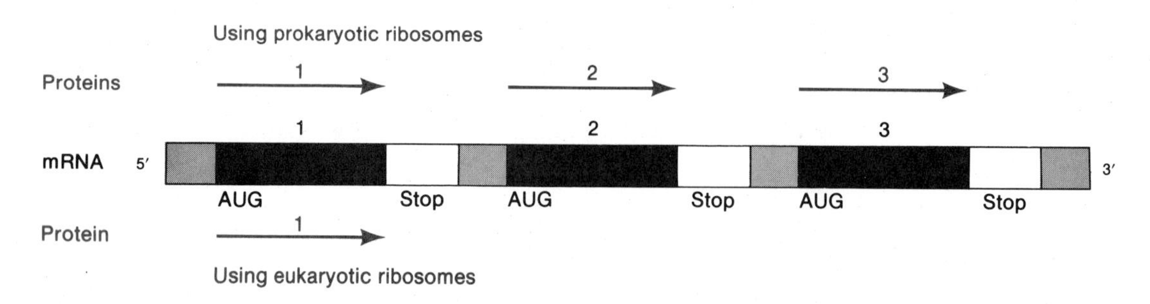

Figure 10-14
Difference in the products translated from a tricistronic mRNA molecule by the ribosomes of prokaryotes and eukaryotes. The prokaryotic ribosome translates all of the cistrons but the eukaryotic ribosome translates only one cistron—the one nearest the 5′ terminus of the mRNA. Translated sequences are in black, stop codons are white, and the leader and spacers are shaded.

COMPLEX TRANSLATION UNITS

The unit of translation is almost never simply a ribosome traversing a mRNA molecule, but is a more complex structure, of which there are several forms. Some of these structures are described in this section.

After about 25 amino acids have been joined in a polypeptide chain, the AUG initiation site of the encoding mRNA molecule is completely free of the ribosome. A second initiation complex then forms. The overall configuration is of two 70S ribosomes moving along the mRNA at the same speed. When the second ribosome has moved along a distance similar to that traversed by the first, a third ribosome is able to attach. This process—movement and reinitiation—continues until the mRNA is covered with ribosomes at a density of about one 70S particle per 80 nucleotides. This large translation unit is called a **polyribosome** or simply a **polysome.** This is the usual form of the translation unit in all cells. An electron micrograph of a polysome and an interpretive drawing are shown in Figure 10-15.

The use of polysomes has a particular advantage to a cell—namely, the overall rate of protein synthesis is increased compared to the rate that would occur if there were no polysomes.

An mRNA molecule being synthesized has a free 5′ terminus; since translation occurs in the 5′→3′ direction, each cistron contained in the mRNA is synthesized in a direction appropriate for immediate translation. That is, the ribosome binding site is transcribed first,

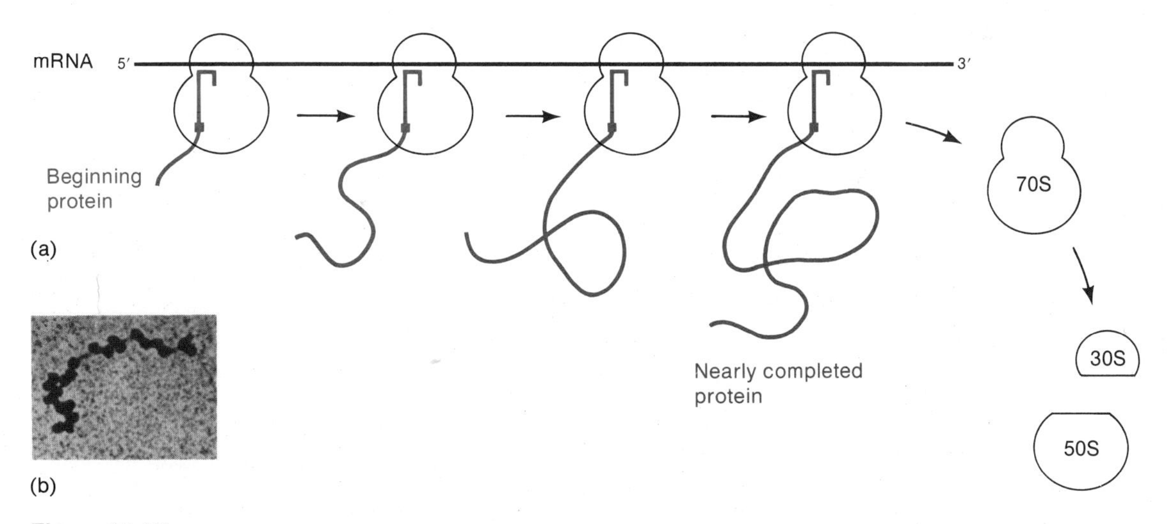

Figure 10-15
Polysomes. (a) Diagram showing relative movement of the 70S ribosome and the mRNA, and growth of the protein chain. (b) Electron micrograph of an *E. coli* polysome. (Courtesy of Barbara Hamkalo.)

followed in order by the AUG codon, the region encoding the amino acid sequence, and finally the stop codon. Thus, in bacteria, in which no nuclear membrane separates the DNA and the ribosome, there is no obvious reason why the 70S initiation complex should not form before the mRNA is released from the DNA. With prokaryotes this does indeed occur; this process is called **coupled transcription-translation.** This coupled activity does not occur in eukaryotes, because the mRNA is synthesized and processed in the nucleus and later transported through the nuclear membrane to the cytoplasm where the ribosomes are located.

Coupled transcription-translation speeds up protein synthesis in the sense that translation does not have to await release of the mRNA from the DNA. Translation can also be started before the mRNA is degraded by nucleases. Figure 10-16(a) shows an electron micrograph of a DNA molecule to which is attached a number of mRNA molecules, each associated with ribosomes. The micrograph is interpreted in panel

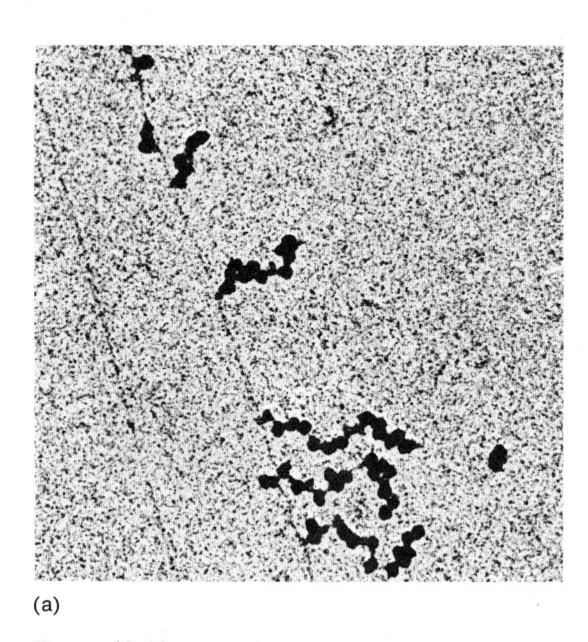

(a)

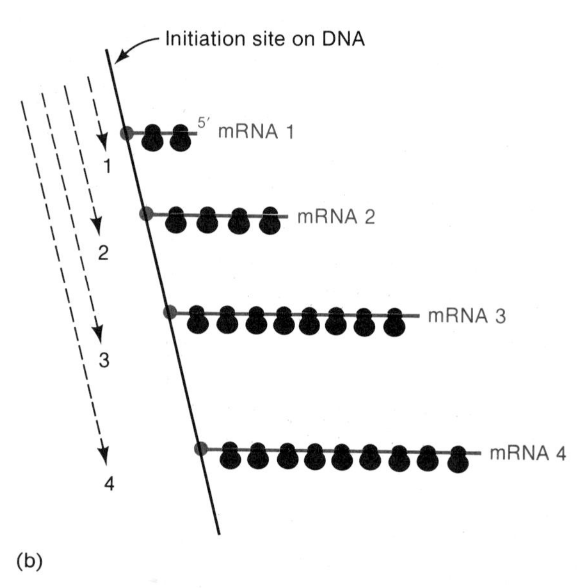

(b)

Figure 10-16
(a) Transcription of a section of the DNA of *E. coli* and translation of the nascent mRNA. Only part of the chromosome is being transcribed. The dark spots are ribosomes, which coat the mRNA. (From O. L. Miller, Barbara A. Hamkalo, and C. A. Thomas. 1977. *Science,* 169, 392.) (b) An interpretation of the electron micrograph of part (a). The mRNA is in red and is coated with black ribosomes. The large red spots are the RNA polymerase molecules; they are actually too small to be seen in the photo. The dashed arrows show the distances of each RNA polymerase from the transcription initiation site. Arrows 1, 2, and 3 have the same length as mRNA 1, 2, 3; mRNA 4 is shorter than arrow 4, presumably because its 5′ end has been partially digested by an RNase.

(b). Note that the lengths of the polysomes increase with distance from the transcription initiation site, because the mRNA is farther from that site and hence longer.

ANTIBIOTICS

Many antibacterial agents (**antibiotics**) have been isolated from fungi. Most of these are inhibitors of protein synthesis. For example, streptomycin and neomycin bind to a particular protein in the 30S particle and thereby prevent binding of $tRNA^{fMet}$ to the P site; the tetracyclines inhibit binding of charged tRNA; lincomycin and chloramphenicol inhibit the peptidyl transferase; and puromycin causes premature chain termination; erythromycin binds to a free 50S particle and prevents formation of the 70S ribosome. A particular antibiotic has clinical value only if it acts on bacteria and not on animal cells; the clinically useful antibiotics usually either fail to pass through the cell membrane of animal cells or do not bind to eukaryote ribosomes, because of some unknown feature of their structure.

Some disease-causing bacteria exert their pathogenic effect because they excrete inhibitors of mammalian protein synthesis. The agent causing diphtheria is an example; it binds to a factor necessary for movement of mammalian ribosomes along the mRNA.

11 Mutagenesis, Mutations, and Mutants

In previous chapters mutants have been encountered repeatedly. In this chapter we examine the biochemical basis of being a mutant. We begin with the definitions of several terms—mutant, mutation, mutagen, and mutagenesis—and a description of several types of mutations.

TYPES OF MUTATIONS

Mutant refers to an organism or a gene that is different from the normal or wild type, as His^- yeast or white-eyed *Drosophila*. However, when referring to biochemical properties and the normal form is − (for example, *E. coli* isolated from nature is unable to metabolize lactose, that is, Lac^-), the + form is called a wild type and any − form is a mutant.

Mutation refers to any change in the base sequence of DNA that gives rise to a mutant phenotype. The most common change is a substitution, addition, or deletion of one or more bases (Figure 11-1).

A **mutagen** is a physical agent or a chemical reagent that causes mutations to occur.

Mutagenesis is the process of producing a mutation. If it occurs in nature without the addition of a mutagen, it is called **spontaneous**

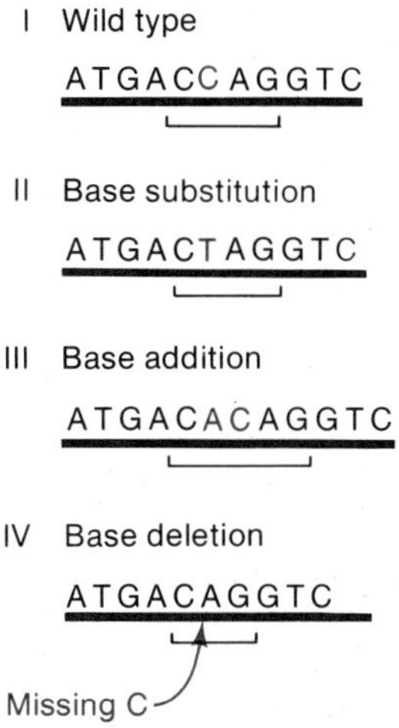

I Wild type

ATGACC AGGTC

II Base substitution

ATGACT AGGTC

III Base addition

ATGACACAGGTC

IV Base deletion

ATGACAGGTC

Missing C

Figure 11-1
Four types of mutations. Only the base sequence in one DNA strand is shown. Changes are shown in red. The horizontal brackets indicate the affected segment.

mutagenesis and the resulting mutants are **spontaneous mutants.** If a mutagen is used, the process is **induced mutagenesis.**

Mutations are classified in several ways. One distinction is based on the nature of the change—specifically, on the number of bases changed. Thus, we may distinguish a **point mutation,** in which there is only a single changed base pair, from a multiple mutation, in which two or more base pairs differ from the wild-type sequence. A point mutation may be a **base substitution,** a **base insertion,** or a **base deletion,** but the term most frequently refers to a base substitution.

A second distinction is based on the consequence of the change in terms of the amino acid sequence that is affected. For example, if there is an amino acid substitution, the mutation is a **missense** mutation.

If the substitution produces a protein that is active at one temperature (typically 30°C) and inactive at a higher temperature (usually 40–42°C), the mutation is called a **temperature-sensitive** or Ts mutation. If the mutation generates a stop codon, protein synthesis will stop, and the mutation is called a **chain termination mutation** or a **nonsense mutation**. Temperature-sensitive and chain termination mutations exhibit the mutant phenotype only under certain conditions and hence are called **conditional mutations**; they are the most versatile and useful mutations available to the molecular biologist.

With microorganisms the phenotype and genotype are written with a capital letter and in lower-case italics, respectively. This convention does not apply to higher organisms. The notation for bacterial mutants is summarized in Table 11-1.

Whether produced spontaneously or by induced mutagenesis, mutants must be selected from large populations. Numerous

Table 11-1
Summary of notation used to designate bacterial mutants and mutations

Phenotype or genotype	*Designation**
Phenotype	
Lacking in or possessing ability to make a substance	Sub$^-$, Sub$^+$
Resistance or sensitivity to an antibiotic	Ant-r, Ant-s
Genotype	
Wild-type gene for making a substance	*sub*$^+$
Mutant gene for making a substance	*sub*$^-$†
Mutant *subA* gene; mutation number 63 in *subA* gene	*subA*$^-$; *subA63*
Gene for resistance or sensitivity to a particular antibiotic	*ant*
Genotype for resistance or sensitivity to an antibiotic	*ant-r, ant-s*

*The arbitrary abbreviations "sub" and "ant" mean substance and antibiotic in this table.
†The notation *sub* for *sub*$^-$ is widespread. Because of possible ambiguity, the superscript minus is used throughout this book.

techniques exist for selecting desired mutants; these techniques are usually applicable to particular organisms and to specific types of mutations. A complete description is beyond the scope of this book; for additional information, see *MB*, pages 342–347 or texts in genetics or microbiology.

BIOCHEMICAL BASIS OF MUTANTS

A mutant may be defined as an organism in which either the base sequence of DNA or the phenotype has been changed. These definitions are the same (except for the case of a silent mutation, which will be discussed shortly) since the base sequence of DNA determines the amino acid sequence of a protein. The chemical and physical properties of each protein are determined by its amino acid sequence, so that a single amino acid change is capable of inactivating a protein.

From the discussion of protein structure in Chapter 4 it is easy to understand how an amino acid substitution can change the structure, and hence, the biological activity, of a protein. For instance, consider a hypothetical protein whose three-dimensional structure is determined entirely by an interaction between one positively charged amino acid (for example, lysine) and one negatively charged amino acid (aspartic acid). A substitution of methionine, which is uncharged, for the lysine would clearly destroy the three-dimensional structure, as would substitution of histidine, which is positively charged, for aspartic acid. Similarly, a protein might be stabilized by a hydrophobic cluster, in which case substitution of glutamine (polar) for leucine (nonpolar) would also be disruptive.

A base substitution does not always yield a mutant phenotype. Because of the redundancy of the code some changes do not alter the amino acid sequence, and some amino acid changes do not significantly affect the structure of a protein. A base change with these properties is said to be **silent.**

The shapes of proteins are determined by such a variety of interactions that sometimes an amino acid substitution is only partially disruptive. For instance, an isoleucine might substitute successfully for leucine and be silent, but replacement with a more bulky amino acid such as phenylalanine might cause subtle stereochemical changes, though a hydrophobic cluster is preserved. This could be manifested as a reduction, rather than a loss, of activity of an enzyme. For example, a bacterium carrying such a mutation in the enzyme that synthesizes adenine might grow very slowly (but it would grow) unless adenine is provided in the growth medium. Such a mutation is called a **leaky**

mutation; these mutations are not particularly useful for most genetic studies.

Generally speaking, the following types of amino acid substitutions are expressed as nonleaky mutations: polar to nonpolar, nonpolar to polar, change of sign of a charge, small side chain to bulky side chain, sulfhydryl to any other side chain, hydrogen-bonding to non-hydrogen-bonding, any change to or from proline (which changes the shape of the polypeptide backbone), and any change in a substrate-binding site.

So far, only amino acid substitution mutations have been discussed. Other mutations that eliminate activity of the protein totally are base deletions, which cause one or more amino acids to be absent from the protein; frameshifts, in which all amino acids starting from the mutant site are different; and chain termination mutants, in which a protein chain is prematurely terminated. These will be discussed in greater detail shortly.

MUTAGENESIS

The production of a mutant requires that a change occur in the base sequence. This can occur spontaneously by replication errors or can be stimulated to occur in five main ways: (1) removal of an incorrectly inserted base is prevented; (2) a base is inserted that tautomerizes and allows a substitution to occur in subsequent replication; (3) a previously inserted base is chemically altered to a base having different base-pairing specificity; (4) one or more bases are skipped during replication; or (5) one or more extra bases are inserted during replication. In the following sections we describe mutagens that act by one or more of these mechanisms and address the question of spontaneous mutagenesis.

Base-Analogue Mutagens

By a base analogue one means a substance other than a standard nucleic acid base which can be built into a DNA molecule by the normal process of polymerization. Such a substance must be able to pair with the base on the complementary strand being copied or the 3′→5′ editing function will remove it. However, if it can tautomerize or if it has two modes of hydrogen-bonding, it will be mutagenic.

The substituted base 5-bromouracil (BU) is an analogue of thymine inasmuch as the bromine has about the same van der Waals radius as the methyl group of thymine (Figure 11-2). In subsequent rounds of

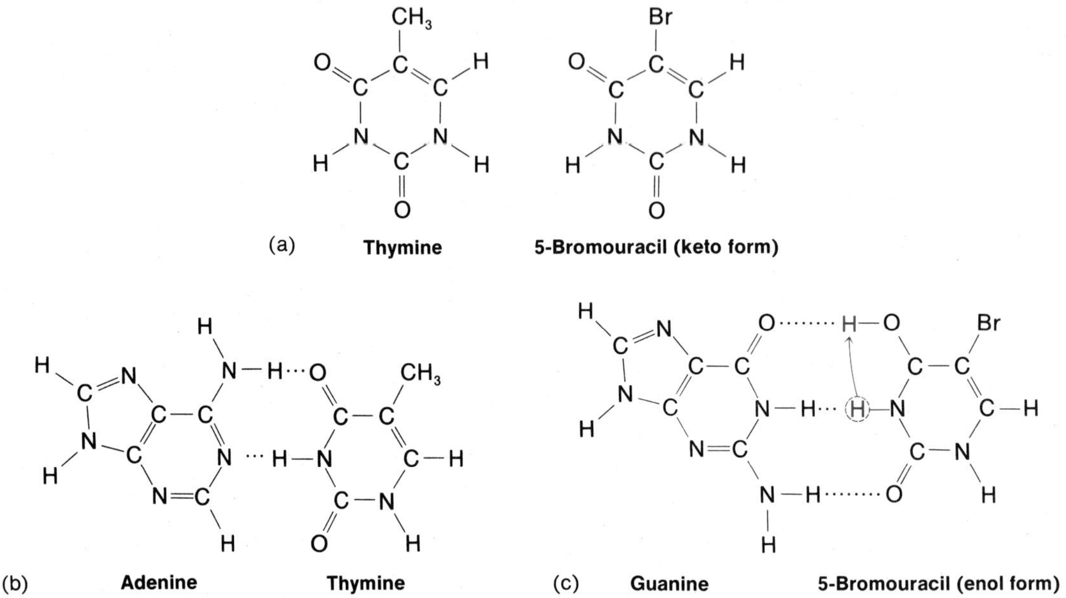

(a) Thymine 5-Bromouracil (keto form)

(b) Adenine Thymine

(c) Guanine 5-Bromouracil (enol form)

Figure 11-2
Mutagenesis by 5-bromouracil. (a) Structural formulas of thymine and 5-bromouracil. (b) A standard adenine-thymine base pair. (c) A base pair between guanine and the enol form of 5-bromouracil. The red H in the dashed circle shows the position of the H in the keto form.

replication BU functions like thymine and primarily pairs with adenine. Thymine can sometimes (but rarely) assume an enol form that is capable of pairing with guanine, and this conversion occasionally gives rise to mutants in the course of replication. The mutagenic activity of 5-bromouracil stems from a shift in the keto-enol equilibrium caused by the bromine atom; that is, the enol form exists for a greater fraction of time for BU than for thymine. Thus, if BU replaces a thymine, in subsequent rounds of replication, it occasionally generates a guanine, which in turn specifies cytosine, resulting in formation of a G • C pair (Figure 11-3).

Recent experiments suggest that 5-bromouracil is also mutagenic in another way. The concentrations of the nucleoside triphosphates in most cells is regulated by the concentration of thymidine triphosphate (TTP). This regulation results in appropriate relative amounts of the four triphosphates for DNA synthesis. One part of this complex regulatory process is the inhibition of synthesis of deoxycytidine triphosphate (dCTP) by excess TTP. The 5-bromouracil nucleoside triphosphate also inhibits production of dCTP. When 5-bromouracil is added to the growth medium, TTP continues to be synthesized by cells at the normal

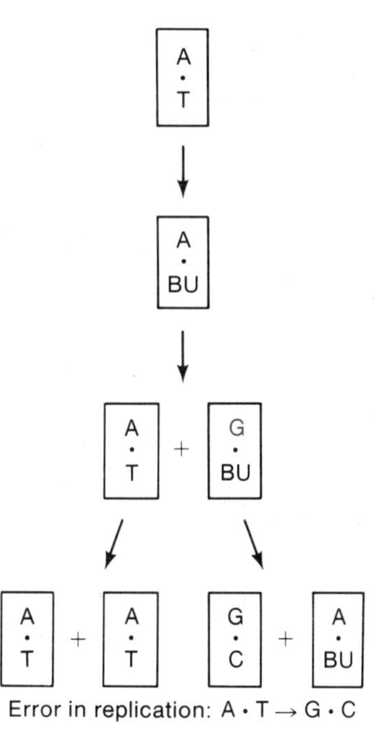

Figure 11-3
A mechanism of 5-bromouracil- (or BU)-induced mutagenesis. (a) During replication, BU, in its usual keto form, substitutes for T and the replica of an initial A · T pair becomes an A · BU pair. In the first mutagenic round of replication the BU, in its rare enol form, pairs with G. In the next round of replication, the G pairs with a C, completing the transition from an A · T pair to a G · C pair.

rate while the synthesis of dCTP is significantly reduced. The ratio of TTP to dCTP then becomes quite high and the frequency of misincorporation of T opposite G increases. The editing function and mismatch repair systems are capable of removing incorrectly incorporated thymine, but in the presence of 5-bromouracil the rate of misincorporation can exceed the rate of correction. An incorrectly incorporated thymine that persists will pair with adenine in the next round of DNA replication, yielding a G·C→A·T transition in one of the daughter molecules. Thus, 5-bromouracil induces transitions in both directions—A·T→G·C by the tautomerization route, and G·C→A·T by the misincorporation route. Thus, mutations that are induced by 5-bromouracil can also be reversed by it.

Both base-pair changes induced by BU maintain the original purine (Pu)-pyrimidine (Py) orientation. That is, the original and the altered base pairs both have the orientation Pu • Py—namely, A • T and G • C. If the original pair was T • A, the altered pair would be C • G—that is, Py • Pu for both the original and the altered pairs. A base change that does not change the Py • Pu orientation is called a **transition.** Base analogue mutations are always transitions. Later we will see changes from Pu • Py to Py • Pu and Py • Pu to Pu • Py; when such a change of orientation occurs, the mutation is called a **transversion.**

Chemical Mutagens

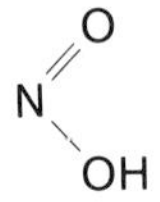

O=N–OH

Nitrous acid

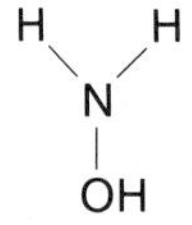

H_2N–OH

Hydroxylamine

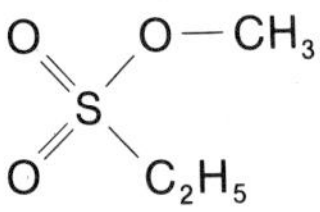

$O_2S(OCH_3)(C_2H_5)$

Ethyl methane sulfonate

Figure 11-4
Structures of three chemical mutagens.

By a chemical mutagen is meant a substance that can alter a base that is already incorporated in DNA and thereby change its hydrogen-bonding specificity. Four very powerful chemical mutagens are nitrous acid (HNO_2), hydroxylamine (HA), and ethylmethane sulfonate (EMS), whose chemical structures are shown in Figure 11-4.

Nitrous acid primarily converts amino groups to keto groups by oxidative deamination. Thus cytosine, adenine, and guanine are converted to uracil (U), hypoxanthine (H), and xanthine (X), respectively. These bases can form the base pairs U • A, H • C, and X • C. Therefore, the changes are G • C→A • T and A • T→G • C as cytosine and adenine respectively are deaminated. Since G and X both pair with C, the change should not be directly mutagenic. However, G • C→A • T transitions have been observed with certain single-stranded DNA phages at sites of a guanine, which suggests that xanthine probably has an undiscovered tautomeric form able to pair with thymine.

Hydroxylamine reacts specifically with cytosine and converts it to a modified base that pairs only with adenine, so that a G • C pair ultimately becomes an A • T pair. The chemistry of the alteration is complex.

EMS and a related substance ethylethane sulfonate (EES) are alkylating agents that have been used extensively in genetic research. Chemical mutagens such as nitrous acid and hydroxylamine, which are exceedingly useful in prokaryotic systems, are not particularly useful as mutagens in higher eukaryotes (because the chemical conditions necessary for reaction are not easily obtained), whereas the alkylating agents are highly effective. Many sites in DNA are alkylated by these agents; of prime importance for the induction of mutations is the addition of an alkyl group to the hydrogen-bonding oxygen of guanine and thymine. These alkylations impair the normal hydrogen-bonding of the bases and cause mispairing of G with T, leading to the transitions A • T→G • C and G • C→A • T. A second important mechanism for mutagenesis by alkylating agents is the induction of error-prone synthesis as in SOS repair; this mechanism is the source of the transversions induced by alkylating agents.

Ultraviolet Irradiation

Ultraviolet light is a fairly potent mutagen. In *E. coli* the number of mutants induced by ultraviolet light can be reduced by exposure to visible light (photoreactivation), which implicates the thymine dimer and other pyrimidine dimers. Bacterial mutants that lack the ability to carry out SOS repair (*lexA*$^-$ and *recA*$^-$ mutants) are not mutagenized by ultraviolet light. These and other experimental results have made it clear that mutagenesis by ultraviolet irradiation is almost exclusively a result of replication errors made during error-prone SOS repair. Error-prone repair of pyrimidine dimers leads to the production of both transitions and transversions.

Mutagenesis by Intercalating Substances

Acridine orange, proflavine, and acriflavine (Figure 11-5), which are substituted acridines, are planar, three-ringed molecules whose dimensions are roughly the same as those of a purine-pyrimidine pair. In aqueous solution, these substances form stacked arrays and are also able to stack with a base pair; this is done by insertion between two base pairs, a process called **intercalation.** Since the thickness of the acridine molecule is approximately that of a base pair and because the two bases of a pair are normally in contact, the intercalation of one acridine molecule causes adjacent base pairs to move apart by a distance equal to that of the thickness of one base pair (Figure 11-6). This has bizarre effects on the outcome of DNA replication, though the mechanism of action of the mutagen is not known. When DNA containing intercalated acridines is replicated, additional bases appear in the sequence (Figure 11-7). The usual addition is a single base, though occasionally two

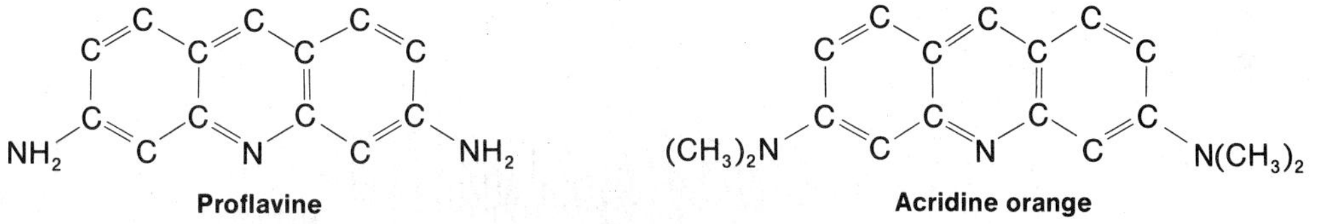

NH_2 Proflavine NH_2

$(CH_3)_2N$ Acridine orange $N(CH_3)_2$

Figure 11-5
Structures of two mutagenic acridine derivatives.

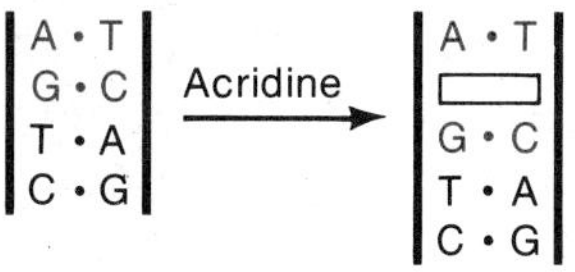

A • T, G • C, T • A, C • G — Acridine → A • T, [ ], G • C, T • A, C • G

Figure 11-6
Separation of two base pairs (shown in red) by an intercalating agent.

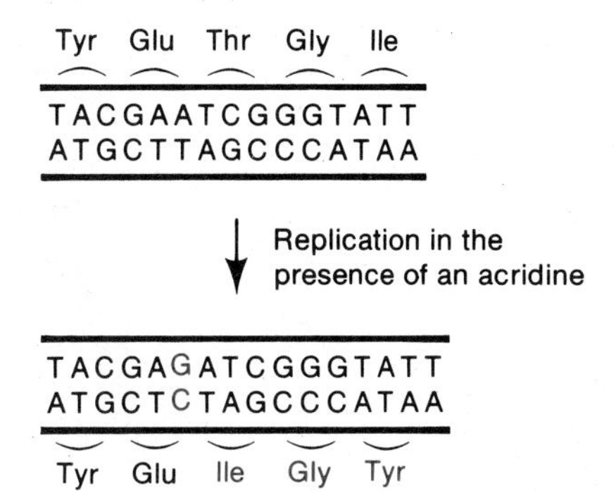

Figure 11-7
A base addition (red) resulting from replication in the presence of an acridine. The change in amino-acid sequence read from the upper strand in groups of three bases is also shown in red.

bases are added. Deletion of a single base also occurs but this is far less common than base addition. Mutations of this sort are called **frameshift mutations.** This is because the base sequence is read in groups of three bases when it is being translated into an amino acid sequence and the addition of a base changes the reading frame (Figure 11-7). This will be discussed in greater detail in the section on reversion.

Mutagenesis by Insertion of Long Segments of DNA (Transposable Elements)

E. coli, and many other organisms as well, contain long DNA segments (hundreds to thousands of base pairs long) that are mobile and that are called **transposable elements.** In a complex way to be described in Chapter 12, a transposable element replicates; one replica remains at the original insertion site and the other replica is inserted in another region of the chromosome. This process of insertion of a replica at a second site is called **transposition.** When transposition occurs, the sequence frequently inserts itself into a bacterial gene, thereby mutating that gene.

Some transposable elements contain sequences for termination of transcription. If such an element inserts between two genes that are transcribed as a polycistronic mRNA or between the promoter and the first gene, all genes downstream from the insertion site will not be transcribed. Such mutations are of a class called **polar mutations.**

Mutator Genes

In *E. coli* there are genes that in a mutant state cause mutations to appear very frequently in other genes throughout the genetic map. These genes are called **mutator genes.** This is a misnomer, because the function of each gene is probably to keep the mutation frequency low; that is, it is only when the product of a mutator gene is itself defective that there is widespread production of mutations.

Of the many mutators that have been observed, four types are understood: (1) a mutant DNA polymerase that reduces or eliminates the $3' \rightarrow 5'$ exonuclease activity of the editing function, (2) a mutant methylating enzyme (the *dam* enzyme) responsible for methylation of the sequences that the mismatch repair system uses to discriminate parental from daughter strands, (3) a mutant enzyme that cannot carry out the excision step in mismatch repair, and (4) mutations in the regulatory circuits that maintain the error-prone SOS repair system in an off state.

MUTATIONAL HOT SPOTS

If several hundred mutations in a single gene are mapped, they are, for the most part, distributed roughly equally over the mutated sites. However, a few sites are represented by as many as 100 times the typical number of mutations; these sites are called **hot spots.**

About 5 percent of the cytosines in a typical DNA molecule are in the methylated form—5-methylcytosine (MeC). The role of MeC (and other methylated bases as well) is not clearly understood in most cases, though one role is described in Chapter 8. At any rate, they are not harmful and do not change the hydrogen-bonding properties of the base—that is, MeC pairs with guanine just as cytosine does. MeC is also subject to alteration by spontaneous deamination, as discussed earlier for cytosine. When cytosine is deaminated, uracil is formed and this is removed by uracil *N*-glycosylase. However, when MeC is deaminated, the result is 5-methyluracil, which is another name for thymine; therefore the G • MeC pair becomes a G • T pair, which in subsequent replication yields an A • T pair. Note that the mismatch repair system is certainly able to convert the G • T pair back to a correct G • C pair. However, spontaneous deamination can occur in nonreplicating DNA (e.g., in a resting cell or in a phage), and both strands will be equally methylated by the A-methylating system (Chapter 8). Thus the mismatch repair system receives no signal indicating that the G • C pair is the correct one and could just as well convert the G • T pair to an A • T pair. Thus the mutation frequency can be very high at a MeC site. A given G • MeC → A • T transition will of course only produce a mutant if the change causes an amino acid substitution that affects the activity of the gene product. Such MeC sites do not occur very often, so hot spots should not be particularly frequent. Direct determination of the base sequence of several genes and of hot spot mutants has shown that, indeed, MeC accounts for most of the hot spots for spontaneous mutagenesis.

REVERSION

So far we have discussed changes from the wild-type to the mutant state. The reverse process, in which the wild-type phenotype is regained, also occurs; this process is called **back mutation, reverse mutation,** or, most commonly, **reversion.** One way that the wild-type phenotype may be restored is to regain the wild-type genotype (that is, the wild-type base sequence). However, this is not always what happens.

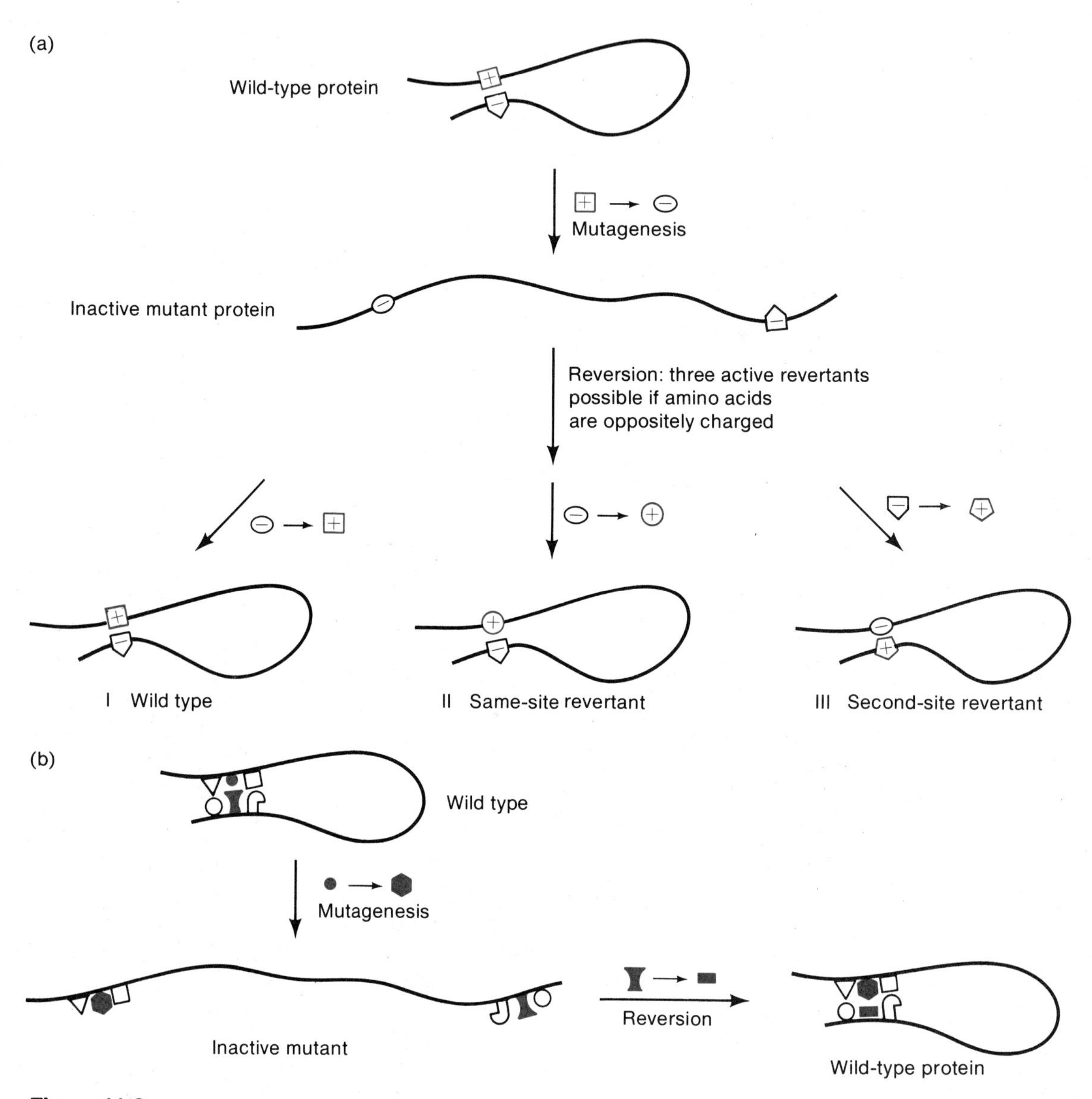

Figure 11-8
Several mechanisms of reversion. In panel (a) the charge of one amino acid is changed and the protein loses activity. The activity is returned by (I) restoring the original amino acid, or (II) by replacing the (−) amino acid by another (+) amino acid, or (III) by reversing the charge of the original (−) amino acid. In each case the attraction of opposite charges is restored. In panel (b) the structure of the protein is determined by interactions between six hydrophobic amino acids. Activity is lost when the small circular amino acid is replaced by the bulky hexagonal one and is restored when space is made by replacing the convex amino acid by the small rectangle.

Intragenic Reversion

The most useful type of revertant in molecular biological studies has been a kind that does not faithfully recreate the wild-type base sequence. In these revertants the reversion does not occur at the site of the mutation but instead entails a mutation at a second site. Such mutations are often called **second-site** or **suppressor mutations.** In some cases the reverse mutation does not even occur within the mutated gene, and we may distinguish between intragenic and intergenic reversion. We first consider the intragenic type.

Consider a hypothetical protein containing 97 amino acids whose structure is determined entirely by an ionic interactior between a positively charged (+) amino acid at position 18 and a negative one (−) at position 64 (Figure 11-8). If the (+) amino acid is replaced by a (−) amino acid, the protein is clearly inactive. Three kinds of reversion events would restore activity (Figure 11-8(a)): (1) The original (+) amino acid could be put back. (2) A different (+) amino acid could be put at position 18. (3) The (−) amino acid at position 64 could be replaced by a (+) amino acid; this second-site mutation would restore the activity of the protein. A possibility which would not generally work but which might work in a specific case is to insert a (+) amino acid at position 17 or 19.

Figure 11-8(b) shows another, but more complicated, example of intragenic reversion. In this case the structure of a protein is maintained by a hydrophobic interaction. The replacement of an amino acid with a small side chain by a bulky phenylalanine changes the shape of that region of the protein. A second amino acid substitution providing space for the phenylalanine could restore the protein structure.

The analysis of second-site amino acid substitution has been an important aid in determining the three-dimensional structure of proteins because the following rule is often obeyed: If a substitution of amino acid A by amino acid X, which creates a mutant, is compensated for by a substitution of amino acid B by amino acid Y, then A and B are either three-dimensional neighbors or are both contained in two interacting regions.

Revertants of frameshift mutations usually occur at a second site. It is of course possible that a particular added base could be removed or a particular deleted base could be replaced by a spontaneous event, but this would not occur very often. Second-site reversion of a frameshift mutation has two requirements illustrated in Figure 11-9: (1) the reverting event must be very near the original site of mutation, so that very few amino acids are altered between the two sites; and (2) the segment of the polypeptide chain in which both changes occur must be able to withstand substantial alterations.

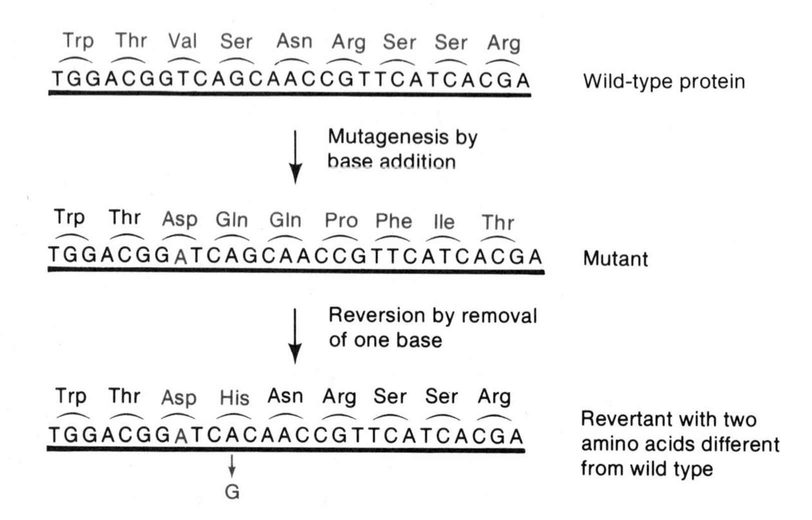

Figure 11-9
Reversion by base deletion from an acridine-induced base-addition mutant.

Intergenic Reversion and Suppression

Intergenic reversion refers to a mutational change in a second gene that eliminates or suppresses the mutant phenotype. One type, which occurs when two proteins interact, is the following. A mutation in the binding site of protein A prevents the protein from interacting with protein B. Another mutation in the binding site of protein B alters this binding site so that the mutant B protein can bind to the mutant A protein; thus, the interaction between the two proteins is restored. The occurrence of intergenic reversion of this kind is an important indicator of the interaction between two proteins and is a splendid example of a genetic result giving information about molecular structure.

A second type of extragenic reversion has the remarkable property that the second-site mutation not only eliminates the effect of the original mutation but also suppresses mutations in many other genes as well. This type of reversion, which is produced by mutations in certain tRNA molecules and aminoacyl synthetases, can be seen most clearly when reversion of chain termination mutations are examined.

Chain termination mutations are common, for they can arise in many ways. For example, a single base change in any of the codons AAG, CAG, GAG, UCG, UUG, UGG, UAC, and UAU can give rise to the chain termination codon UAG. If such a mutation occurs within a gene, a mutant protein with little or no function will result because no tRNA molecule exists whose anticodon is complementary to UAG. Thus, only a fragment of the wild-type protein is produced and this usually fails to function unless the mutation is very near the carboxyl terminus of the protein.

In certain bacterial strains the presence of a chain termination mutation is not sufficient to stop polypeptide synthesis. For example, a phage may have acquired a UGA codon in a gene encoding a critical protein; when the phage infects most host bacteria, no phage progeny will be produced. However, in a particular bacterial strain the mutant phage may grow normally, indicating that the mutation is made silent by some element in the bacterium. Such a bacterium is said to be able to **suppress** the mutation and to contain a **suppressor.** A bacterium able to suppress a particular type of chain termination mutation—in this example, a UAG codon—is usually able to do so with a large number of mutations of that class, whether the mutation is in a phage or the bacterium itself. In general, other types of chain termination mutations, for example, UGA, will not be suppressed. The explanation for this phenomenon is that the bacterium (called a suppressor mutant) contains an altered tRNA molecule that can respond to a particular stop codon. For UAG, the altered tRNA molecule might contain the anticodon CUA, which can pair with that codon. Such a tRNA molecule is called a **suppressor tRNA,** and a mutation on which it can act is said to be suppressor-sensitive.

What has been mutated in the production of a suppressor tRNA? Clearly it must be a normal tRNA gene. Therefore, in the example just given, a $tRNA^{Lys}$ molecule whose anticodon is CUU, has been altered to have the anticodon CUA, which can hydrogen-bond to the codon UAG.

Inasmuch as a single base change is sufficient to alter the complementarity of an anticodon and a codon, there are (at most) eight tRNA molecules having a complementary anticodon that, with a single changed base, will also suppress a UAG codon. Thus, the following amino acids (whose codons are also indicated) can be put at the site of a chain termination codon: Lys (AAG), Gln (CAG), Glu (GAG), Ser (UCG), Trp (UGG), Leu (UUG), and Tyr (UAC and UAU). Note that these are the same amino acid codons that can be altered by mutation to form a UAG site. Suppressors also exist for chain termination mutants of the UAA and UGA type. These too are mutant tRNA molecules whose anticodons are altered by a single base change.

In conventional notation, suppressors are given the genetic symbol *sup* followed by a number (or occasionally a letter) that distinguishes one suppressor from another. A cell lacking a suppressor is designated *sup0* and *sup*$^-$.

Several features of nonsense suppression should be recognized:

1. Not every UAG suppressor can restore a functional protein by suppressing each UAG chain termination mutation. Thus, a UAG codon

produced by mutating the leucine UUG codon might be suppressed by a suppressor tRNA that inserts tyrosine, serine, or tryptophan, but might not be able to tolerate a substitution by the electrically charged amino acids lysine, glutamine, or glutamic acid.

2. Suppression may be incomplete in that the activity of the suppressed mutant protein may not be as great as that of the wild-type protein and the stop codon may not always be read as a sense codon.

3. A cell can survive the presence of a suppressor only if the cell also contains two or more copies of the tRNA gene. Clearly if a $tRNA^{Ser}$ molecule that reads the UCG codon is mutated, then the UCG can no longer be read as a sense codon. This will lead to chain termination wherever UCG occurs and a cell harboring such a mutant tRNA molecule will fail to terminate virtually every protein made by the cell. However, as was mentioned earlier, there are multiple copies of most tRNA molecules; moreover, there are also minor tRNA molecules having the same anticodons as the major molecules. Thus, in any living cell containing a suppressor tRNA, there must always be an additional copy of a wild-type tRNA that can function in normal translation.

If a cell contains an UAG suppressor, then proteins terminated by a single UAG codon will not be completed and the existence of a suppressor tRNA should be lethal. There are two ways that this problem can be avoided:

1. Protein factors active in termination (see Chapter 10) respond to chain termination codons even though a tRNA molecule that recognizes the codon is present, i.e., suppression is weak. For example, if the probability of recognition is only x percent, then only x percent of the prematurely terminated mutant proteins would be completed and only x percent of normally terminated proteins would be ruined by lack of termination.

2. Normal chain termination often uses pairs of distinct termination codons such as the sequence UAG–UAA. Thus, the existence of a UAG suppressor would not prevent termination of a doubly terminated protein.

Suppression of missense mutations also occurs. For example, a protein in which valine (nonpolar) has been mutated to aspartic acid (polar), resulting in loss of activity, is restored to the wild-type phenotype by a missense suppressor that substitutes alanine (nonpolar) for aspartic acid. Such a substitution can occur in three ways: (1) a mutant tRNA molecule may recognize two codons, possibly by a change in the anticodon loop; (2) a mutant tRNA molecule can be recognized by a noncognate aminoacyl synthetase and be misacylated; and (3) a mutant synthetase can charge a noncognate tRNA molecule. Examples of each type of suppressor are known. Suppression of missense mutations is necessarily inefficient. If a suppressor that substitutes alanine for aspartic acid worked with, say, 20 percent efficiency, then in virtually every protein molecule synthesized by the cell at least one aspartic acid would be replaced, which is a situation that a cell could not possibly survive. The usual frequency of missense suppression is about 1 percent.

REVERSION AS A MEANS OF DETECTING MUTAGENS AND CARCINOGENS

In view of the increased number of chemicals present as environmental contaminants and because most cancer-causing agents (carcinogens) are also mutagens, tests for the mutagenicity of these substances have become important. One simple method for screening large numbers of substances for mutagenicity is a reversion test using nutritional mutants of bacteria. In the simplest reversion test, known numbers of a mutant bacterium are plated on a growth medium containing a potential mutagen, and the number of revertant colonies that arise is counted. If the substance is a mutagen, the number of colonies will be greater than that obtained in the absence of the compound. However, such simple tests fail to demonstrate the mutagenicity of many carcinogens; these substances are not directly mutagenic (or carcinogenic), but are converted to active compounds by enzymatic reactions that occur in the liver of animals and have no counterpart in bacteria. The normal function of these enzymes is to protect the organism from various noxious substances that occur naturally by enzymatically converting them to nontoxic substances. However, when the enzymes encounter certain manmade and natural compounds, they convert these substances, which may not be themselves directly harmful, to mutagens or carcinogens. The enzymes are contained in the microsomal fraction of liver cells. Addition of the microsomal fraction to the growth medium enables these substances to show up as mutagens and is the basis of the **Ames test** for carcinogens.

A set of histidine-requiring (His^-) mutants of the bacterium *Salmonella typhimurium*, which contain either a base substitution or a frameshift mutation, are used for tests of reversion to His^+. The frequency of spontaneous His^+ revertants is low in this mutant but revertants are readily produced in one or more of these mutants by most known mutagens. Agar is prepared containing (1) a very small amount of histidine, sufficient to initiate growth of individual cells but not enough for colony formation (needed because two rounds of replication are required for a mutant to be expressed, as was discussed earlier), and (2) a known carcinogen or substance to be tested. A small amount of an extract of rat liver and about 10^8 His^- cells are spread on the agar (plate A). The same number of cells is applied to another plate (B) which lacks the carcinogen; the number of colonies appearing on plate B is usually about 5 to 10 and these are the spontaneous revertants. When a known carcinogen is present, more colonies are found on plate A. The number of colonies on plate A depends on the concentration of the substance being tested and, for a known carcinogen, correlates roughly with its known effectiveness as a carcinogen.

The Ames test has now been used with thousands of substances and mixtures (such as industrial chemicals, food additives, pesticides, hair dyes, etc.) and numerous unsuspected substances have been found to stimulate reversion in this test. This does not mean that the substance is definitely a carcinogen but only that it has a high probability of being so. **As a result of these tests, many industries have reformulated their products: for example, the cosmetic industry has changed the formulation of many hair dyes and cosmetics to render them nonmutagenic. Ultimate proof of carcinogenicity is determined from testing for tumor formation in laboratory animals. The Ames test and several other microbiological tests are used to reduce the number of substances that have to be tested in animals since to date only a few percent of more than 300 substances known from animal experiments to be carcinogens failed to increase the reversion frequency in the Ames test.**

12 Plasmids and Transposable Elements

Plasmids are extrachromosomal circular DNA molecules found in most bacterial species and in some species of eukaryotes. Under normal circumstances a particular plasmid is dispensable to its host cell; for example, sometimes at the time of cell division a plasmid-free daughter cell is formed and such a cell is almost always viable. However, many plasmids contain plasmid genes that may be essential in certain environments. For example, the R plasmids carry genes that confer resistance to numerous antibiotics so that in nature a cell containing such a plasmid can survive in the presence of an antibiotic, whether humanly administered or produced by a fungus.

Transposable elements (or transposons, as they are called in bacteria) are another type of accessory DNA molecule. They are chromosomal segments that differ from other regions of the chromosome in that they are able to relocate (transpose) to another part of the chromosome or to a plasmid, virus, or separate chromosome. Their movement, which occurs infrequently and usually at random in time, is mediated by enzymes encoded in the element; other enzymes, for instance, some conferring drug resistance, are contained in some elements.

Currently, major interest in plasmids centers on their practical value in genetic engineering. Transposable elements are responsible for many genetic phenomena in both bacteria and eukaryotes; in the latter, they are involved in the regulation of the activity of some genes.

PLASMID DNA

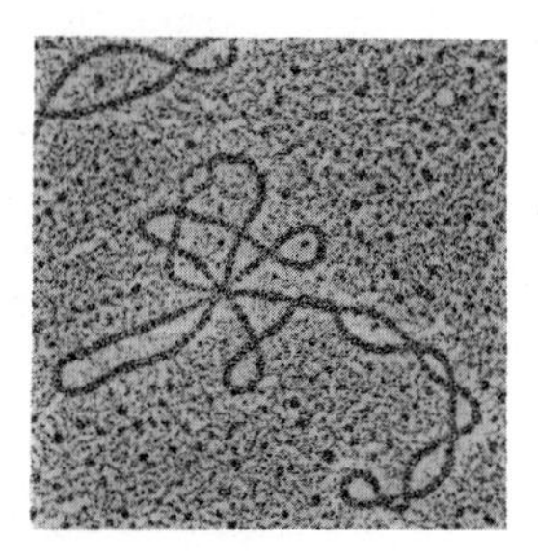

Figure 12-1
A supercoiled plasmid.

With only a single exception (the killer-plasmid of yeast, which is an RNA molecule) all known plasmids are supercoiled circular DNA molecules (Figure 12-1). The molecular weights of the DNA range from about 10^6 for the smallest plasmid to slightly more than 10^8 for the largest one. The number of copies of a plasmid per cell ranges from 1–2 (low-copy number plasmids) to 10–60 (high-copy-number plasmids).

Plasmid DNA can usually be isolated from bacteria in a simple way. A culture of plasmid-containing bacteria is lysed by adding a detergent and then the lysate is centrifuged. The bacterial chromosome complex, which contains protein and RNA, is very large and compact and moves rapidly to the bottom of the centrifuge tube; the smaller plasmid DNA remains in the supernatant. CsCl and the fluorescent dye ethidium bromide are then added to the supernatant and the solution is centrifuged to equilibrium. The CsCl forms a density gradient. In the presence of ethidium bromide supercoiled molecules have a higher density than the linear fragments of chromosomal DNA (which is broken during isolation); thus, at equilibrium the supercoiled DNA is located in a different region of the centrifuge tube than the linear molecules. The fluorescence of the ethidium bromide makes both DNA fractions visible, so the supercoils can be easily identified and removed from the tube.

TRANSFER OF PLASMID DNA

The bacterium *E. coli* possesses two mating types: donors or **males,** and recipients or **females.** The determinant of maleness is the **F** or **sex plasmid;** as donor, a male cell is designated F^+. A female cell lacks the F plasmid and is designated F^-. When a culture of males is mixed with a culture of females, male-female pairs form (this is called **conjugation**), each pair joined by a conjugation bridge (Figure 12-2). This pairing induces looped rolling circle replication of F, and one copy of F is transferred to the female in about one minute (Figure 12-3). In contrast with other sexual systems, the female is converted to a male inasmuch as, after the mating, the recipient cell contains the F plasmid.

Note that during transfer DNA synthesis occurs in both the donor and recipient cells. In the donor the synthesis replaced the single strand that is transferred; synthesis in the recipient converts the transferred single strand to double-stranded DNA. The mechanism of recirculariza-tion of the transferred DNA is not well understood. It is believed that when one cycle of looped rolling circle replication is completed, the

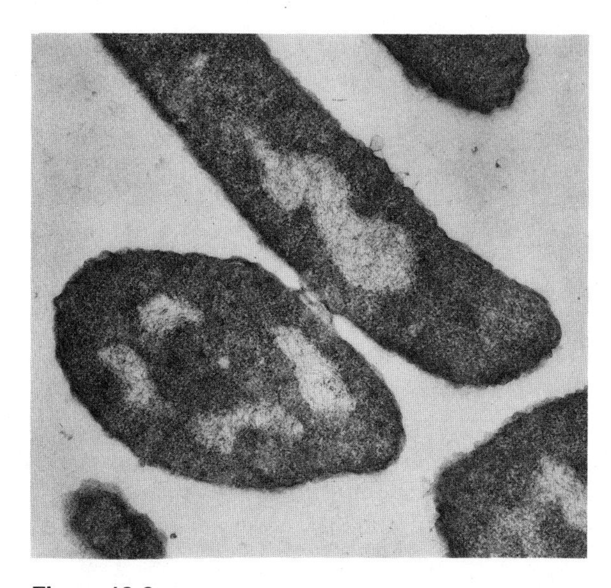

Figure 12-2
Electron micrograph of two *E. coli* cells during conjugation. (Courtesy of Lucien Caro.)

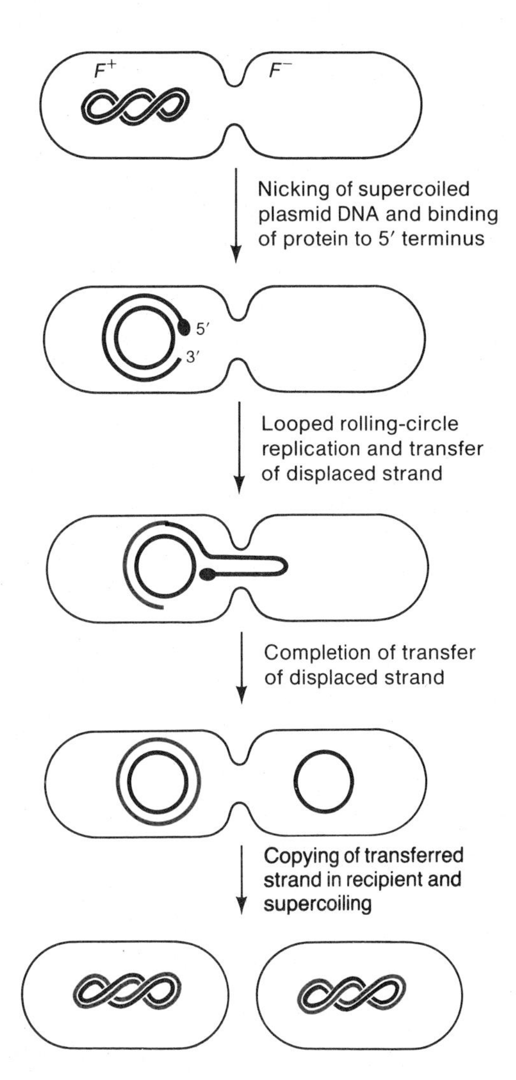

Figure 12-3
A model for transfer of F plasmid DNA from an F^+ cell by a looped rolling-circle mechanism. The displaced single strand is transferred to the F^- recipient cell, where it is converted to double-stranded DNA. Chromosomal DNA is omitted for clarity.

protein, shown in Figure 12-3, that binds to the 5′ terminus makes a cut at the replication origin and brings the 3′ and 5′ termini together in preparation for a joining event.

Many, but not all, classes of plasmids are capable of transfer. Plasmids unable to transfer usually either are unable to form stable pairs or lack a system for transfer replication.

Another property possessed by the F plasmid is the ability to integrate into the bacterial chromosome. (The mechanism for this will be discussed later in the chapter.) When this occurs, the chromosome remains a single, circular DNA molecule and F behaves as if it were part of this chromosome, increasing the size of the chromosome. The integration process happens rarely but it is possible to purify a culture of cells that are the progeny of the cell in which integration occurred. The cells in such a culture are called **Hfr males.** (Hfr is an acronym for *h*igh *f*requency of recombination.) When a culture of Hfr cells is subsequently mixed with an F^- culture, conjugation also occurs as described earlier, though the material transferred is slightly different from that in an $F^+ \times F^-$ mating (Figure 12-4). This time, under the influence of F, DNA replication begins in the Hfr cell and a replica is transferred to the F^- cell; however, the direction of replication is such that a small portion of F is transferred first and the major portion is transferred last. Moreover, because bacteria are very small and in constant motion by being bombarded by solvent molecules (Brownian motion), and because it takes 100 minutes to transfer an entire chromosome, the mating pair usually breaks apart before transfer is completed. Thus, the female receives both a large fragment of the male chromosome, which may contain hundreds or thousands of genes, and a small functionless fragment of F. Because of this, the exconjugant female remains a female in an Hfr $\times$ F^- mating.

The presence of the new chromosomal fragment in the female sets in motion a recombination system that causes genetic exchanges to occur, so that a recombinant F^- cell often results. Thus, in a mating between an Hfr leu^+ culture and an F^-leu^- culture, F^-leu^+ cells form.

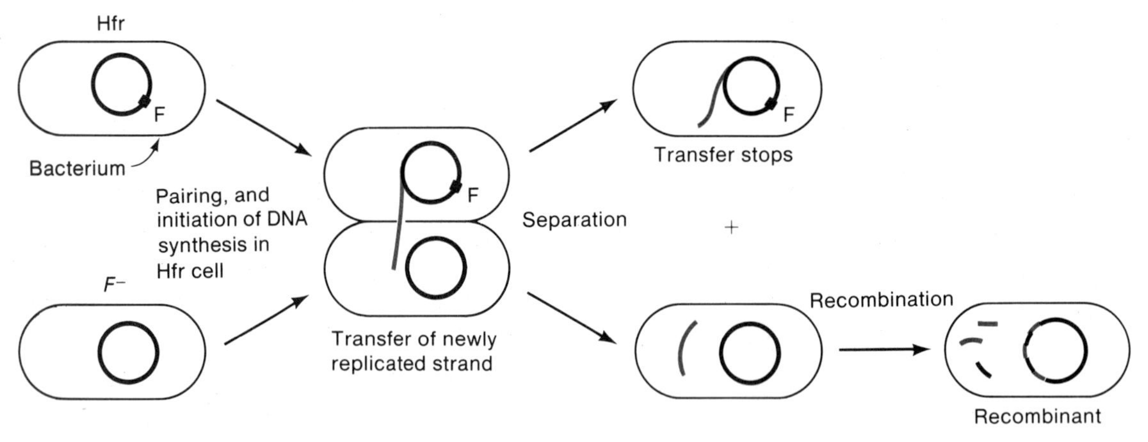

Figure 12-4
A diagram showing mating between an Hfr male bacterium and an F^- female bacterium. The two bacteria form a pair. Then, under the influence of a unit called F, DNA replication begins in the male, adjacent to F, and a replica of F is transferred to the female. By random motion the mating cells break apart. At a later time a portion of the transferred DNA exchanges with the corresponding piece in the female. The intact female chromosome then replicates; the fragments do not replicate and are ultimately lost in the course of cell division.

Hfr $\times$ F^- matings are powerful matings for creating bacteria with desired combinations of mutations. Discussions of this technique can be found in *MB*, pages 48–51 and in most books on genetics.

An Hfr is produced when F integrates stably into the chromosome, as we have already stated. At a low frequency, F can excise itself. When this happens, the excised circular DNA is sometimes found to contain genes that were adjacent to F in the chromosome (Figure 12-5). A plasmid containing both F genes and chromosomal genes is called an **F′ plasmid.** It is usual to describe an F′ plasmid by stating the genes it is known to possess—for example, F′*lac pro* contains the genes for lactose utilization and proline synthesis. F′ plasmids can also be transferred from an F′ male to a female. This occurs sufficiently rapidly that the entire F′ is usually transferred before the mating pair breaks apart. Thus the female recipient is converted to an F′ male in an F′ $\times$ F^- mating.

F′ plasmids have been useful in the production of partial diploid bacteria. For example, if an F′lac^+/*str-s* male is mated with a lac^-*str-r* female and a Lac^+ Str-r colony is isolated, the cells in the colony will carry two copies of the *lac* gene—the lac^+ version brought to the female in the F′ and the lac^- version already present in the female chromosome. To denote this, the genotype of the cell is written F′lac^+/lac^- *str-r,* as is usual with partial diploid cells. By convention, genes carried on the F′ plasmid are written at the left of the diagonal line.

F′ plasmids have been exceedingly useful in the analysis of regulation of bacterial systems. An example of this will be seen in Chapter 14 when the system responsible for metabolism of lactose is examined.

Often it is desirable to transfer a plasmid that is incapable of conjugal transfer, for example, the Col plasmids. When this is required, the plasmid can be transferred, albeit at much lower efficiency, simply by mixing free plasmid DNA with recipient cells suspended in a cold $CaCl_2$ solution, a procedure known as **$CaCl_2$ transformation.** The main requirement for successful transfer is that the recipient bacterium

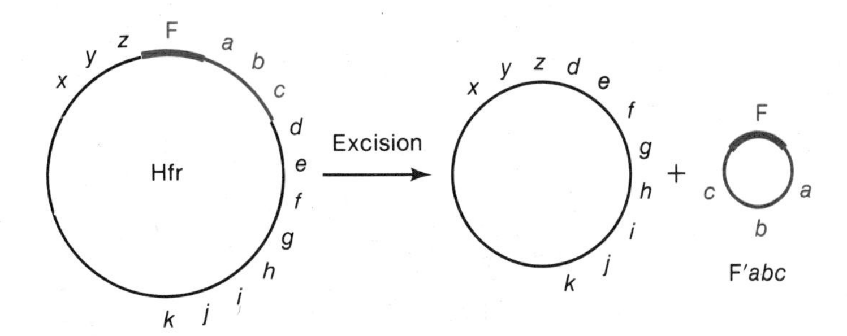

Figure 12-5
Diagram showing production of an F′ plasmid by aberrant excision from an Hfr chromosome.

possess replication enzymes that are active on the plasmid, which is invariably the case if the donor and recipients are of the same species; sometimes, interspecific transformation is also successful. This procedure is important in genetic engineering, as will be seen in Chapter 13.

PROPERTIES OF PARTICULAR PLASMIDS

We have just seen some of the properties of the F plasmid. In this section we briefly describe some of the more commonly studied plasmids. These have been used primarily as means of studying DNA replication (see later section) and as a cloning vehicle in genetic engineering (Chapter 13).

The drug-resistance, or R, plasmids were first isolated from the bacterium *Shigella dysenteriae* during an outbreak of dysentery in Japan and have since been found in *E. coli* and various species of *Salmonella, Vibrio, Bacillus, Pseudomonas,* and *Staphylococcus*. Their defining characteristics are that they confer resistance on their host cell to a variety of fungal antibiotics and are usually self-transmissible. Most R plasmids consist of two contiguous segments of DNA (Figure 12-6). One of these segments is called **RTF (resistance transfer factor);** it carries genes regulating DNA replication and copy number, the transfer genes, and sometimes the gene for tetracycline resistance *(tet),* and has a molecular weight of 11×10^6. The other segment, sometimes called the **r determinant,** is variable in size (from a few million to more than 100×10^6 molecular weight units) and carries other genes for antibiotic resistance. Resistance to the drugs penicillin (Pen), ampicillin (Amp), chloramphenicol (Cam), streptomycin (Str), kanamycin (Kan), and sulfonamide (Sul), in combinations of one or more, appears commonly.

Col plasmids are *E. coli* plasmids encoding colicins, proteins that are capable of preventing growth of a bacterial strain that does not

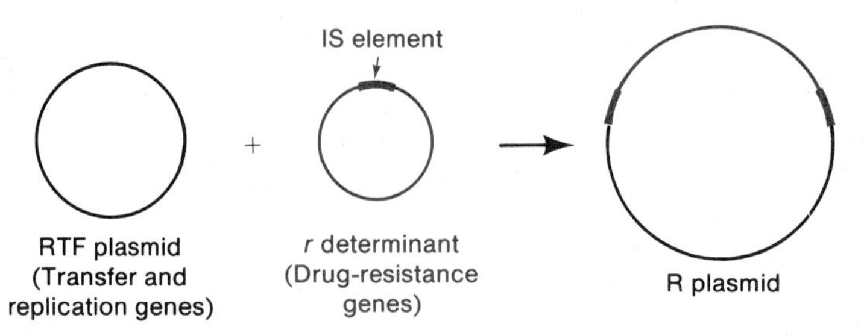

Figure 12-6
The components of an infectious R plasmid. The IS element is a transposon responsible for merging of RTF and *r* by the process of replicon fusion described later in the chapter.

contain a Col plasmid.* There are many types of colicins, each designated by a letter (e.g., colicin B) and each having a particular mode of inhibition of sensitive cells. The best-studied Col plasmid is ColE1, whose molecular weight is 4.2×10^6. It is used extensively in recombinant DNA research (Chapter 13) and in an *in vitro* DNA replication system.

A crown gall tumor found in many dicotyledonous plants is caused by the bacterium *Agrobacterium tumefaciens.* The tumor-causing ability resides in a plasmid called Ti. In an infected plant some of the bacteria enter and grow within the plant cells and lyse there, releasing their DNA in the cell, and from this point on, the bacteria are no longer necessary for tumor formation. By an unknown mechanism a small fragment of the Ti plasmid, containing the genes for replication, becomes integrated into the plant cell chromosomes. The integrated fragment breaks down the hormonally regulated system that controls cell division and the cell is thereby converted to a tumor cell. This plasmid has recently become very important in plant breeding because specific genes can be inserted into the Ti plasmid by recombinant DNA techniques; sometimes these genes can become integrated into the plant chromosome, thereby permanently changing the genotype and phenotype of the plant. It is believed that new plant varieties having desirable and economically valuable characteristics derived from unrelated species can be developed in this way.

Plasmids are known in yeast, a unicellular eukaryote. One of the more intriguing ones is the **killer particle,** which is a double-stranded RNA molecule of molecular weight about 15×10^6. It is the only known plasmid that does not contain DNA. This particle contains ten genes for replication and several others for synthesis of the killer substance, a colicinlike material.

The best-studied yeast plasmid is the so-called 2μm plasmid, a DNA molecule having a length of 2μm and a molecular weight of 4×10^6. There are about sixty copies of this plasmid per cell and all are found in the yeast nucleus. This location is consistent with the fact that for replication the 2μm plasmid uses the same enzymes needed for replication of the yeast chromosomes. Furthermore, the plasmid DNA is coated with histones, as is all nuclear DNA. The plasmid DNA apparently does not integrate into the host DNA.

The 2μm plasmid constitutes a useful system for studying DNA replication in yeast. However, at present its greatest value is as a vector for cloning foreign genes in yeast by the recombinant DNA techniques described in Chapter 13.

*The Col plasmids are one class of a general type of plasmid called a bacteriocinogenic plasmid, which produce bacteriocins in many bacterial species. Bacteriocins, of which colicins are one example, are proteins that bind to the cell wall of a sensitive bacterium and inhibit one or more essential processes such as replication, transcription, translation, or energy metabolism.

PLASMID REPLICATION

A plasmid can only replicate within a host cell; one might therefore expect that all plasmids native to the same host species would have the same mode of replication. However, just as with phages, which also replicate only within a host cell, there is enormous variation in both the enzymology and mechanics of plasmid DNA replication. The major types of variability are the following:

1. *Reliance on host enzymes.* Some plasmids use host enzymes exclusively. Others encode some of their own enzymes.
2. *Identity of polymerase.* Most *E. coli* plasmids use polIII for chain growth and polI for synthesis of precursor fragments, as in replication of the *E. coli* chromosome. However, others (ColE1 is an example) use polI for chain growth.
3. *Directionality.* Both purely unidirectional and purely bidirectional replication have been observed. In addition, there are plasmids in which both modes are present. For example, the plasmid RK6 replicates first in one direction and then later in the opposite direction from the same origin.
4. *Termination.* In unidirectionally replicating plasmids termination necessarily occurs at the origin after one cycle of replica-

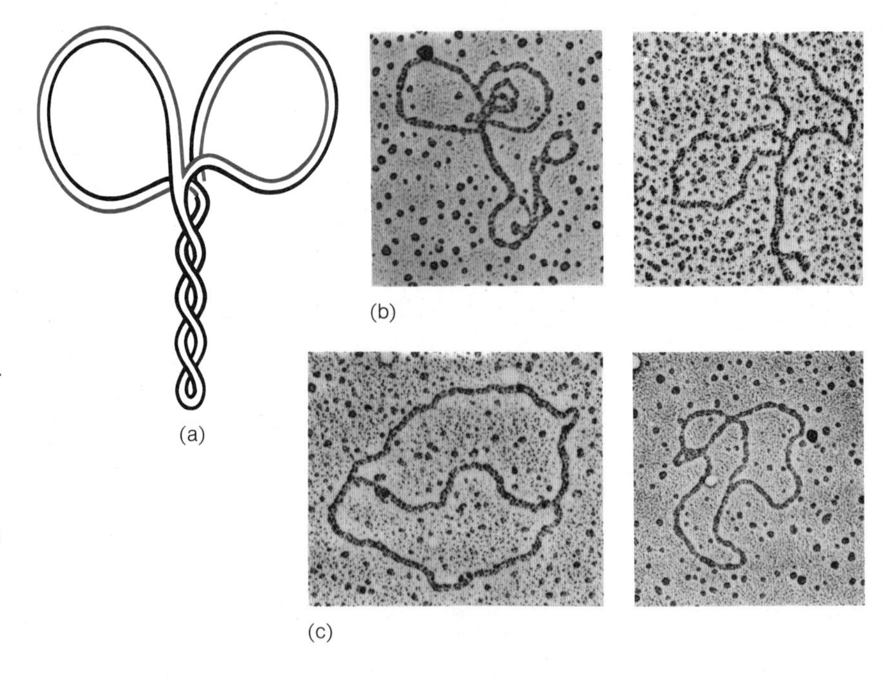

Figure 12-7
Replication of ColE1 DNA. (a) Diagram of a butterfly molecule. The newly synthesized strands are red. (b) Electron micrographs of butterfly molecules. (c) Nicked butterfly molecules, showing that a nick converts a butterfly molecule to a θ molecule. (Courtesy of Donald Helinski.)

tion. Bidirectionally replicating molecules are of two types though, both of which have been observed. In one type, termination occurs when the growing forks reach the same region. Others have a fixed termination site that is sometimes reached by one growing fork before the other fork reaches it. The signal for termination is unknown.

5. *Replicating form.* In the most carefully studied plasmids, replication occurs by the so-called *butterfly* mode, first observed for animal viruses (Figure 12-7). In a partially replicated molecule the replicated portions are untwisted, as is usually the case in θ replication, but the unreplicated portion is supercoiled. When the replication cycle is completed, one of the circles must be cleaved (possibly by DNA gyrase). The result after one round of replication is one nicked molecule and one supercoiled molecule. The nicked molecule is then sealed and, somewhat later, supercoiled. Whether this is a general mechanism for plasmid replication is not known.

Control of Copy Number

Some plasmids are present in cells in low-copy number—one or a few per cell—whereas others exist in large numbers—from 10 to 100 per cell. **The number is regulated by controlling the rate of initiation of DNA synthesis.**

The generally accepted explanation for the regulation of copy number is that there is a plasmid-encoded inhibitor that binds to the replication origin and inhibits replication. Let us first see how this idea explains maintenance of copy number from generation to generation. In the current theory it is assumed that the inhibitor is active only as a multisubunit protein and that the monomer-multimer equilibrium is very dependent on concentration. Thus, as a cell grows (enlarges), the inhibitor concentration drops and inactive monomers form; replication is thereby derepressed, and the number of plasmid DNA molecules doubles. At this point there will exist twice the initial number of inhibitor genes; therefore, by protein synthesis the inhibitor concentration also doubles. This results in the formation of active inhibitor and replication stops. A similar sequence of events would occur if there were initially only one copy of a high-copy-number plasmid–that is, replication would continue until there is sufficient inhibitor to turn off synthesis. How can this explanation account for differences in copy number? The most likely possibility is that for the high-copy-number plasmids the association of monomers to form a multimeric active inhibitor requires a higher concentration of monomers than is the case for the low-copy-number plasmids. Thus, only when the number of plasmids per cell is high is the "gene dosage" high enough to produce active inhibitor.

STRUCTURE OF TRANSPOSABLE ELEMENTS

Transposable elements were first detected in *E. coli* as polar mutations in the *gal* operon. In order to understand these mutations, fragments of DNA containing such mutations were isolated and their nucleotide sequences were determined. It was found that each polar mutation was associated with a segment of DNA that had somehow been inserted in the gene. These segments were called **insertion** or **IS elements** and later shown to be members of a class of sequences called **tranposable elements** or, in bacteria, **transposons.**

A variety of different IS elements, ranging in length from a few hundred to a few thousand nucleotide pairs, have been isolated and sequenced (Table 12-1). A feature of all IS elements is the presence of inverted-repeat sequences at the termini (Figure 12-8). The polar effects are the result of either transcription termination signals or stop codons (in all reading frames) in the main body of the element. IS elements are found in various regions of the *E. coli* chromosome, in plasmids, and in

Table 12-1
Properties of some *E. coli* insertion elements

Element	*Number of copies and location*	*Size, base pairs*
IS1	5–8 in chromosome	768
IS2	5 in chromosome, 1 in F	1327
IS3	5 in chromosome, 2 in F	Approximately 1400
IS4	1 or 2 in chromosome	Approximately 1400
IS5	Unknown	1250
$\gamma\delta$	1 or more in chromosome, 1 in F	5700

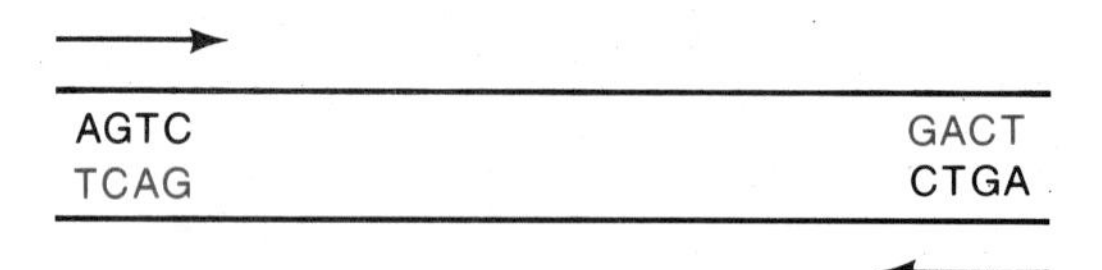

Figure 12-8
An example of a terminal inverted repeat in a DNA molecule. The arrows indicate the inverted base sequences. Note that the sequences AGTC and CTGA are *not* in the same strand. In a direct repeat, the sequence in the upper strand would be AGTC . . . AGTC.

some phages (Table 12-1). They have also been found at different sites within a single gene, inserted in both orientations; having both orientations is not unexpected in view of the symmetry of the termini. The feature that has aroused greatest interest in IS elements is their ability to move from one site to another, which is the origin of the term transposable element. This movement, called **transposition,** is the defining characteristic of transposons and will be described shortly.

Since the discovery of the IS elements numerous transposons carrying genes that are easily identifiable—in particular, antibiotic-resistance genes—have been found in several bacterial species. These transposons have been studied extensively since their presence can be ascertained merely by noting whether the host cell can form a colony on a growth medium containing a particular antibiotic. The transposons containing accessory genes are of two sorts—the composite and the Tn3 families.

1. *The composite family.* Transposons in this family consist of two identical IS elements that flank the antibiotic-resistance segment (Figure 12-9). The IS elements in these composite units can be in an inverted or direct-repeat configuration. Note that since the two ends of an IS element are themselves inverted repeats, the relative orientation (inverted or direct) of the flanking IS elements of a composite transposon does not alter its terminal sequences—they remain the same in the inverted-repeat array.

2. *The Tn3 transposon family.* The Tn3 family of transposons consists of quite large elements (about 5000 base pairs). Each transposon carries three genes, one encoding β-lactamase (which confers resistance to ampicillin) and two others needed for transposition (which will be discussed later). All Tn3-like transposons contain short (38-base-pair) inverted repeats and none are flanked by IS-like elements. A map of Tn3 will be seen later in the chapter in Figure 12-14.

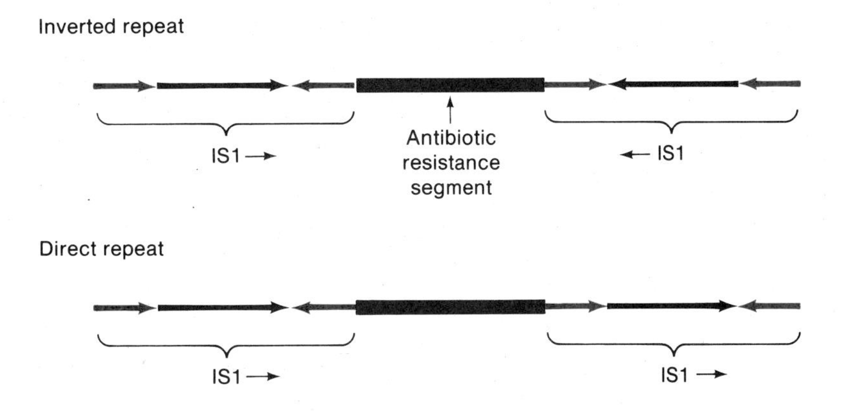

Figure 12-9
Two composite transposons flanked by IS1 in either inverted or direct repeat.

TRANSPOSITION

Consider a culture of a particular bacterium that contains a single transposon at some site in the chromosome. If this culture is grown for many generations, rare mutant bacteria will result in which the transposon is present within genes at sites different from the initial site. This apparent movement exemplifies the phenomenon of transposition, "apparent" because genetic and physical analysis usually show that such mutant cells contain two copies of the transposon—one at the original site and another at the new site (this will be discussed shortly). On continued growth of a cell containing two copies, another rare transposition event may occur, leading to a cell with three copies. Ultimately, all cells would contain an enormous number of copies of a particular transposon were it not for the fact that transposons occasionally are excised and lost.

Transposition is usually not detected by mutant formation, for technical reasons and because insertion does not always occur within genes. Instead, one normally works with transposons carrying antibiotic-resistance markers and looks for their apparent movement from one DNA molecule to another—for example, from the chromosome to a plasmid or from a plasmid to a phage. Since transposition is a recombination event, bacterial mutants defective in normal genetic recombination are used—for example, bacterial *recA*$^-$ mutants; in this way, one can study movement that is controlled only by the transposon itself—namely, transposition.

In the experiment illustrated in Figure 12-10, an R plasmid containing an *amp-r* transposon is transferred by conjugation to a *recA*$^-$ recipient cell that carries another transposon having the *kan-r* (kanamycin-resistance) gene. On continued growth of a culture containing donors, recipients, and conjugated cells in the presence of both antibiotics (so that neither donor nor recipient can mutiply), only cells that have received the R plasmid can grow. In these cells transposition of the *kan-r* transposon from the recipient chromosome to the R factor occasionally occurs, yielding an R plasmid carrying both antibiotic-resistance markers. One can demonstrate that the *kan-r* transposon is carried by the R factor by mating these cells with a Kan-s Amp-s culture (which carries another antibiotic-resistance marker so that donor and recipient cells can be distinguished), allowing conjugation to occur, and plating on a medium containing all three antibiotics. Only recipient cells that have obtained both the *kan-r* and *amp-r* markers together on the R plasmid can form colonies.

When transposition occurs, a particular transposon can usually be inserted at one of a large number of positions. For example, if an *amp-r* transposon is transferred to *lac*$^+$ *leu*$^+$ *amp-s* cells, in time both *lac*$^-$

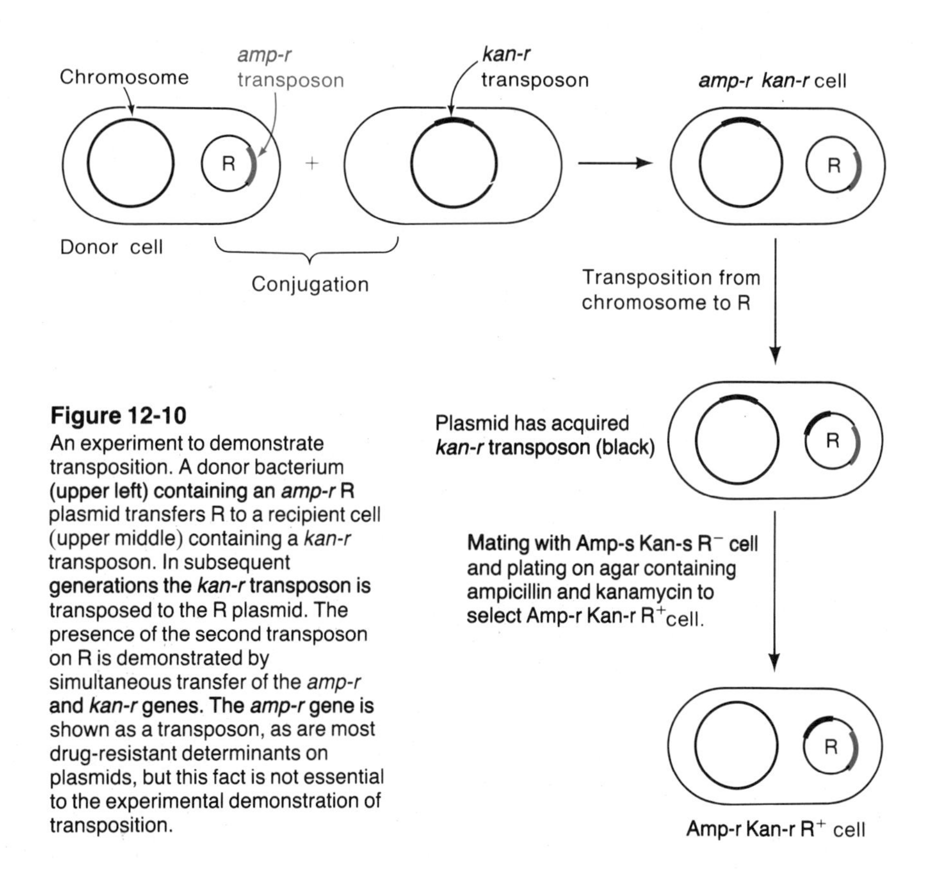

Figure 12-10
An experiment to demonstrate transposition. A donor bacterium (upper left) containing an *amp-r* R plasmid transfers R to a recipient cell (upper middle) containing a *kan-r* transposon. In subsequent generations the *kan-r* transposon is transposed to the R plasmid. The presence of the second transposon on R is demonstrated by simultaneous transfer of the *amp-r* and *kan-r* genes. The *amp-r* gene is shown as a transposon, as are most drug-resistant determinants on plasmids, but this fact is not essential to the experimental demonstration of transposition.

leu$^+$ *amp-r* and *lac*$^+$ *leu*$^-$ *amp-r* cells will accumulate in the culture. However, the insertion sites are not randomly distributed at the nucleotide level, and different transposons have various degrees of selectivity for particular sites. At one extreme is IS4, for which 20 independent insertion events have been observed in the *E. coli galT* gene at a single site. In transposition of the element Tn9 into the terminal 160 bases of the *E. coli lacZ* gene, 28 independent transpositions were detected; 16 positions were observed but 5 of these were represented by multiple occurrences of insertion. The least selective element is Mu, which apparently can insert at any site. The cause of selectivity in some cases is not known.

THE MECHANISM OF TRANSPOSITION

The end result of the transposition process is the insertion of a transposon between two base pairs in a recipient DNA molecule. Base sequence analysis of many transposons and their insertion sites reveals

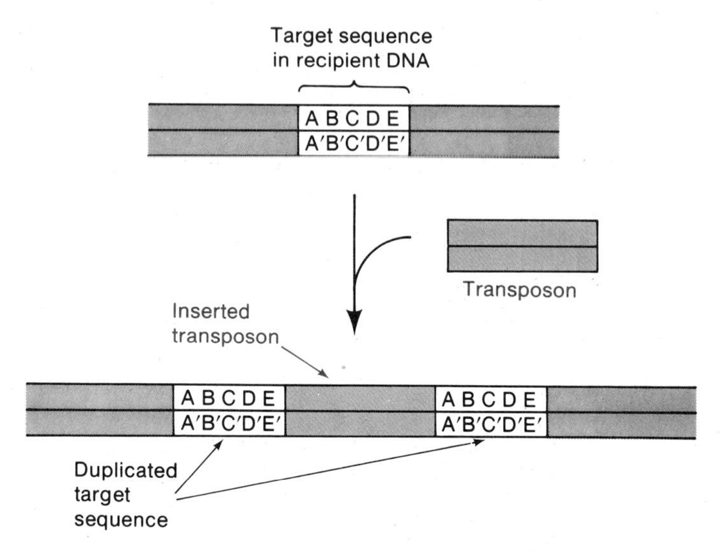

Figure 12-11
Insertion of a transposon generates a duplication of the target sequence.

that there is no sequence homology, which is consistent with the lack of a requirement for the *E. coli recA* system. However, the sequences of the regions in which the inserted element joins the recipient DNA has yielded several surprises, which we document in this section.

A characteristic of the insertion process is that insertion of a transposon always involves the duplication of a short base sequence (3–12 base pairs long) in the recipient DNA molecule, called the **target sequence,** and the inserted transposon is sandwiched between the repeated bases. This arrangement is shown in Figure 12-11. We repeat, emphatically, that *only one copy of the duplicated target sequence is present in the recipient DNA prior to insertion of the transposon and it is not present in the transposon itself.*

The length of the target sequence varies from one transposon to the next but *is the same for all insertions of a particular transposon.* For example, an inserted IS1 is always flanked by a nine-base-pair sequence, whereas Tn3 is always flanked by a duplication of five base pairs. Note that *for a particular transposon, the target sequence is different for each insertion site; only the length of the duplicated sequence is constant.*

As mentioned earlier, in bacterial transposition one copy of the transposon remains at the original site. This can be demonstrated by the arrangement shown in Figure 12-12, in which transposition between plasmids is observed. After many generations of growth of a bacterium

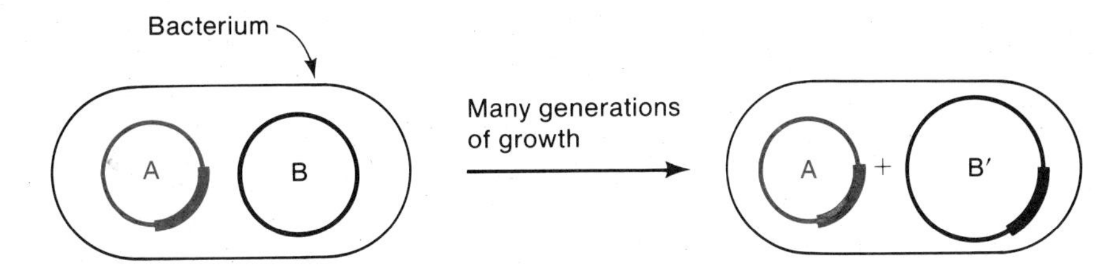

Figure 12-12
Transposition of a transposon from plasmid A to a second plasmid B.

containing both plasmid A, which has a transposon, and plasmid B, which lacks a transposon, cells are produced in which plasmid B (now called B′) also possesses the transposon. This observation is significant in understanding the mechanism of transposition, because it indicates that the original transposon is duplicated in the transposition process. Thus, *both* the target sequence and the transposon are duplicated, which means that *transposition is a replicative process.*

Some insight into transposition is obtained from the following phenomenon. Consider a cell with two plasmids, one of which contains a transposon. At a frequency of about 10^{-7} events per cell generation, the two plasmids fuse to form a single plasmid called a cointegrate. (This process is sometimes also called **replicon fusion.**) This hydrid plasmid is not simply a fusion of the two plasmids, because it contains two (not just one) copies of the transposon. Both copies are in the same orientation (that is, in direct repeat) and precisely at the junction between the donor and recipient plasmid sequences (Figure 12-13). As far as is known, all transposons are capable of mediating cointegration, though the cointegrates are not always stable.

The formation of cointegrates is significant for two reasons. First, cointegrates are thought to be an essential intermediate in the transposition process and, second, the fact that the cointegrate has two copies of

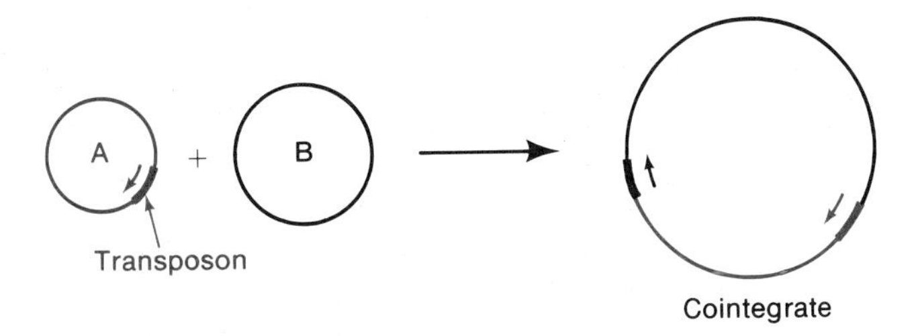

Figure 12-13
Formation of a cointegrate by transposon-mediated fusion of two plasmids A and B. Two copies of the transposon are generated in the process. The arrows denote an arbitrary direction to show that both copies of the transposon in the cointegrate are in direct repeat.

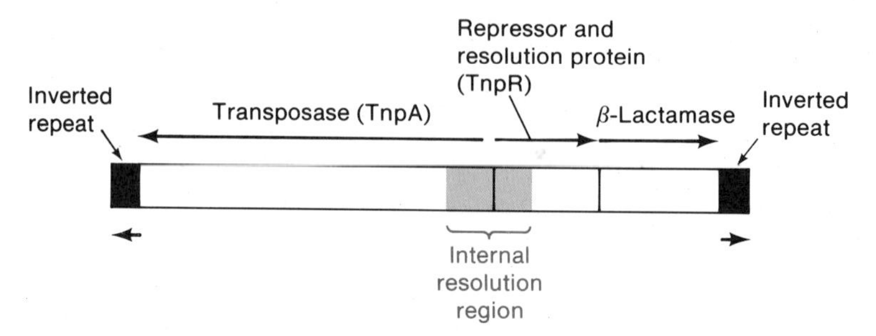

Figure 12-14
The physical map of transposon Tn3. There is a total of 4957 base pairs; each inverted repeat contains 38 base pairs. The three genes are indicated above the upper arrows, which show the direction of transcription. The internal resolution region is discussed in the text.

the transposon, whereas only one copy was present before cointegration occurred, again indicates that DNA *replication of the transposon has occurred.*

The cointegrate as an intermediate is suggested by certain features of Tn3, whose structure is shown in Figure 12-14. Tn3 has two inverted repeats, like all transposons, and these flank three genes. The three gene products have been isolated and from their size and the base sequence of Tn3, we know that the genes utilize all of the bases between the inverted repeats. The leftmost gene encodes a gigantic protein (1015 amino acids), denoted TnpA, which is a transposase and which is responsible for formation of cointegrates. The rightmost gene encodes β-lactamase, an enzyme that inactivates ampicillin. The central gene encodes a small protein (185 amino acids) called TnpR, which has two functions—negative regulation or repression of the synthesis of TnpA, and promotion of a site-specific exchange that resolves cointegration in a second step of the transposition process. Near the boundary of the genes encoding TnpA and TnpR is a sequence of DNA called the **internal resolution site** and it includes the base sequence at which TnpR acts.

Genetic experiments have yielded the following information. All *tnpA*$^-$ mutants are unable either to transpose or to form cointegrates. Also, *tnpR*$^-$ mutants and deletions of the internal resolution site form cointegrates yet are unable to transpose. That is, in a cell containing a plasmid with a *tnpR*$^-$ copy of Tn3 and a second plasmid lacking Tn3, the process shown in Figure 12-13 can occur, but the process shown in Figure 12-14 does not occur. These observations suggest that Tn3-mediated transposition occurs by a two-step process—first, TnpA induces formation of a cointegrate and then TnpR promotes a site-specific exchange at the internal resolution site. A schematic diagram of this sequence of events is shown in Figure 12-15.

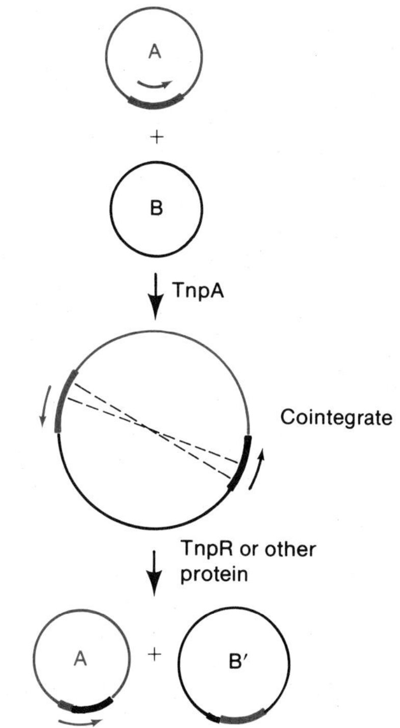

Figure 12-15
A model for transposition utilizing a cointegrate intermediate.

Some transposons do not carry a TnpR-like site-specific recombination system but are still able to transpose. These systems presumably use a homology-dependent exchange system since, once a cointegrate has formed, there are two identical sequences–namely, copies of the transposon—so that any homology-dependent process could carry out the exchange shown in Figure 12-15.

Numerous molecular models for transposition have been proposed. The model for which there is greatest evidence is the Shapiro model, which is explained in *MB*, pages 786–787.

TRANSPOSABLE ELEMENTS IN EUKARYOTES

The first evidence for the existence of transposable elements came not from bacteria but from ingenious genetic studies of maize by Barbara McClintock. Her observations indicated that the activities of certain genetic loci in corn are disturbed by a controlling element that inhibits the activity only when the element is adjacent to the locus; change in level of gene activity seemed to occur by movement of the element to and from its inhibiting position. The controlling element, called *Dissociation*, has been found to be a transposable element, and its movement is regulated by a second transposable element called *Activator*. Both of these have been isolated and their base sequences determined.

Transposable elements have been observed in other eukaryotes. The yeast *Saccharomyces cerevisiae* contains several different transposable elements of which the best-studied are the IS-like element δ and the composite element Ty1 (for which about 35 copies are present per haploid cell). In the fruit fly *Drosophila melanogaster* 5–10 percent of the DNA consists of 30–40 distinct families of transposable elements. The best-understood ones are copia and the P elements. Transposition of copia affects the activity of certain genes, presumably by movement of regulatory sites, such as promoters and other protein-binding sites.

A number of genetic results with *Drosophila* can be explained by the existence of transposons. For example, there are mutations that inhibit nearby recombination events, are polar, and have been shown to be insertions.

An important possibility is that some (or perhaps all) of the RNA tumor viruses are themselves transposons; however, the only evidence for this view at present is that several integrated tumor viruses are flanked by 600-base-pair sequences in direct repetition and, in addition, a duplication of a short target sequence.

GENETIC PHENOMENA MEDIATED BY TRANSPOSONS

Transposable elements mediate a variety of genetic phenomena, such as gene rearrangements (for example, inversions), deletions, and fusion of DNA molecules (as in cointegrate formation). Transposable elements can also serve as switches turning genes on and off (for example, *Dissociation* in maize) or by inverting sequences containing promoters. These phenomena are direct consequences of transposition, and some have already been discussed; hypotheses for the mechanisms for producing inversion and deletions can be found in *MB*, pages 790–793.

Several processes in bacteria are known to be the result of physical exchange (recombination) between multiple copies of transposons. Two of these, formation of Hfr cells and gene amplification, are described in this section.

Hfr cells are formed by integration of F, as has already been stated. Critical to understanding how this occurs are two facts: (1) F contains several IS elements (Figure 12-16) and (2) in an Hfr an integrated F is always flanked by two copies, in direct repeat, of one of the IS elements of F. The primary mechanism for formation of Hfr cells is a reciprocal exchange between an IS element in F and the same element in the chromosome (Figure 12-17). Since the IS elements are found at various sites in the chromosome and since they are capable of relocating by

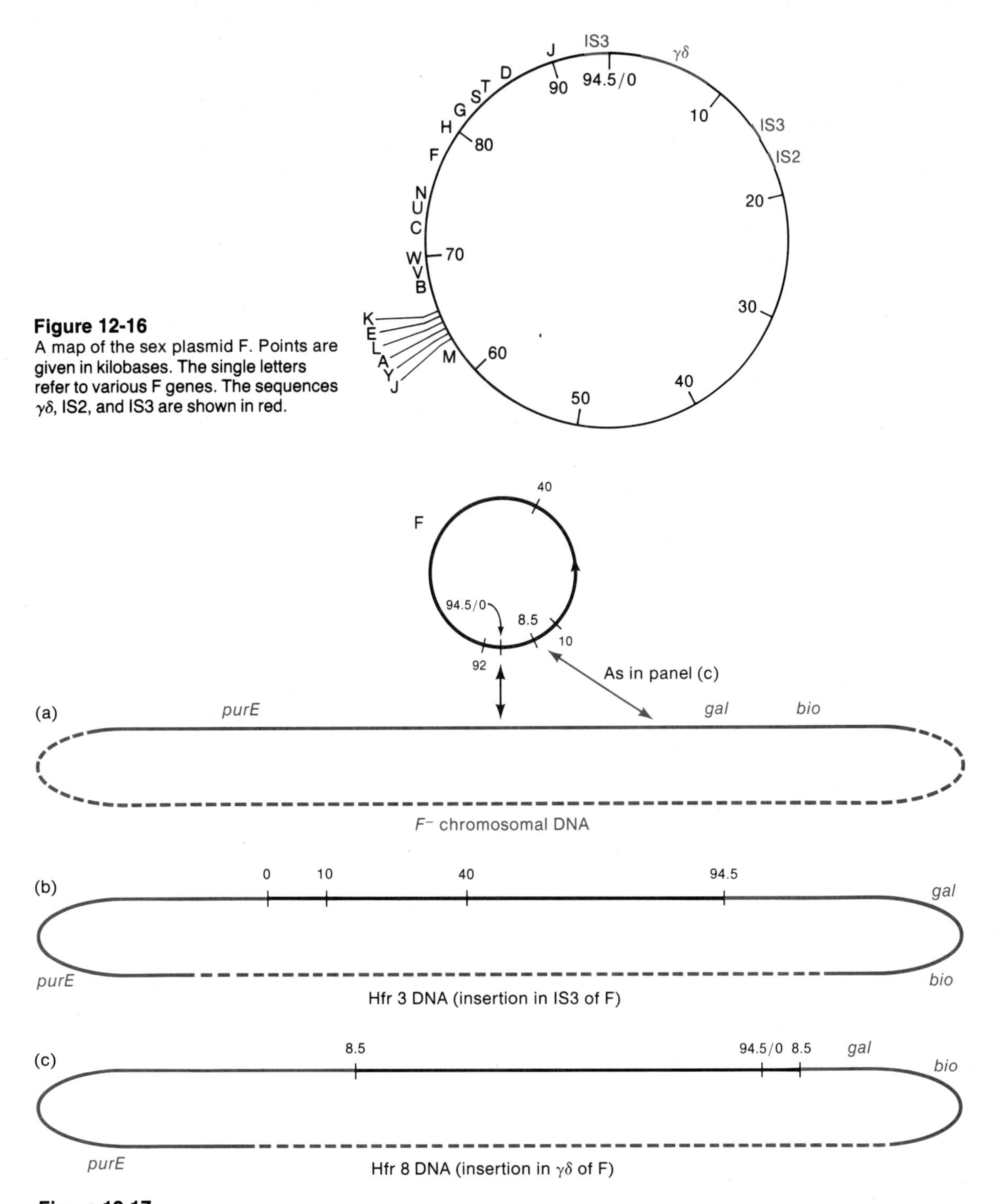

Figure 12-16
A map of the sex plasmid F. Points are given in kilobases. The single letters refer to various F genes. The sequences $\gamma\delta$, IS2, and IS3 are shown in red.

Figure 12-17
Formation of two different Hfr strains. (a) The F^- chromosome and F. The exchange in both molecules occurs at the points indicated by the black double-headed arrow. (b) The result of the exchange shown in panel (a). (c) The result of the exchange indicated in panel (a) by the double-headed red arrow.

transposition, numerous integration sites are possible. Furthermore, since transposed IS elements in the chromosome can be in two different orientations, F can be integrated both clockwise and counterclockwise with respect to the genetic map of *E. coli*. This explanation is sufficient to understand the origin of the many different Hfr cell lines and the fact that Hfr strains that transfer DNA in both directions with respect to the *E. coli* map are known (Figure 12-18).

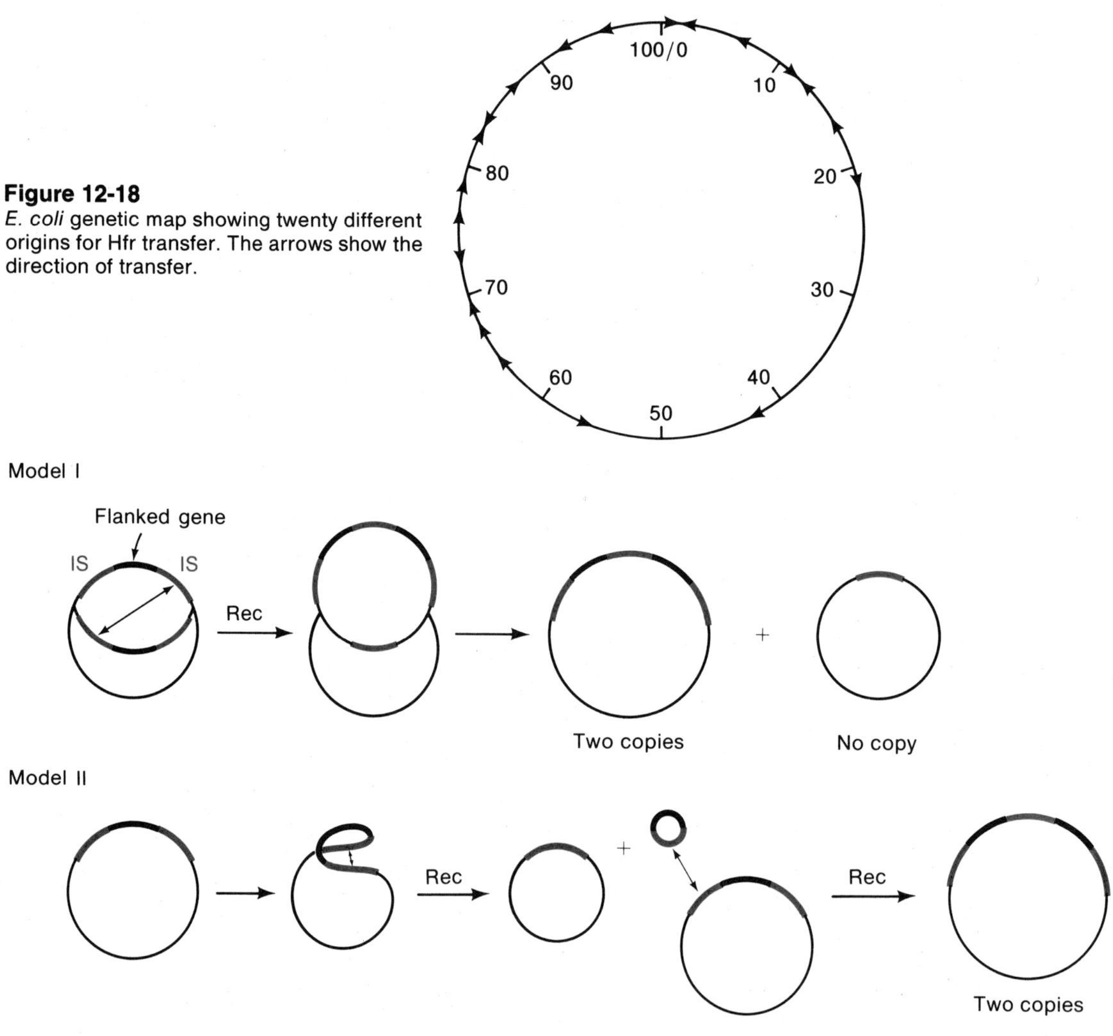

Figure 12-18
E. coli genetic map showing twenty different origins for Hfr transfer. The arrows show the direction of transfer.

Figure 12-19
Two possible models for gene amplification. Rec-mediated recombination occurs between IS elements as shown by the two-headed arrow. In Model I recombination occurs during replication. In Model II there are two Rec-mediated events. The second event is between an excised minicircle and another copy of the original plasmid.

Another phenomenon in which homologous recombination between multiple copies of transposons plays a role is gene amplification in bacteria. Some R plasmids have the property that if a bacterium containing the plasmid is exposed to an antibiotic concentration considerably lower than the maximum concentration tolerated by the cell, over a period of many generations the cell becomes resistant to ever-increasing concentrations of the antibiotic. This results from a gradually increasing number of antibiotic-resistance segments in the R plasmid. That is, the R plasmid increases in size owing to repeated tandem duplications of the antibiotic-resistance genes. This process requires an active bacterial Rec system, showing that duplication is not a transposition process. Based on the fact that the antibiotic-resistance segment of R plasmids are flanked by transposons, two models for gene amplification shown in Figure 12-19 have been developed. (There are other models but these two are especially simple.) Note that the two basic requirements of the models are that the genes to be amplified are flanked by identical sequences and that the Rec system acts effectively on pairs of homologous regions. There is no proof for these models at present.

13 Recombinant DNA and Genetic Engineering

Technical developments often are responsible for great steps forward in science. The most recent development in biology has been a technique for joining DNA molecules together *in vitro*, namely, the recombinant DNA technology, or **genetic engineering.** This technique and its many variants have revolutionized the study of eukaryotes and have provided sources of particular proteins in quantities hitherto considered to be nearly impossible to obtain. The technology changes continually; in this chapter we will examine only its principles.

Circular plasmid DNA is especially important in genetic engineering. When DNA is isolated from plasmid-containing bacteria or yeast, the sample obtained will contain both chromosomal and plasmid DNA, which must be separated. A large part of the chromosomal DNA can be removed by centrifugation at low speed. Circular plasmid DNA molecules are then purified by a procedure that is based on (1) the circularity of the plasmid DNA and (2) the considerable fragmentation (and hence linearization) of the circular bacterial chromosome that occurs during isolation. Plasmid DNA is quite resistant to fragmentation because it is so much smaller than chromosomal DNA. The technique used is equilibrium centrifugation (Section 4.6) in CsCl solutions containing **ethidium bromide.** This molecule binds very tightly to DNA by slipping in between the DNA base pairs (*intercalation*), causing the DNA to unwind; the lower density of the ethidium bromide molecules compared to that of a nucleotide causes a decrease in the density of the DNA by approximately 0.15 g/cm^3. The density shift is greater for linear DNA molecules (for example, the chromosome fragments) than for covalent circles (the plasmid DNA), and forms the basis of the separation. A covalently closed, circular DNA molecule has no free ends; thus, for simple topological reasons, unwinding the individual polynucleotide strands is accompanied by a twisting of the entire molecule in the opposite direction of the unwinding motion. For example, a circular molecule that has bound enough ethidium bromide molecules to produce one clockwise-unwinding turn will twist in the counterclockwise direction, yielding a molecule shaped like the figure 8. As additional ethidium bromide molecules intercalate, the 8-shaped molecule becomes more twisted; ultimately, the DNA molecule is unable to twist further, so no additional ethidium bromide molecules can be bound. (Actually, plasmid DNA molecules are negatively supercoiled at the outset, so the supercoiling is first removed by binding ethidium bromide and then reversed in direction—that is the DNA becomes positively supercoiled.) A linear DNA molecule does not have the topological constraint that a covalent circle has and therefore can bind a greater number of ethidium bromide molecules. Because of the decrease in density of DNA caused by the binding of ethidium bromide molecules and the lesser binding to a covalent circle than to a linear molecule, a covalent circle has a *higher* (less reduced) density at concentrations of ethidium bromide that saturate the covalent circle. Therefore, by equilibrium centrifugation in a density gradient, covalent circles (plasmids) can be separated from linear molecules (chromosomal fragments), as shown in Figure 4-27, *MB*, page 134.

In genetic engineering the immediate goal of an experiment is usually to insert a *particular* fragment of chromosomal DNA into a plasmid or a viral DNA molecule. This is facilitated by techniques for breaking DNA molecules at specific sites and for isolating particular DNA fragments. These techniques are described in the following section.

ISOLATION AND CHARACTERIZATION OF PARTICULAR DNA FRAGMENTS

In the recombinant DNA methodology DNA fragments are usually obtained by treatment of DNA samples with a specific class of nucleases. Many nucleases have been isolated from a variety of organisms, and most produce breaks at random sites within a DNA sequence (endonucleases) or remove nucleotides only from the termini (exonucleases), as discussed in Chapter 3. However, the class of nucleases called **restriction endonucleases**, or, more simply, **restriction enzymes**, consists of sequence-specific enzymes. Most restriction enzymes recognize only one short base sequence in a DNA molecule and make two single-strand breaks, one in each strand, generating 3′-OH and 5′-P groups at each position. Several hundred of these enzymes have been purified from hundreds of species of microorganisms.

The sequences recognized by restriction enzymes are **palindromes** —that is, the sequence has symmetry of the form

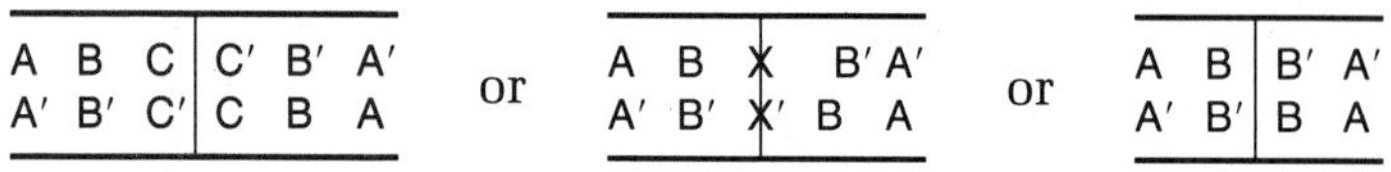

```
A  B  C  | C′ B′ A′          A  B  X  B′ A′          A  B  | B′ A′
A′ B′ C′ | C  B  A     or    A′ B′ X′ B  A     or    A′ B′ | B  A
```

in which the capital letters represent bases, a ′ indicates a complementary base, X is any base, and the vertical line is the axis of symmetry. Most of these sequences have 4–6 bases.

One of the most exciting events in the study of restriction enzymes was the observation by electron microscopy that fragments produced by many restriction enzymes spontaneously circularize. These circles could be relinearized by heating, but if after circularization they were also treated with *E. coli* DNA ligase, which joins 3′-OH and 5′-P groups, circularization became permanent. This observation was the first evidence for three important features of restriction enzymes:

1. Restriction enzymes make breaks in symmetric sequences.
2. The breaks are usually not directly opposite one another.
3. The enzymes generate DNA fragments with complementary ends.

These properties are illustrated in Figure 13-1.

Examination of a very large number of restriction enzymes showed that the breaks are usually in one of two distinct arrangements: (1) staggered, but symmetric around the line of symmetry (forming **cohesive ends**) or (2) both at the center of symmetry (forming **blunt ends**). Two types of enzymes produce cohesive ends—those yielding a single-stranded extension with a 5′-P terminus and those yielding a 3′-OH extension. These arrangements and their consequences are shown in Figure 13-2. Table 13-1 lists the sequences and cleavage sites for several restriction enzymes, some of which generate cohesive sites and others of which yield blunt ends. Fragments having blunt ends cannot circularize spontaneously.

Most restriction enzymes recognize one base sequence without regard to the source of the DNA. Thus, *fragments obtained from a DNA molecule from one organism have the same cohesive ends as the frag-*

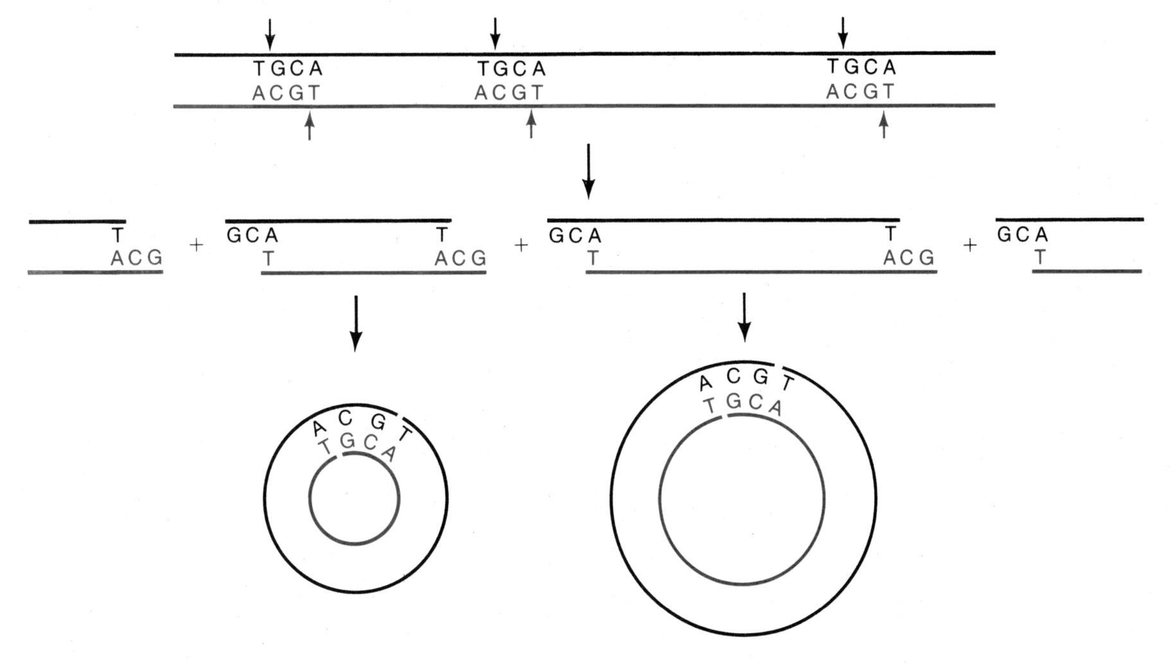

Figure 13-1
An experiment showing that restriction fragments can circularize. Arrows indicate cleavage sites.

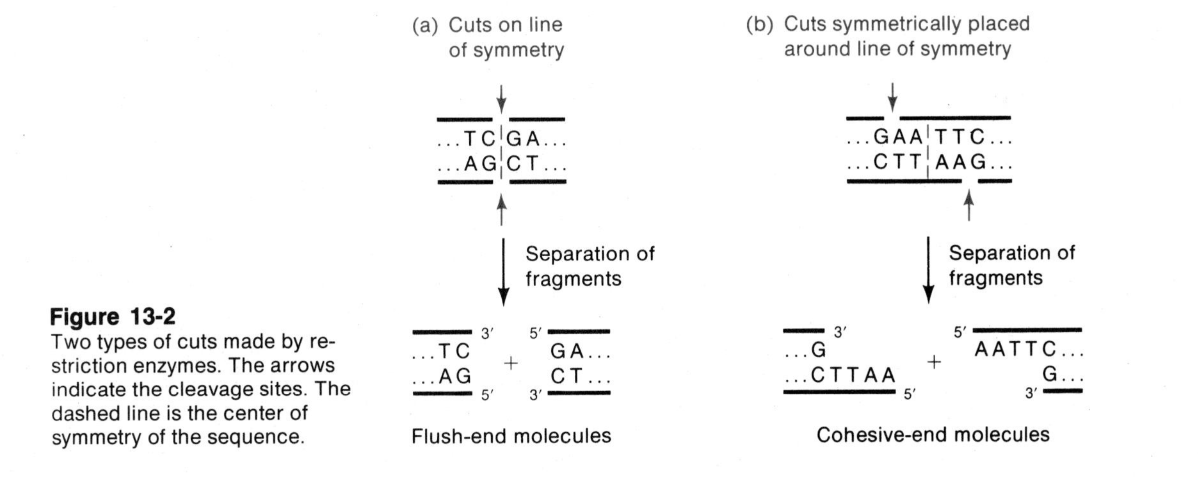

Figure 13-2
Two types of cuts made by restriction enzymes. The arrows indicate the cleavage sites. The dashed line is the center of symmetry of the sequence.

ments produced by the same enzyme acting on DNA molecules from another organism. This point will be seen to be one of the foundations of the recombinant DNA technology.

Since most restriction enzymes recognize a unique sequence, *the number of cuts made in the DNA from an organism by a particular*

Table 13-1
Some restriction endonucleases and their cleavage sites

Microorganism	*Name of enzyme*	*Target sequence and cleavage sites*
Generates cohesive ends		
E. coli	EcoRI	G↓A A ¦ T T C C T T ¦ A A↑G
Bacillus amyloliquefaciens H	BamHI	G↓G A ¦ T C C C C T ¦ A G↑G
Haemophilus influenza	HindIII	A↓A G ¦ C T T T T C ¦ G A↑A
Generates flush ends		
Brevibacterium albidum	BalI	T G G↓¦C C A A C C↑¦G G T

Note: The vertical dashed line indicates the axis of dyad symmetry in each sequence. Arrows indicate the sites of cutting.

enzyme is limited. A typical bacterial DNA molecule, which contains roughly 3×10^6 base pairs, is cut into several hundred to several thousand fragments, and nuclear DNA of mammals is cut into more than a million fragments. These numbers are large but still small compared to the number of sugar-phosphate bonds in an organism. Of special interest are the smaller DNA molecules, such as viral or plasmid DNA, which may have only 1–10 sites of cutting (or even none, for particular enzymes). Plasmids having a single site for a particular enzyme are especially valuable, as we will see shortly.

Because of the sequence specificity, *a particular restriction enzyme generates a unique set of fragments for a particular DNA molecule.* Another enzyme will generate a different set of fragments from the same DNA molecule. Figure 13-3(a) shows the sites of cutting of *E. coli* phage λ DNA by the enzymes EcoRI and BamHI. A map showing the unique sites of cutting of the DNA of a particular organism by a single enzyme is called a **restriction map.** The family of fragments generated by a single enzyme can be detected easily by gel electrophoresis of enzyme-treated DNA (Figure 13-3(b)), and particular DNA fragments can be isolated by cutting out the portion of the gel containing the fragment and removing the DNA from the gel.

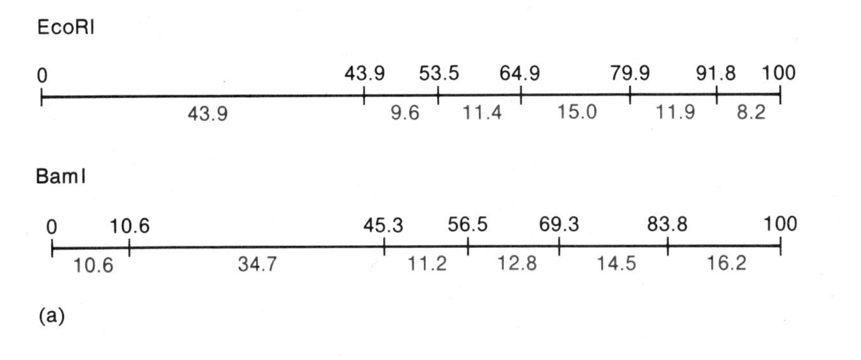

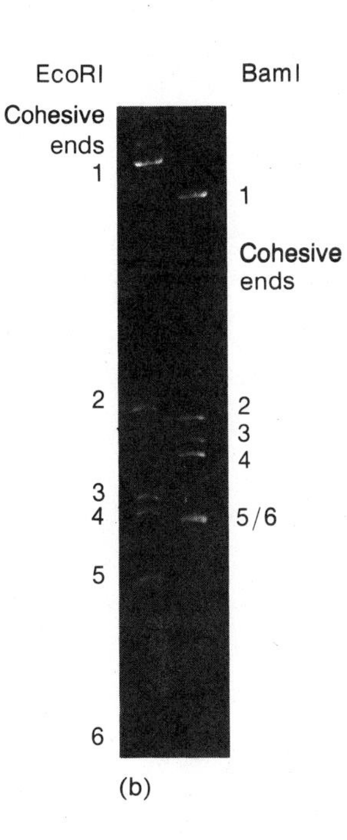

Figure 13-3
(a) Restriction maps of λ DNA for EcoRI and BamI nucleases. The vertical bars indicate the sites of cutting. The black numbers indicate the percentage of the total length of λ DNA measured from the gene-*A* end of the molecule. The red numbers are the lengths of each fragment, again expressed as percentage of total length. (b) A gel electrophoregram of EcoRI and BamI restriction-enzyme digests of λ DNA. The bands labeled cohered ends contain molecules consisting of the two terminal fragments joined by the normal cohesive ends of λ DNA. Numbers indicate fragments in order from largest (1) to smallest (6). Bands 5 and 6 of the BamI digest are not resolved. (Courtesy of Dennis Anderson and Lynn Enquist.)

THE JOINING OF DNA MOLECULES

In genetic engineering a particular DNA segment of interest is joined to a small, but essentially complete, DNA molecule that is able to replicate—a **vector** or **cloning vehicle.** By a transformation procedure, described below, this recombinant molecule is placed in a cell in which replication can occur (Figure 13-4). When a stable transformant has been isolated, the genes or DNA sequences on the donor segment are said to be **cloned.** In this section several types of vectors and a few procedures used for joining molecules are described.

A vector is a DNA molecule in which a DNA fragment can be cloned; to be useful, it must have three properties:

1. A means of introducing vector DNA into a cell must be available.
2. The vector must have a replication origin—that is, be able to replicate.
3. Transformants must be selectable by a straightforward assay, preferably by growth of a host cell on a solid medium.

The types of vectors most commonly used at present are plasmids and *E. coli* phages λ and M13; several animal viruses are also gaining in use.

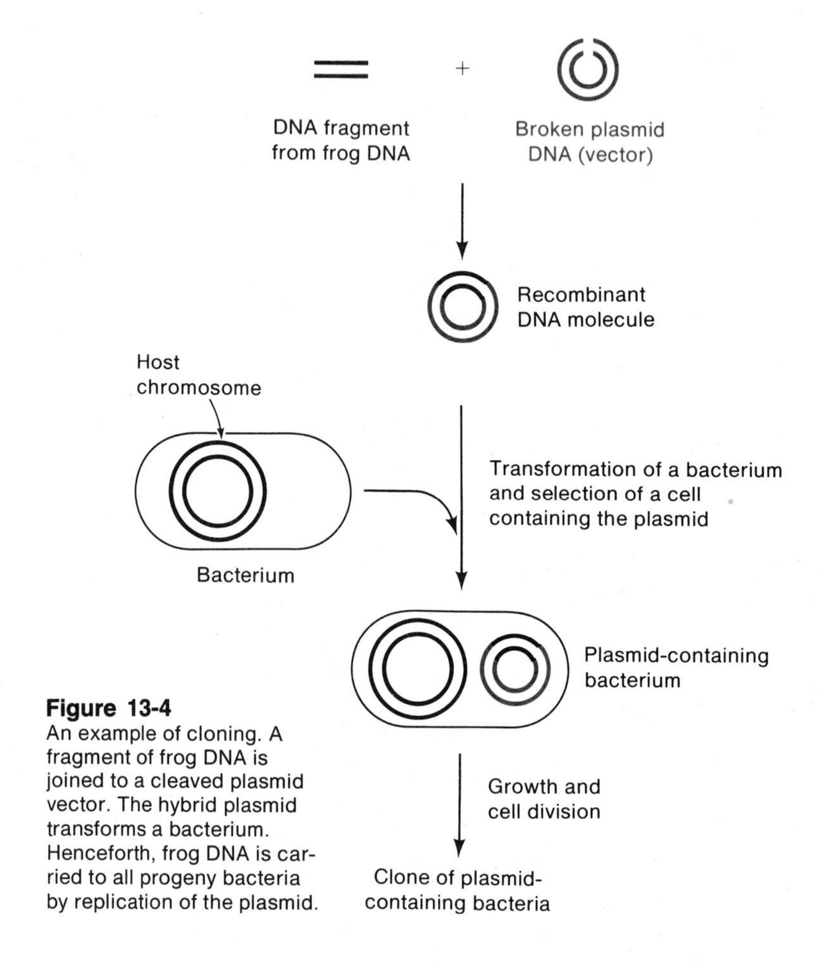

Figure 13-4
An example of cloning. A fragment of frog DNA is joined to a cleaved plasmid vector. The hybrid plasmid transforms a bacterium. Henceforth, frog DNA is carried to all progeny bacteria by replication of the plasmid.

These cloning vehicles can be detected in host cells by means of genetic features or particular markers made evident during colony or plaque formation. Plasmid and phage DNA can be introduced into cells by the **$CaCl_2$ transformation procedure,** as described in Chapter 12 (page 193).

Retroviruses are RNA-containing animal viruses that have an unusual life cycle. These viruses, which contain the enzyme **reverse transcriptase** in their protein coats, use this enzyme to synthesize a double-stranded DNA copy of the viral RNA shortly after infection. As a normal part of the viral life cycle, this double-stranded DNA becomes inserted into the animal-cell chromosome, apparently at one of an enormous number of potential sites, and only after insertion can transcription occur. In certain conditions the infected host cell survives the infection, yet retains the retroviral DNA in its chromosome. One of the best-understood retroviruses is **Rous sarcoma virus,** which causes tumors in chickens. Retrovirus DNA, either isolated from a cell or prepared synthetically, is a useful vector with animal cells.

Earlier circularization of fragments having complementary terminal bases was described; similarly, two DNA fragments can join together by the pairing of their complementary cohesive ends. Therefore, because a particular restriction enzyme produces fragments with the *same* cohesive ends, without regard for the source of the DNA, fragments from **DNA molecules isolated from two *different* organisms (for example, a bacterium and a frog) can be joined, as shown in Figure 13-5. Furthermore,** if DNA is treated with DNA ligase to seal the joint after base-pairing, the fragments will be joined permanently, and a molecule will be generated that may never have existed before. The ability to join a DNA fragment of interest to a vector is the basis of the recombinant DNA technology.

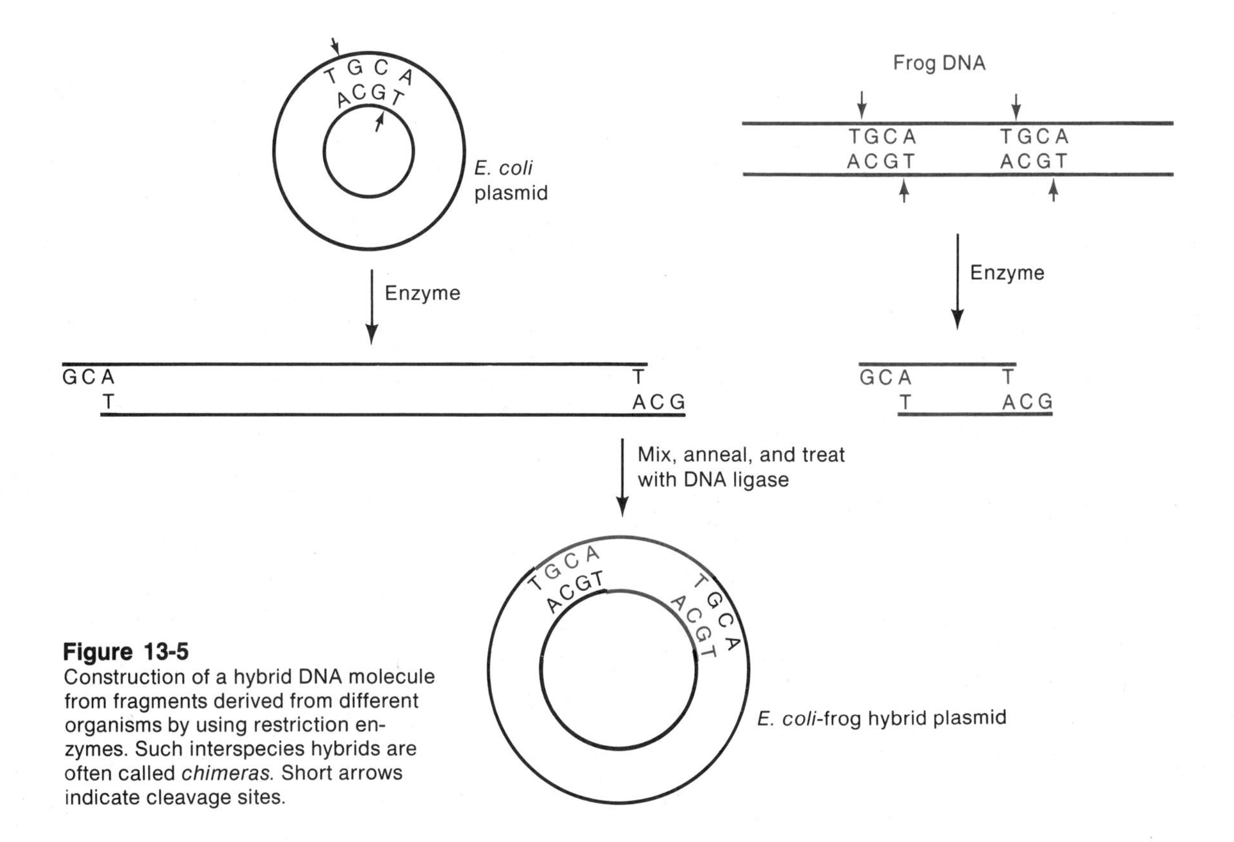

Figure 13-5
Construction of a hybrid DNA molecule from fragments derived from different organisms by using restriction enzymes. Such interspecies hybrids are often called *chimeras*. Short arrows indicate cleavage sites.

Joining cohesive ends does not always produce a DNA sequence that has functional genes. For example, consider a linear DNA molecule that is cleaved into four fragments—A, B, C, and D—whose original sequence in the molecule was ABCD. Reassembly of the fragments can yield the original molecule, but since B and C have the same pair of cohesive ends, molecules with the fragment arrangements ACBD or BADC can also form with the same probability as ABCD. The problem of the scrambling of vector fragments is minimized by using a vector having only one cleavage site for a particular restriction enzyme. Plasmids of this type are available (most have been genetically engineered); usually they have sites for several different restriction enzymes, but only one enzyme is used at a time.

DNA molecules lacking cohesive ends can also be joined. A direct method uses the DNA ligase made by *E. coli* phage T4. This enzyme differs from other DNA ligases in that it not only seals single-strand breaks in double-stranded DNA but can also join blunt-ended molecules. Another method uses the enzyme **terminal deoxynucleotidyl transferase,** an unusual DNA polymerase obtained from animal tissue. This enzyme adds nucleotides (by means of deoxynucleoside triphosphate precursors) to the 3′-OH group of an extended single-stranded segment of a DNA chain—*without the need to copy a template strand.* In order to produce an extended single strand one need only treat a DNA molecule with a 5′-specific exonuclease to remove a few terminal nucleotides. If a mixture is prepared containing exonuclease-treated DNA, terminal deoxynucleotidyl transferase, and a single kind of nucleoside triphosphate—for example, deoxyadenosine triphosphate (dATP)—a DNA extension having adenine as the only base will form at both 3′-OH termini (Figure 13-6). Such extended segments are called **poly(dA) tails.** If instead thymidine triphosphate (dTTP) were provided, the DNA molecule would have poly(dT) tails. Since a poly(dA) tail is complementary to a poly(dT) tail, any two DNA molecules can subsequently be joined if a poly(dA) tail is put on one molecule and a poly(dT) tail on the other, as shown in the figure. Gaps in the joined molecule can be filled by treatment with a DNA polymerase (*E. coli* polymerase I is commonly used), and the remaining single-stranded interruptions are again sealed by DNA ligase. This method, which is called **homopolymer tail-joining,** is a general method for joining any pair of DNA molecules.

Often at a later time one will want to isolate a cloned fragment from a vector. This is straightforward when cohesive-end joining has been used in that the restriction enzyme used originally to produce the cohesive termini can excise the cloned DNA by making a cut in the new joint. However, if T4 DNA ligase had been used to join blunt-ended fragments that had not been formed by the same restriction enzyme or if homopolymer-tail joining had been used, no enzyme would be avail-

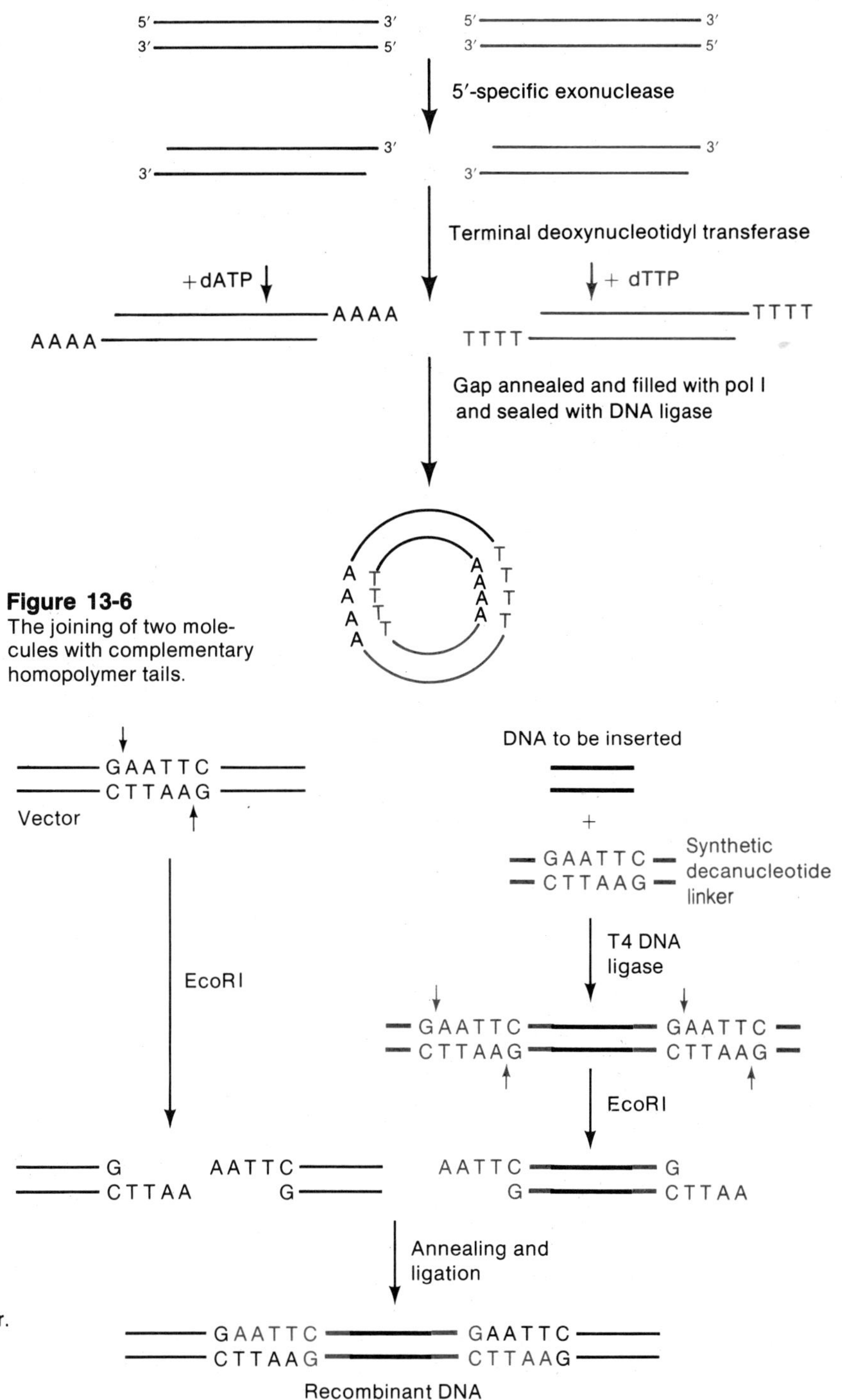

Figure 13-6
The joining of two molecules with complementary homopolymer tails.

Figure 13-7
Formation of recombinant DNA through use of a linker. The short arrows indicate the sites of cutting by the EcoRI enzyme.

able to perform the specific cleavage. In those cases, the use of **linkers** (Figure 13-7) provides the necessary sequences. With the linker method a synthetic short DNA segment, which contains a known restriction site, is coupled by T4 DNA ligase to each blunt end of both the fragment of interest and the vector. Then, both DNA molecules are treated with the restriction enzyme whose site is contained in the linker. In this way, the same complementary single-stranded termini are produced on both the fragment to be inserted and on the vector, and joining of cohesive termini becomes possible.

INSERTION OF A PARTICULAR DNA MOLECULE INTO A VECTOR

In the cloning procedure that we have described so far, a collection of fragments obtained by digestion with a restriction enzyme can be made to anneal with a cleaved vector molecule, yielding a large number of hybrid vectors containing different fragments of foreign DNA. However, if a particular DNA segment or gene is to be cloned, the vector possessing that segment must be isolated from the set of all vectors possessing foreign DNA. The kinds of selections used will be described shortly.

For many genes, simple selection techniques are adequate for recovery of a vector containing that gene. For example, if the gene to be cloned were a bacterial *leu* gene, one would use a Leu^- host and select a Leu^+ colony on a medium lacking leucine. However, often the clone of interest is either so rare or so difficult to detect that it is preferable if, prior to joining, the fragment containing the DNA segment can be purified. One method has already been discussed: if the DNA of interest (for example, a viral gene) is known to be contained in a particular restriction fragment, that fragment can be isolated from an agarose gel after electrophoresis and joined to an appropriate vector. However, eukaryotic cells contain about a million cleavage sites for a typical restriction enzyme, so direct isolation of a eukaryotic gene from a mixture of fragments separated by electrophoresis is not feasible. In this section another procedure for cloning a particular DNA molecule is described.

In both prokaryotes and eukaryotes, selection of a vector containing a particular gene from a collection of vectors containing foreign DNA is not always straightforward. In theory, a gene is most easily detected by its expression—that is, the phenotype conferred by the gene, as mentioned previously for the *E. coli leu* gene. However, all prokaryotic genes do not produce such easily detected phenotypes and for eukaryotes, a particular gene that has been inserted into a bacterial DNA vector may, for a variety of reasons, fail either to be transcribed or to be correctly translated into a functional protein. By the technique described in this section the insertion of any coding sequence whose mRNA can be isolated in almost pure form—which is possible for many eukaryotic genes—can be simply done.

Let us assume that the chicken gene for ovalbumin (an eggwhite protein) is to be cloned into a bacterial plasmid, in order to determine the base sequence of the gene. A recombinant plasmid could be formed by treating both chicken cellular DNA and the *E. coli* plasmid DNA with the same restriction enzyme, mixing the fragments, annealing, and finally transforming *E. coli* with the DNA mixture containing the recombinant plasmid. However, identification of a bacterial colony containing this plasmid could not be carried out by a test for the presence of ovalbumin, because a bacterium containing the intact ovalbumin gene will not synthesize ovalbumin. Introns will be present in the gene (Chapter 9), and bacteria lack the enzymes needed to remove these noncoding sequences. Other tests might be performed to find the desired cell, such as hybridization of bacterial DNA with either the primary transcript or the mRNA of the ovalbumin gene isolated from chicken cells, but since only about one plasmid-containing colony in 10^5 would contain the ovalbumin gene, this method would be very tedious. A more convenient procedure is clearly desirable.

Some specialized animal cells, such as those producing ovalbumin, make only one or a very small number of proteins. In these cells the specific mRNA molecules, whose introns will have been removed by processing enzymes *if the mRNA is isolated from the cytoplasm*, constitute a large fraction of the total mRNA synthesized in the cell; consequently, mRNA samples can usually be obtained that consist predominantly of a single mRNA species—in this case, ovalbumin mRNA. If genes of this class—that is, those whose gene products are the major cellular proteins—are to be cloned, the purified mRNA of each type of cell can serve as a starting point for creating a collection of recombinant plasmids many of which should contain only the gene of interest.

Many RNA-containing animal tumor viruses contain reverse transcriptase in the virus particle. This enzyme can use a single-stranded RNA molecule (such as mRNA) as a template and, by an incompletely understood mechanism, synthesize a double-stranded DNA copy, called **complementary DNA** or **c-DNA.** If the template RNA molecule is an mRNA molecule (that is, if the introns have been removed from the primary transcript), the corresponding full-length c-DNA will contain an uninterrupted coding sequence. *This sequence will not be that of the original eukaryotic gene;* however, if the purpose of forming the recombinant DNA molecule is to synthesize a eukaryotic gene product in a bacterial cell, and if such processed RNA can be isolated, then c-DNA formed from processed mRNA is the material of choice to be inserted.

Joining of c-DNA to a vector can be accomplished by any of the procedures for joining blunt-ended molecules; the linker method is most commonly used.

DETECTION OF RECOMBINANT MOLECULES

When a vector is cleaved by a restriction enzyme and renatured with a mixture of all restriction fragments from a particular organism, many types of molecules result—some examples are a self-joined vector that has not acquired any fragments, a vector with one or more fragments, and a molecule consisting only of many joined fragments. Molecules of the third class cannot be established in a bacterium; they usually lack a DNA replication origin and replication genes, and are of no consequence. However, to facilitate the isolation of a vector containing a particular gene, some means is needed to ensure, first, that a vector established after $CaCl_2$ transformation does possess an inserted DNA fragment, and, second, that it is the DNA segment of interest. In this section several useful procedures for detecting plasmid vectors and recombinant vectors will be described.

In using the $CaCl_2$-transformation procedure to establish a plasmid in a bacterium, the initial goal is to isolate bacteria that contain the plasmid from a mixture of plasmid-free and plasmid-containing colonies. A common procedure is to use a plasmid possessing an antibiotic-resistance marker and to grow the transformed bacteria on a medium containing the antibiotic: only cells in which a plasmid has become established will form a colony. A useful plasmid is pBR322. It is small, consisting of 4362 nucleotide pairs of known sequence, and has two different antibiotic-resistance markers—resistance to tetracycline (*tet*-r) and to ampicillin (*amp*-r). Thus, plasmid-containing transformants are easily detected by growth of a transformed culture on medium containing either one of these antibiotics. Also, pBR322 is very useful because it contains only one copy of each of seven different types of restriction-enzyme cleavage sites at which DNA can be inserted, so the position of inserted DNA is always known.

In addition to a screening procedure for identifying plasmid-containing cells, a method is needed to identify plasmids in which DNA has been inserted. Having two antibiotic-resistance markers, such as are available in pBR322, allows the use of a procedure for detecting insertion called **insertional inactivation;** this is carried out as follows. In pBR322 the *tet* gene contains sites for cutting by the restriction enzymes BamHI and SalI. Thus, insertion at either of these sites will yield a plasmid that is *amp-r tet-s*, because insertion interrupts and hence inactivates the *tet* gene. If wildtype (Amp-s Tet-s) cells are transformed with a DNA sample in which the cleaved pBR322 and restriction fragments have been joined, and the cells are plated on a medium containing ampicillin, all surviving colonies must be Amp-r and hence must possess the plasmid. Some of these colonies will be Tet-r and some Tet-s, and these can be identified by plating onto a medium containing tetracycline. Because unaltered pBR322 carries the *tet*-r allele, an Amp-r colony will also be Tet-r unless the *tet-r* allele has been inactivated by insertion of foreign DNA. Thus, an Amp-r Tet-s cell must contain not only pBR322 DNA but donor DNA as well.

The proportion of the transformed cells in which insertion has occurred can be enriched by growing the transformed cells, prior to plating, in a liquid medium containing both cycloserine and tetracycline. Cycloserine kills growing cells, but Tet-s cells are merely inhibited, not killed, by tetracycline. Thus, in this growth medium Tet-r cells (which grow) are killed, and Tet-s cells (which are inhibited) survive. Plating of cells treated in this way on medium containing ampicillin yields Amp-r Tet-s colonies; these all possess pBR322 containing a donor DNA fragment. One would, of course, always retest for the Tet-s character by replica plating, because some Tet-r cells may have survived the treatment. The insertional inactivation procedure is illustrated in Figure 13-8.

Other useful plasmids contain one antibiotic-resistance marker and the *E. coli lac* genes. The Lac^+ and Lac^- phenotypes can be distinguished by plating on an indicator medium on which each phenotype yields colonies of a particular color.

Other procedures can also be used to select against reconstituted plasmids lacking foreign DNA. For example, the use of homopolymer tail-joining automatically selects against re-formation of the plasmid. The two termini of the cleaved plasmid will both have the same homopolymers, so restoration of circularity of the plasmid, which is essential for plasmid replication, is not possible.

When *E. coli* phage λ is the vector, two procedures are used for selecting for the presence of inserted genes. Both are based on the fact that λ has several centrally located genes that are not needed for lytic growth. A λ variant, one of several that have been constructed by recombinant DNA techniques, has two restriction sites for the enzyme EcoRI near the termini of this nonessential region. To use this vector requires cleaving the DNA with the EcoRI enzyme and separating the three fragments by gel electrophoresis. The terminal fragments are isolated and the central fragment is discarded. The terminal fragments can be joined via the cohesive termini produced by EcoRI, but the resulting DNA molecule, whose length is 72 percent that of wildtype λ DNA, will be noninfective. This is because the minimum length of DNA that can be packaged in a λ phage head is 77 percent of the wildtype length.

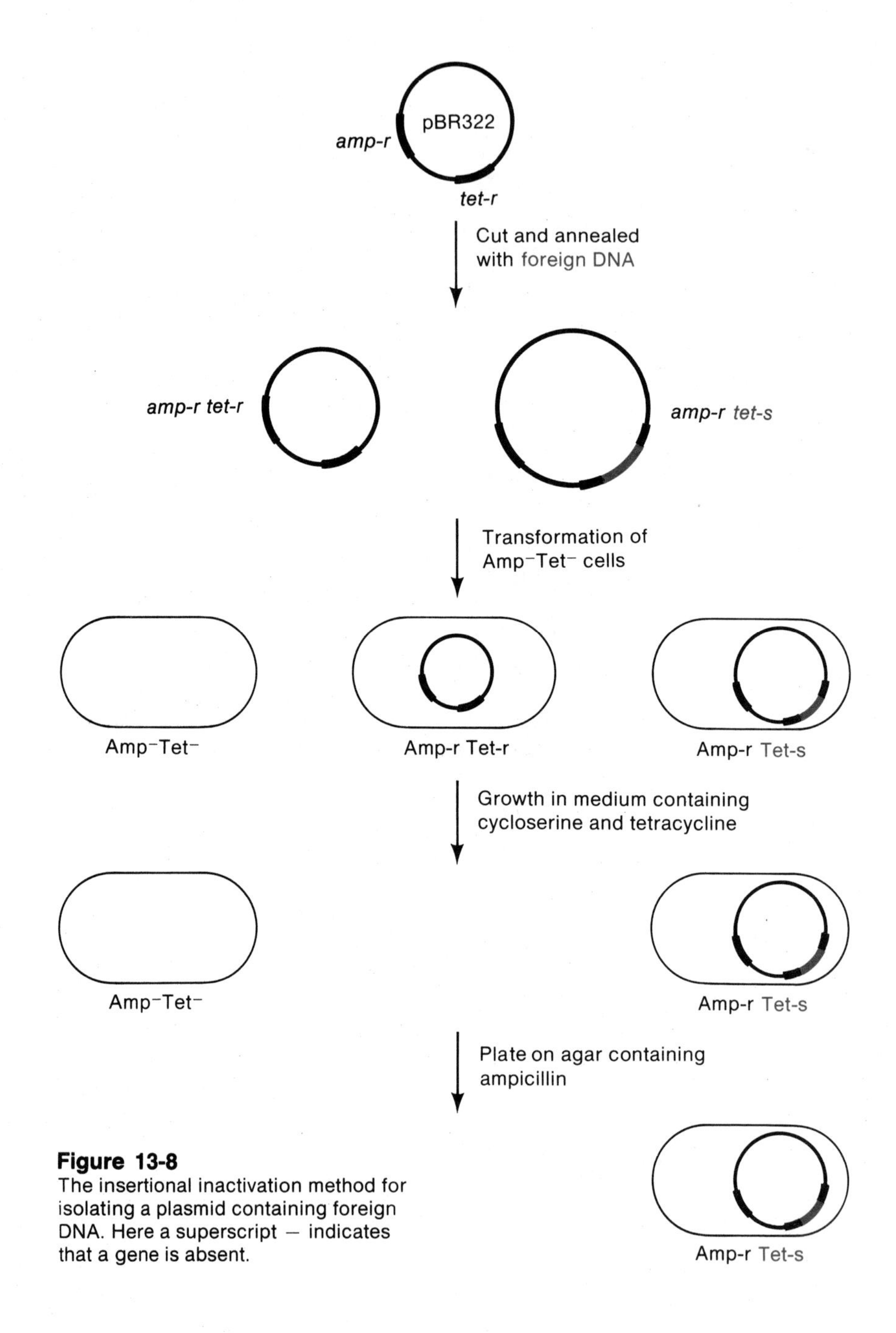

Figure 13-8
The insertional inactivation method for isolating a plasmid containing foreign DNA. Here a superscript – indicates that a gene is absent.

However, if foreign DNA whose size is 5–35 percent of the size of wild-type λ DNA is inserted, the resulting recombinant DNA molecule will be packageable and hence able to produce progeny phage. Thus, if an *E. coli* culture is infected (by the $CaCl_2$ transformation procedure) with DNA obtained by annealing a mixture of the two terminal fragments and foreign DNA, any phage that are produced will contain foreign DNA.

SCREENING FOR PARTICULAR RECOMBINANTS

As described earlier, the simplest procedure for detecting a cloned gene is complementation of a bacterial gene. Some eukaryotic genes can also be detected in this way. In the first such experiment reported, a cloned *his* (histidine synthesis) gene of yeast was detected by transforming a particular His$^-$ *E. coli* mutant and selecting for growth on a medium lacking histidine. However, this method is not generally successful with animal or plant DNA cloned in bacteria, because either the genes are not expressed or there is no corresponding bacterial gene. The following two procedures described in this section are more generally applicable and are commonly used for eukaryotic genes.

The **colony** or ***in situ*** **hybridization assay** allows detection of the presence of any gene for which radioactive mRNA is available (Figure 13-9). Colonies to be tested are replica-plated from a solid medium onto filter paper. A portion of each colony remains on the medium, which constitutes the reference plate. The paper is treated with NaOH, which simultaneously opens the cells and denatures the DNA. The paper is then saturated with ^{32}P-labeled mRNA, complementary to the gene being sought, and DNA-RNA renaturation occurs. After washing to remove unbound [^{32}P]mRNA, the positions of the bound radioactive phosphorus, usually detected by autoradiography, locate the desired colonies. A similar assay is done with phage vectors.

If the protein product of a gene of interest is synthesized, immunological techniques allow the protein-producing colony to be identified.

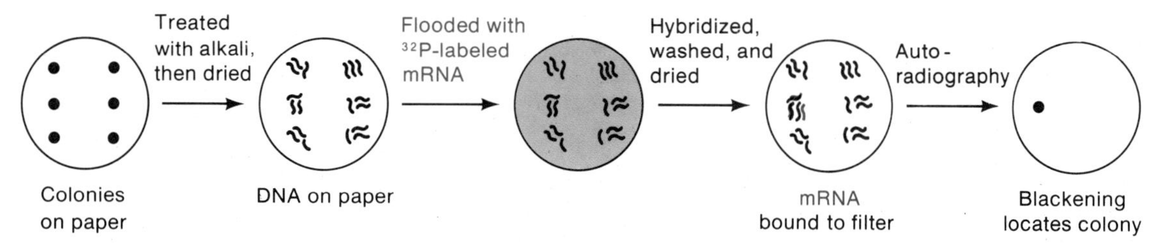

Figure 13-9
Colony hybridization. The reference plate, from which the colonies on paper were obtained, is not shown.

In one test the colonies are transferred, as in colony hybridization, and the transferred copies are exposed to a radioactive antibody directed against the particular protein. Colonies to which the radioactivity adheres are those containing the gene of interest. The radioactivity is generally detected by autoradiography.

APPLICATIONS OF GENETIC ENGINEERING

Recombinant DNA technology has revolutionized biology in the past decade. At present, its main uses are (1) facilitating the production of useful proteins, (2) creating bacteria capable of synthesizing economically important molecules, (3) supplying DNA and RNA sequences as a research tool, (4) altering the genotype of organisms such as plants, and (5) potentially correcting genetic defects in animals (gene therapy). Some examples of these applications follow.

Commercial Possibilities

Bacteria can carry out an enormous number of chemical reactions, but in each species the number is limited. However, by genetic engineering, organisms can be created that combine the features of several bacteria. For example, several genes from different bacteria have been inserted into a single plasmid that has then been established in a marine bacterium to yield an organism capable of metabolizing petroleum; this organism has been used to clean up oil spills in the oceans. Furthermore, bacteria are being designed that can synthesize industrially important chemicals—for example, ethyl alcohol—and that can simplify the large-scale manufacture of expensive antibiotics, which has the potential for reducing the cost of these drugs. A human insulin, synthesized in *E. coli,* is already commercially available.

For decades, an important role of genetics has been in plant breeding—that is, in the development of new strains of plants having desirable characteristics. For the most part, plant breeders have had to rely on selection of mutants and on hybridization between different varieties of the same species. Creating new varieties of plants by direct alteration of genotype is now becoming an important application of genetic engineering. Of great use is the bacterium *Agrobacterium tumefaciens* and its plasmid Ti, which produces crown gall tumors in dicotyledonous plants. These tumors result from disruption of the bacterium inside plant cells, release of bacterial DNA, and integration of a segment of the plasmid DNA into the plant chromosome. By genetic engineering, genes obtained from one plant can be introduced into this plasmid and then, via infection of a second plant with the bacterium carrying a recombinant plasmid, the genes of the first plant can be transferred to the second plant. The tumor-causing genes of Ti have been removed by genetic engineering and transformation systems have been established by which purified DNA rather than bacteria can be used to produce the alterations. Some attempts to perform plant breeding in this way have recently been partially successful, but the technique must still be considered to be in its infancy.

Uses in Research

Recombinant DNA technology is extraordinarily useful in basic research. In Chapter 14 several examples are given of mutant bacteria in which particular genetic systems have been altered in an effort to make them more amenable to study (for example, the numerous mutants in the *lac* operon). Such mutants have usually been derived through standard genetic techniques (for example, mutagenesis and genetic recombination). For simple mutants this procedure is straightforward, but for mutants required to have many genetic markers (which may be very closely linked) the frequency of production of mutants can be so low that their isolation becomes very tedious. Recombinant DNA techniques can simplify mutant construction, since fragments containing desired genetic markers can be purified, altered, and combined in a test tube, and then introduced into another cell. This saves time, labor, and often enables mutants to be constructed that cannot in practice be formed in any other way. An example is the formation of double mutants of animal viruses, which undergo crossing over at such a low frequency that mutations can rarely be recombined by genetic crosses.

The greatest impact of the new technology on basic research has been in the study of eukaryotes, in particular, of eukaryotic regulation. Experiments of the type considered in Chapter 14, which study the regulation of operons in bacteria, have been made possible by the use of mutations in promoters, operators, and structural genes. However, this approach has not been feasible with most eukaryotes, because they are diploid and, hence, mutants are difficult to isolate. Furthermore, except for the unicellular eukaryotes such as yeast, there is no simple and rapid way to do multiple genetic manipulations with eukaryotic cells.

Cloning techniques have made it possible to study regulation of gene expression in eukaryotes by direct assays of mRNA molecules produced by particular genes. The approach is based on the success with which it has become possible to understand gene activity in bacteria by studying the synthesis of mRNA in a variety of conditions. An excellent assay is DNA-RNA hybridization, in which either a primary transcript or mRNA is detected by hybridization to a DNA sample that has been enriched for the gene being studied.

In microbial experiments specialized transducing particles (Chapter 15) would be the source of the DNA—for example, *E. coli gal* mRNA is assayed by hybridization of radioactively labeled intracellular RNA with DNA of λ *gal* transducing particles. However, transducing particles do not exist for eukaryotic systems, so recombinant DNA techniques have been used to clone a gene whose regulation is to be studied. The usual procedure is to clone the gene first in plasmid (or a phage) vector and then allow the cell containing the plasmid to multiply. Purified plasmid DNA provides the researcher with a large supply of the DNA of that gene. The vector containing that DNA sequence is called a DNA **probe,** because its DNA, in denatured form, can be added to a cell extract containing mRNA, to probe for a particular mRNA by hybridization. Since no natural genes of the vector are present in eukaryotic cells, the vector DNA can be regarded as a source of pure DNA copies of a particular eukaryotic gene because no other genes in the vector will participate in the hybridization.

Production of Eukaryotic Proteins

One of the most valuable applications of genetic engineering is the production of large quantities of particular proteins that are otherwise difficult to obtain (for example, proteins of which only a few molecules are normally present in a cell). The method is simple in principle. The gene encoding the desired protein is coupled to a vector containing a bacterial promoter, and tests are performed to ensure that the gene is oriented such that its coding strand is linked to the strand containing the promoter. A plasmid for which many copies are present in each cell, or occasionally an actively replicating phage (such as λ), is used as a vector; in both cases, cells can be prepared that contain several hundred copies of the gene, and this can result in synthesis of a gene product to reach a concentration of about 1 to 5 percent of the cellular protein. In practice, production of large quantities of a prokaryotic protein in a bacterium is straightforward. However, if the gene is from a eukaryote, special problems exist: (1) Eukaryotic promoters are not usually recognized by bacterial RNA polymerases. (2) The mRNA transcribed from eukaryotic genes may not be translatable on bacterial ribosomes (3) Introns may be present, and bacteria are unable to excise eukaryotic introns. (4) The protein itself often must be processed (for example, insulin); and bacteria cannot recognize processing signals from eukaryotes. (5) Eukaryotic proteins are often recognized as foreign material by bacterial protein-digesting enzymes and are broken down. Several approaches to solving these problems have been taken with some degree of success. One procedure uses a yeast plasmid (the 2-μm plasmid) as a cloning vehicle for eukaryotic genes. Recent studies suggest that when a recombinant 2-μm plasmid containing eukaryotic genes is put into yeast, which is a eukaryote, some of the problems just described no longer exist.

Another approach uses a plasmid containing the *E. coli lac* region, cleaved in the *lacZ* gene (which encodes the enzyme β-galactosidase) by a restriction enzyme making blunt ends. Either c-DNA or a synthetic DNA molecule whose sequence is known from the amino acid sequence of the protein product is inserted into the *lacZ* gene, so that the eukaryotic protein is synthesized as the terminal region of β-galactosidase, from which it can be cleaved. The first example of this approach resulted in a synthetic gene capable of yielding a 14-residue polypeptide hormone—somatostatin—which *in vivo* is synthesized in the mammalian hypothalamus. The procedure, applicable to most short polypeptides, is described in *MB*, pages 829–834.

Genetic Engineering with Animal Viruses

Genetic engineering with retroviruses allows the possibility of altering the genotype of an animal cell. Since a wide variety of retroviruses are known, including two that can grow on human hosts, genetic defects may be correctable by these procedures in the future. However, many retrovirus species contain a gene that produces uncontrolled growth of a cell containing the retrovirus, thereby causing tumors in animals. If a retrovirus vector is to be used to change a genotype, the tumor-causing ability must be removed. This is possible by removal of the tumor-causing gene, which also provides the space needed for incorporation of foreign DNA. Other retroviruses lack tumor-forming genes and produce tumors by virtue of their insertion next to particular cellular genes whose activation leads to uncontrolled cell division. This type of virus has also been modified by removal of its terminal sequences, which are necessary for integration, thereby eliminating the tumorigenicity. The recombinant DNA procedure employed with retroviruses consists of synthesis in the laboratory of double-stranded DNA from the viral RNA, through use of reverse transcriptase. The DNA is then cleaved with a restriction enzyme and, by the techniques already described, foreign DNA is inserted and selected. Treatment of mutant cells in culture with the recombinant DNA, and application of a transformation procedure suitable for animal cells, yields cells in which recombinant retroviral DNA has been permanently inserted. In this way, the genotype of the cells can be altered. Experiments have been done in which human cells deficient in the synthesis of purines have been obtained from patients with Lesch-Nyhan syndrome and grown in culture; these cells have been converted to normal cells by transformation with recombinant DNA. The exciting potential of this technique lies in the possibility of correcting genetic defects—for example, restoring the ability of a diabetic individual to make insulin or correcting immunological deficiencies. This technique has been termed **gene therapy.** However, it must be recognized that retroviruses are not well understood and are potentially dangerous.

Gene therapy is not yet a practical technique, for major problems exist; for example, there is no reliable way to ensure that a gene is inserted in the appropriate target cell or target tissue. In addition, some means is needed to regulate the expression of the inserted genes.

Animal proteins can also be made in large quantities by animal cells that contain many copies of a gene. For example, in a preliminary study to test the feasibility of cloning in an animal virus vector the rabbit β-globin gene (globin is a protein subunit of hemoglobin) has been cloned in SV40 virus, a virus that grows in cultures of monkey cells and produces thousands of intracellular DNA molecules; when this hybrid virus infects monkey kidney cells, rabbit β-globin is synthesized in large quantities.

Other exciting possibilities are being investigated. For example, vaccinia virus, the virus used in smallpox vaccination, has been modified so that some of its coat proteins are those of viruses that are difficult to grow and dangerous to handle, such as hepatitis virus. Individuals immunized with the hybrid vaccinia virus will also become immune to hepatitis virus. This new technique for antigen production is being developed by numerous pharmaceutical and bioengineering companies.

14 Regulation of Gene Activity in Prokaryotes

The number of protein molecules produced per unit time by active genes varies from gene to gene, satisfying the needs of a cell and sometimes also avoiding wasteful synthesis. The different rates result mainly from different efficiencies of either recognition of a promoter by RNA polymerase or initiation of translation. However, the flow of genetic information is regulated in other ways also. For example, many gene products are needed only on occasion, and regulatory mechanisms of an on-off type exist that enable such products to be present only when demanded by external conditions. More subtly regulated systems can adjust the intracellular concentration of a particular protein in response to needs imposed by the environment. In general, the synthesis of particular gene products is controlled by mechanisms collectively called **gene regulation.**

The regulatory systems of prokaryotes and eukaryotes are somewhat different from each other. Prokaryotes are generally free-living unicellular organisms that grow and divide indefinitely as long as environmental conditions are suitable and the supply of nutrients is adequate. Thus, their regulatory systems are geared to provide the maximum growth rate in a particular environment, except when such growth would be detrimental. This strategy seems to apply to the free-living unicells such as yeast, algae, and protozoa, though less information is available about these organisms than for bacteria.

The requirements of tissue-forming eukaryotes are different from those of prokaryotes. In a developing organism—for example, in an embryo—a cell must not only grow and produce many progeny cells but also must undergo considerable change in morphology and biochemistry and then maintain the changed state. Furthermore, during the growth and cell-division phases of the organism, these cells are challenged less by the environment than are bacteria in that the composition and concentration of their growth media do not change drastically with time. Some examples of such media are blood, lymph, or other body fluids, or, in the case of marine animals, sea water. Finally, in an adult organism, growth and cell division in most cell types have stopped, and each cell needs only to maintain itself and its properties. Many other examples could be given; the main point is that because a typical eukaryotic cell faces different contingencies than a bacterium does, the regulatory mechanisms of eukaryotes and prokaryotes are not the same.

In this chapter we consider regulation in prokaryotes. Eukaryotes are examined in Chapter 16.

PRINCIPLES OF REGULATION

The best-understood regulatory mechanisms are those used by bacteria and by phages. In these organisms an on-off regulatory activity is obtained by controlling transcription—that is, synthesis of a particular mRNA is allowed when the gene product is needed and inhibited when the product is not needed. In bacteria, few examples are known of switching a system completely off. When transcription is in the off state, a basal level of gene expression almost always remains, often consisting of only one or two transcriptional events per cell generation; hence, very little synthesis of the gene product occurs. For convenience, when discussing transcription, the term "off" will be used, but it should be kept in mind that usually what is meant is "very low." In only one case in bacteria, namely, in spores, is expression of most genes totally turned off. In eukaryotes complete turning off of a gene is quite prevalent. Regulatory mechanisms other than the on-off type are also known in both prokaryotes and eukaryotes; for example, the activity of a system may be modulated from fully on to partly on, rather than to off.

In bacterial systems, when several enzymes act in sequence in a single metabolic pathway, usually either all or none of these enzymes are produced. This phenomenon, which is called **coordinate regulation,** results from control of the synthesis of a single polycistronic mRNA molecule encoding all of the gene products. This type of regulation does not occur in eukaryotes because eukaryotic mRNA is usually monocistronic, as discussed in Chapter 9.

Isopropylthiogalactoside (IPTG)

Several mechanisms of regulation of transcription are common; the particular one used often depends on whether the enzymes being regulated act in degradative or synthetic metabolic pathways. For example, in a multistep degradative system the availability of the molecule to be degraded frequently determines whether the enzymes in the pathway will be synthesized. In contrast, in a biosynthetic pathway the final product is often the regulatory molecule. Even in a system in which a single protein molecule (not necessarily an enzyme) is translated from a monocistronic mRNA molecule, the protein may be **autoregulated**—that is, the protein itself may inhibit initiation of transcription and high concentrations of the protein will result in less transcription of the mRNA that encodes the protein. The molecular mechanisms for each of the regulatory patterns vary quite widely but usually fall in one of two major categories—**negative regulation** and **positive regulation** (Figure 14-1). In a negatively regulated system, an inhibitor is present in the cell and prevents transcription. An antagonist of the inhibitor, generally called an **inducer,** is needed to allow initiation of transcription. In a positively regulated system, an effector molecule (which may be a protein, a small molecule, or a molecular complex) activates a promoter; no inhibitor must be overridden. Negative and positive regulation are not mutually exclusive, and some systems are both positively and nega-

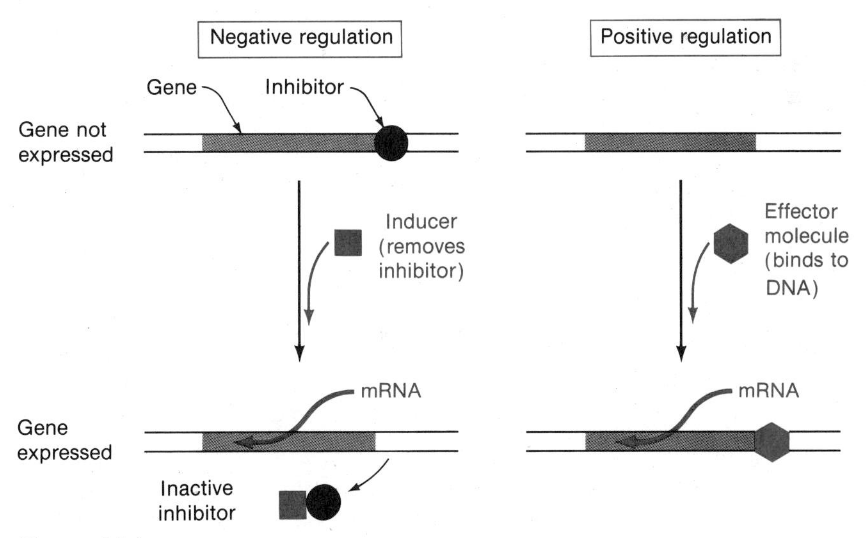

Figure 14-1

The distinction between negative and positive regulation. In negative regulation an inhibitor, bound to the DNA, must be removed before transcription can occur. In positive regulation an effector molecule must bind to the DNA. A system may also be regulated both positively and negatively; then, the system is "on" when the positive regulator is bound to the DNA and the negative regulator is not bound to the DNA.

tively regulated, utilizing two regulators to respond to different conditions in the cell.

A degradative system may be regulated either positively or negatively. In a biosynthetic pathway, the final product usually negatively regulates its own synthesis; in the simplest type of negative regulation, absence of the product increases its synthesis and presence of the product decreases its synthesis.

In both prokaryotic and eukaryotic systems enzyme *activity* rather than enzyme *synthesis* may also be regulated. That is, the enzyme may be present but its activity is turned on or off. When this occurs, the product of the enzymatic reaction or, in the case of a biosynthetic pathway, the end product of a sequence of reactions usually inhibits the enzyme. This mode of regulation is called **feedback inhibition.** Small molecules other than reaction products are also frequently used either to activate or inhibit a particular enzyme. Such a molecule is termed an **allosteric effector molecule.**

THE *E. COLI* LACTOSE SYSTEM AND THE OPERON MODEL

In *E. coli* two proteins are necessary for the metabolism of lactose—the enzyme **β-galactosidase,** which cleaves lactose (a β-galactoside) to yield galactose and glucose, and a carrier, **lactose permease,** which is required for the entry of lactose into a cell. The existence of two different proteins in the lactose-utilization system was first shown by a combination of genetic experiments and biochemical analysis.

First, hundreds of mutants unable to use lactose as a carbon source—Lac^- mutants—were isolated. Some of the mutations were in the *E. coli* chromosome and others were in F′*lac*, a plasmid carrying the genes for lactose utilization. By performing $F' \times F^-$ matings partial diploids having the genotypes F′lac^-/lac^+ or F′lac^+/lac^- were constructed, as described in Chapter 12. (The genotype of the plasmid is given to the left of the diagonal line and that of the chromosome to the right.) It was observed that these diploids always produced a Lac^+ phenotype (that is, they made β-galactosidase); none produced an inhibitor that prevented functioning of the *lac* gene. Other partial diploids were then constructed in which both the F′*lac* plasmid and the chromosome carried lac^- genes; these were tested for the Lac^+ phenotype, with the result that all of the mutants initially isolated could be placed into two groups, *lacZ* and *lacY*. The partial diploids F′$lacY^-lacZ^+/$ $lacY^+lacZ^-$ and F′$lacY^+lacZ^-/lacY^-lacZ^+$ had a Lac^+ phenotype, producing β-galactosidase, but the genotypes F′$lacY^-lacZ^+/lacY^-lacZ^+$ and F′$lacY^+lacZ^-/lacY^+lacZ^-$ had the Lac^- phenotype. The existence of two groups was good evidence that the *lac* system consisted of at least two genes ("at least," because mutations had not yet been obtained in other genes).

Further experimentation was needed to establish the precise function of each gene. Experiments in which cells were placed in a medium containing ^{14}C-labeled (radioactive) lactose showed that no [^{14}C]lactose would enter a *lacY*$^-$ cell, whereas it readily penetrated a *lacZ*$^-$ mutant. Treatment of a *lacY*$^-$ cell with lysozyme, an enzyme that destroys part of the cell wall and makes it permeable, enabled radioactive lactose to enter a *lacY*$^-$ cell, indicating that the *lacY* gene is probably concerned with transport of lactose through the cell membrane into the cell and is the structural gene for *lac* permease. Enzymatic assays showed that β-galactosidase is present in *lac*$^+$ but not *lacZ*$^-$ cells; these results provided the initial evidence that the lacZ *gene is the structural gene for β-galactosidase.* A final important result—that the *lacY* and *lacZ* genes are adjacent—was obtained by genetic mapping.

The on-off nature of the lactose-utilization system is evident in the following observations:

1. If a culture of Lac$^+$ *E. coli* is growing in a medium lacking lactose or any other β-galactoside, the intracellular concentrations of β-galactosidase and permease are exceedingly low—roughly one or two molecules per bacterium. However, if lactose is present in the growth medium, the number of each of these molecules is about 10^5-fold higher.
2. If lactose is added to a Lac$^+$ culture growing in a lactose-free medium (also lacking glucose, a point that will be discussed shortly), both β-galactosidase and permease are synthesized nearly simultaneously, as shown in Figure 14-2. Analysis of the total mRNA present in the cells before and after addition of lactose shows that no *lac* mRNA (the mRNA that encodes β-galactosidase and permease) is present before lactose is added and that the addition of lactose triggers synthesis of *lac* mRNA. The analysis is done by growing cells in a radioactive medium in which newly synthesized mRNA is radioactive, then isolating the mRNA, and finally allowing it to renature with the DNA of a λ *lac*-transducing particle. Since the only radioactive mRNA that will renature to the *lac*-containing DNA is *lac* mRNA, the amount of radioactive DNA-RNA hybrid molecules is a measure of the amount of *lac* mRNA.

These two observations led to the view that the lactose system is **inducible** and that lactose is an **inducer.**

Lactose itself is rarely used in experiments to study the induction phenomenon for a variety of reasons; one important reason is that the

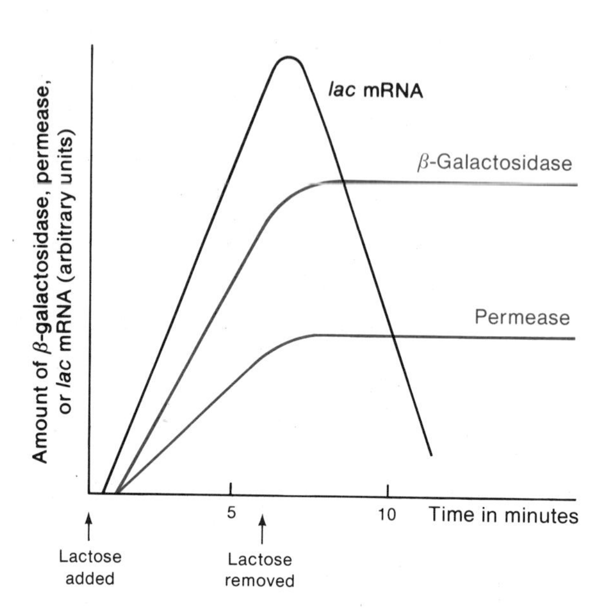

Figure 14-2
The "on-off" nature of the *lac* system. *Lac* mRNA appears very soon after lactose is added; β-galactosidase and permease appear at the same time but are delayed with respect to mRNA synthesis because of the time required for translation. When lactose is removed, no more *lac* mRNA is made and the amount of *lac* mRNA decreases owing to the usual degradation of mRNA. Both β galactosidase and permease are stable. Their concentration remains constant even though no more can be synthesized. A third protein of the *lac* system, β-galactoside transacetylase, is synthesized coordinately with β-galactosidase and permease. This protein, the product of the *a* gene, will be discussed later.

β-galactosidase that is synthesized catalyzes the cleavage of lactose and results in a continual decrease in lactose concentration, which complicates the analysis of many types of experiments (for example, kinetic experiments). Instead, a sulfur-containing analogue of lactose is used—either isopropylthiogalactoside (IPTG) or thiomethylgalactoside (TMG); these analogues are inducers but are not substrates of β-galactosidase.

Mutants have also been isolated in which *lac* mRNA is synthesized (hence also β-galactosidase and permease) in *both* the presence and the absence of an inducer. These mutants provided the key to understanding induction because they eliminated regulation; they were termed **constitutive.** Complementation tests—again with partial diploids carrying two constitutive mutations, one in the chromosome and the other in a plasmid—showed that the mutants fall into two groups termed *lacI* and $lacO^c$. The characteristics of the mutants are shown in Table 14-1. The $lacI^-$ mutants are recessive (entries 3, 4). In the absence of an inducer a $lacI^+$ cell fails to make *lac* mRNA, whereas this mRNA is made by a $lacI^-$ mutant. Thus, the *lacI* gene is apparently a regulatory gene *whose product is an inhibitor that keeps the system turned off.* A $lacI^-$ mutant lacks the inhibitor and hence is constitutive. Wildtype copies of the *lacI*-gene product are present in a $lacI^+/lacI^-$ partial diploid, so the system is inhibited. The *lacI*-gene product, a protein molecule, is called the ***lac* repressor.** Genetic mapping experiments place the *lacI* gene adjacent to the *lacZ* gene and establish the gene order *lacI lacZ lacY*. How the *lacI* repressor prevents synthesis of *lac* mRNA will be explained shortly.

Table 14-1
Characteristics of partial diploids having several combinations of *lacI* and *lacO* alleles

Genotype	Constitutive or inducible synthesis of lac mRNA
1. F′$lacO^clacZ^+/lacO^+lacZ^+$	Constitutive
2. F′$lacO^+lacZ^+/lacO^clacZ^+$	Constitutive
3. F′$lacI^-lacZ^+/lacI^+lacZ^+$	Inducible
4. F′$lacI^+lacZ^+/lacI^-lacZ^+$	Inducible
5. F′$lacO^clacZ^+/lacI^-lacZ^+$	Constitutive
6. F′$lacO^clacZ^-/lacO^+lacZ^+$	Inducible
7. F′$lacO^clacZ^+/lacO^+lacZ^-$	Constitutive

The $lacO^c$ *mutations are dominant* (entries 1, 2, and 5, Table 14-1), but the dominance is evident only in certain combinations of *lac* mutations, as can be seen by examining the partial diploids shown in entries 6 and 7. Both combinations are Lac^+, because a functional *lacZ* gene is present. However, in the combination shown in entry 6, synthesis of β-galactosidase is inducible even though a $lacO^c$ mutation is present. The difference between the two combinations in entries 6 and 7 is that in entry 6 the $lacO^c$ mutation is carried on a DNA molecule that also has a $lacZ^-$ mutation, whereas in entry 7, $lacO^c$ and $lacZ^+$ are *carried on the same DNA molecule*. Thus, a $lacO^c$ mutation causes constitutive synthesis of β-galactosidase only when the $lacO^c$ and $lacZ^+$ alleles are located *on the same DNA molecule;* the $lacO^c$ mutation is said to be ***cis*-dominant,** since only genes *cis* to the mutant are expressed in dominant fashion. Confirmation of this conclusion comes from an important biochemical observation: the mutant enzyme (encoded in the $lacZ^-$ sequence) is synthesized constitutively in a $lacO^clacZ^-/lacO^+\ lacZ^+$ partial diploid (entry 6), whereas the wildtype enzyme (encoded in the $lacZ^+$ sequence) is synthesized only if an inducer is added. All $lacO^c$ mutations are located between the *lacI* and *lacZ* genes; thus, the gene order of the four elements of the *lac* system is

lacI lacO lacZ lacY

An important feature of all $lacO^c$ mutations is that they cannot be complemented (a feature of all *cis*-dominant mutations). That is, a $lacO^+$ allele cannot alter the constitutive activity of a $lacO^c$ mutation. Thus, *lacO* does not encode a diffusible product and must define a *site* or a noncoding region of the DNA rather than a gene. This site determines whether synthesis of the product of the adjacent *lacZ* gene is inducible or constitutive. The *lacO* region is called the **operator.**

The regulatory mechanism of the *lac* system was first explained by the **operon model,** which has the following features (Figure 14.3):

1. The lactose-utilization system consists of two kinds of components—*structural genes* needed for transport and metabolism of lactose, and *regulatory elements* (the *lacI* gene, the *lacO* operator, and the *lac* promoter). Together these components comprise the ***lac* operon.**
2. The products of the *lacZ* and *lacY* genes are encoded in a single polycistronic mRNA molecule. (This mRNA molecule contains a third gene, denoted *lacA*, which encodes the enzyme trans-

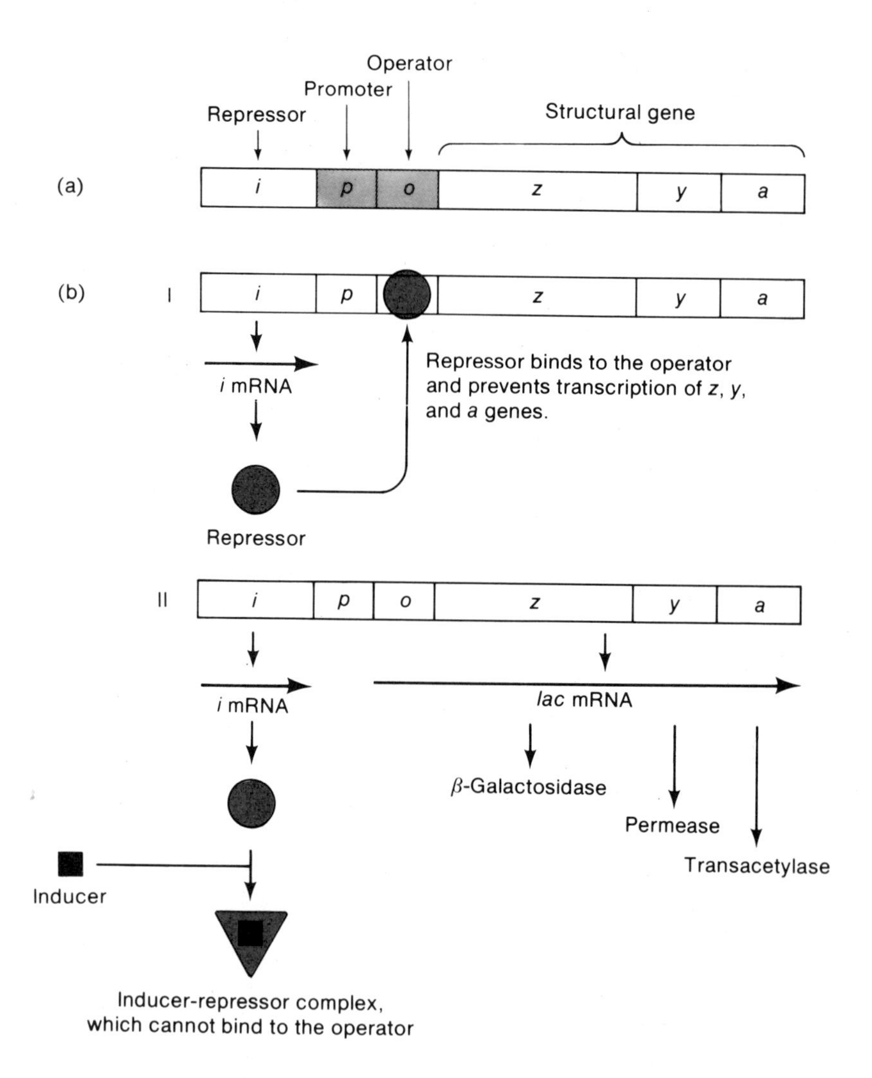

Figure 14-3
(a) Genetic map of the *lac* operon, not drawn to scale: the *p* and *o* sites are actually much smaller than the genes.
(b) Diagram of the *lac* operon in (I) repressed and (II) induced states. The inducer alters the shape of the repressor, so the repressor can no longer bind to the operator. The symbols *i, p, o, z, y, a* represent *lac I*, *lac P*, etc.

acetylase. This enzyme is used in the metabolism of certain β-galactosides other than lactose and will not be of further concern.)

3. The promoter for the *lacZ lacY lacA* mRNA molecule is immediately adjacent to the *lacO* region. This location has been substantiated by the isolation and mapping of promoter mutants (*lacP*$^-$) that are completely incapable of making either β-galactosidase or permease, because no *lac* mRNA is made.
4. The *lacI*-gene product, the repressor, binds to a unique sequence of DNA bases, namely, the operator.
5. When the repressor is bound to the operator, initiation of transcription of *lac* mRNA by RNA polymerase is prevented.
6. Inducers stimulate mRNA synthesis by binding to and inactivating the repressor, a process called **derepression.** Thus, in the presence of an inducer the operator is unoccupied, and the promoter is available for initiation of mRNA synthesis.

Note that regulation of the operon requires that the *lacO* operator be adjacent to the structural genes of the operon (*lacZ, lacY, lacA*), but proximity of the *lacI* gene is not necessary, because the *lacI* repressor is a soluble protein and is therefore diffusible throughout the cell.

The operon model is supported by a wealth of experimental data and explains many of the features of the *lac* system as well as numerous other negatively regulated genetic systems. One aspect of the regulation of the *lac* operon—the effect of glucose—has not yet been discussed. Examination of this feature indicates that the *lac* operon is also subject to positive regulation.

The function of β-galactosidase in lactose metabolism is to form glucose by cleaving lactose. (The other cleavage product, galactose, is also ultimately converted to glucose by the enzymes of the galactose operon.) Thus, if both glucose and lactose are present in the growth medium, activity of the *lac* operon is not needed, and indeed, no β-galactosidase is formed until virtually all of the glucose in the medium is consumed. The lack of synthesis of β-galactosidase is a result of lack of synthesis of *lac* mRNA. No *lac* mRNA is made in the presence of glucose, because in addition to an inducer to inactivate the *lacI* repressor, another element is needed for initiating *lac* mRNA synthesis; the activity of this element is regulated by the concentration of glucose. However, the inhibitory effect of glucose on expression of the *lac* operon is quite indirect.

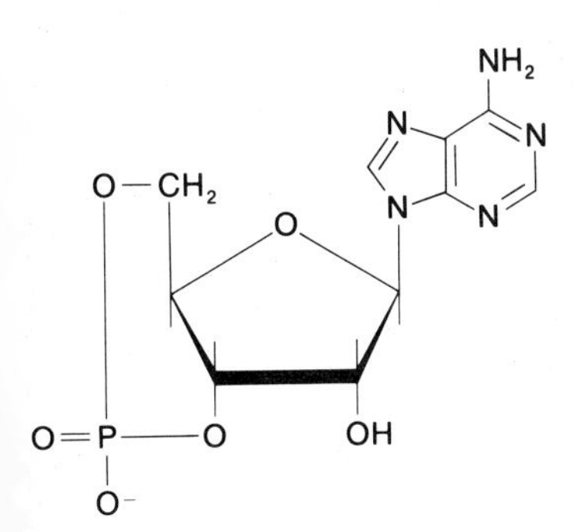

Figure 14-4
Structure of cyclic AMP.

The small molecule **cyclic AMP (cAMP)** is universally distributed in animal tissues, and in multicellular eukaryotic organisms it is important in regulating the action of many hormones (Figure 14-4). It is also present in *E. coli* and many other bacteria. Cyclic AMP is synthesized enzymatically by **adenyl cyclase,** and its concentration is regulated

Table 14-2
Concentration of cyclic AMP in cells growing in media having the indicated carbon sources

Carbon source	*cAMP concentration*
Glucose	Low
Glycerol	High
Lactose	High
Lactose + glucose	Low
Lactose + glycerol	High

indirectly by glucose metabolism. When bacteria are growing in a medium containing glucose, the cAMP concentration in the cells is quite low. In a medium containing glycerol or any carbon source that cannot enter the biochemical pathway used to metabolize glucose (the glycolytic pathway), or when the bacteria are otherwise starved of an energy source, the cAMP concentration is high (Table 14-2). The mechanism by which glucose controls the cAMP concentration is poorly understood; the significant point is that *cAMP regulates the activity of the* lac *operon* (and other several other operons as well).

E. coli (and many other bacterial species) contain a protein called the **catabolite activator protein (CAP)**, which is encoded in a gene called *crp*. Mutants of either *crp* or the adenyl cyclase gene are unable to synthesize *lac* mRNA, indicating that both CAP function and cAMP are required for *lac* mRNA synthesis. CAP and cAMP bind to one another, forming a unit denoted **cAMP-CAP**, which is an active regulatory element in the *lac* system. The requirement for cAMP-CAP is independent of the *lacI* repression system since *crp* and adenyl cyclase mutants are unable to make *lac* mRNA even if a $lacI^-$ or a $lacO^c$ mutation is present. *The cAMP-CAP complex must be bound to a base sequence in the DNA in the promoter region in order for transcription to occur* (Figure 14-5). Thus, *cAMP-CAP is a positive regulator,* in contrast

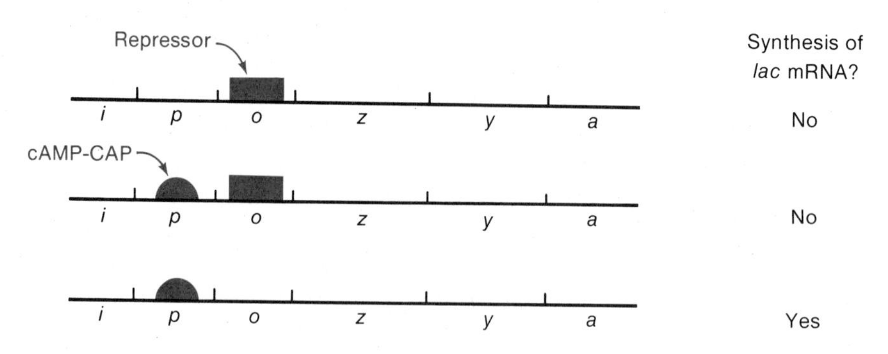

Figure 14-5
Three states of the *lac* operon showing that *lac* mRNA is made only if cAMP-CAP is present and repressor is absent. The symbols *i, p, o,* etc. represent *lac I, lac P, lac O,* etc.

with the repressor, and the *lac* operon is independently regulated both positively and negatively.

The precise mechanism by which cAMP-CAP stimulates and the repressor inhibits transcription is not known in detail. However, by using mixtures containing purified *lac* DNA, *lac* repressor, cAMP-CAP, and RNA polymerase two points have been established:

1. In the absence of cAMP-CAP, RNA polymerase binds only weakly to the promoter, but its binding is stimulated when cAMP-CAP is also bound to the DNA. The weak binding rarely leads to initiation of transcription, because the correct interaction between RNA polymerase and the promoter does not occur.
2. If the repressor is first bound to the operator, RNA polymerase cannot stably bind to the promoter.

These results are sufficient to explain how lactose and glucose function to regulate transcription of the *lac* operon.

The ratios of the number of copies of β-galactosidase, permease, and transacetylase (the third structural gene of the system), are 1.0:0.5:0.2. These differences, which are examples of **translational regulation,** are achieved in two ways:

1. The *lacZ* gene is translated first (Figure 11-6). Frequently, the *lac* mRNA molecule detaches from its translating ribosome following chain termination. The frequency with which this occurs is a function of the probability of reinitiation at each subsequent AUG codon. Thus, there is a gradient in the amount of poly-

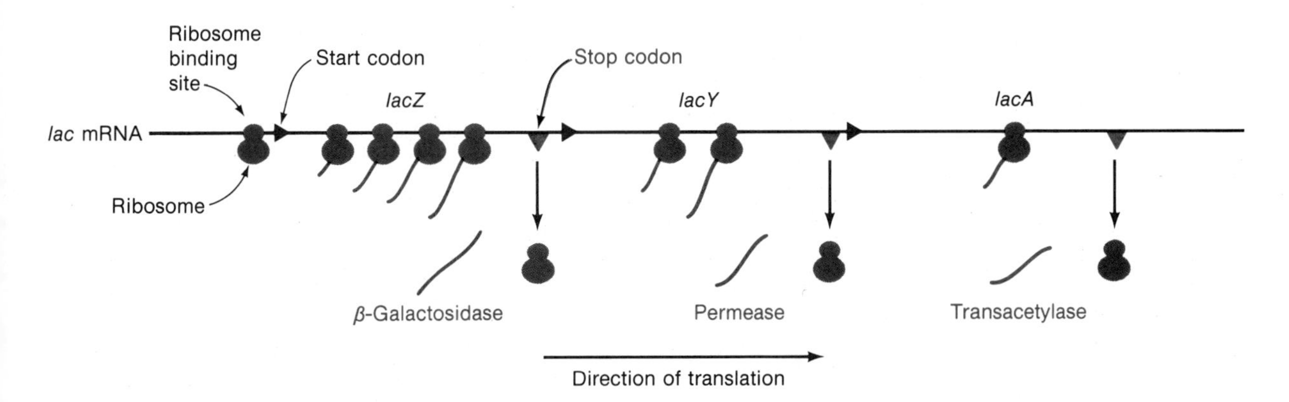

Figure 14-6
One explanation for polarity in the *lac* operon. All ribosomes attach to the mRNA at the ribosome binding site. At each stop codon some ribosomes detach, so the number of ribosomes translating each gene decreases for each subsequent gene.

peptide synthesis from the 5′ terminus to the 3′ terminus of the mRNA molecule, an effect called **polarity** that occurs with most polycistronic mRNA molecules.

2. Degradation of *lac* mRNA is initiated more frequently in the *lacA* gene than in the *lacY* gene and more often in the *lacY* gene than in the *lacZ* gene. Hence, at any given instant, there are more complete copies of the *lacZ* gene than of the *lacY* gene, and more copies of the *lacY* gene than of the *lacA* gene. In prokaryotes this mode of regulation occurs repeatedly:

> The overall expression of activity of an operon is regulated by controlling transcription of a polycistronic mRNA, and the relative concentrations of the proteins encoded in the mRNA are determined by controlling the frequency of initiation of translation of each cistron.

However, the mechanism by which transcription is regulated varies from one system to the next. An inducer-repressor system is common to many operons responsible for degradative metabolism and cAMP-CAP is an element in many carbohydrate-degrading systems. However, particular features of the regulatory mechanisms differ; for example, two promoters are present in the galactose operon, and the arabinose operon is mainly positively regulated. These operons are described in *MB*, pages 563–573.

THE TRYPTOPHAN OPERON, A BIOSYNTHETIC SYSTEM

The tryptophan (*trp*) operon of *E. coli* is responsible for the synthesis of the amino acid tryptophan. Regulation of this operon occurs in such a way that when tryptophan is present in the growth medium, the *trp* operon is not active. That is, when adequate tryptophan is present, transcription of the operon is inhibited; however, when the supply is insufficient, transcription occurs. The *trp* operon is quite different from the *lac* operon in that tryptophan acts directly in the repression system rather than as an inducer. Furthermore, since the *trp* operon encodes a set of biosynthetic rather than degradative enzymes, neither glucose nor cAMP-CAP functions in operon activity.

A simple on-off system, as in the *lac* operon, is not optimal for a biosynthetic pathway; a situation may arise in nature in which some tryptophan is available, but not enough to allow normal growth if synthesis of tryptophan were totally shut down. Tryptophan starvation when the supply of the amino acid is inadequate is prevented by a

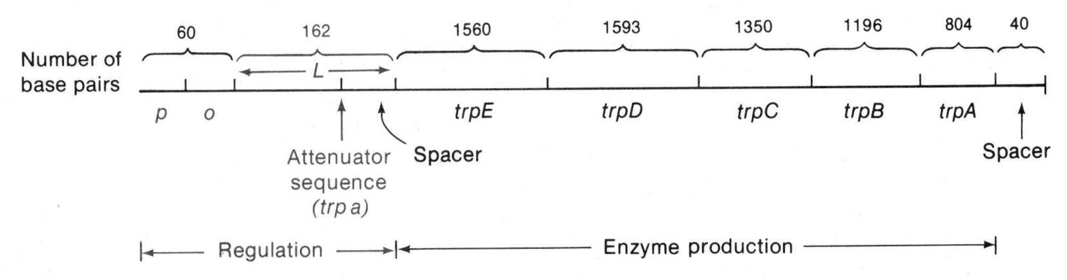

Figure 14-7
The *E. coli trp* operon. For clarity, the regulatory region is enlarged with respect to the coding region. The proper size of each region is indicated by the number of base pairs. *L* is the leader. The regulatory elements are shown in red.

modulating system in which *the amount of transcription in the derepressed state is determined by the concentration of tryptophan.* This mechanism is found in many operons responsible for amino acid biosynthesis.

Tryptophan is synthesized in five steps, each requiring a particular enzyme. In the *E. coli* chromosome the genes encoding these enzymes are adjacent to one another in the same order as their use in the biosynthetic pathway; they are translated from a single polycistronic mRNA molecule and are called *trpE, trpD, trpC, trpB,* and *trpA*. The *trpE* gene is the first one translated. Adjacent to the *trpE* gene are the promoter, the operator, and two regions called the **leader** and the **attenuator,** which are designated *trpL* and *trp a* (not *trp A*), respectively (Figure 14-7). The repressor gene *trpR* is located quite far from this gene cluster.

The regulatory protein of the repression system of the *trp* operon is the *trpR*-gene product. Mutations either in this gene or in the operator cause constitutive initiation of transcription of *trp* mRNA, as in the *lac* operon. This protein, which is called the *trp* **aporepressor,** does not bind to the operator unless tryptophan is present. The aporepressor and the tryptophan molecule join together to form the active *trp* repressor, which binds to the operator. The reaction scheme is:

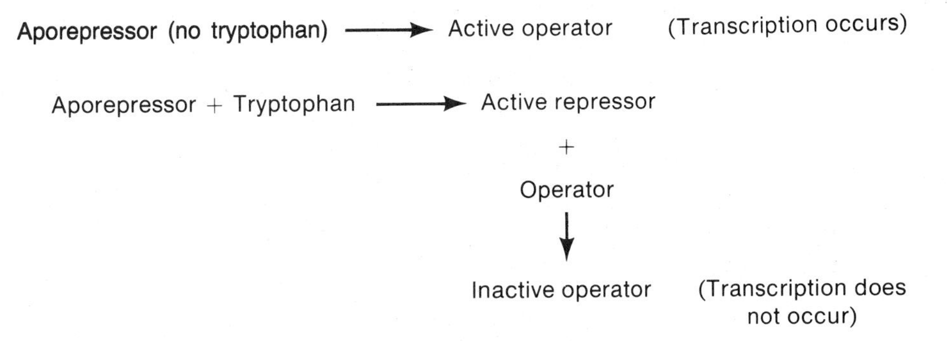

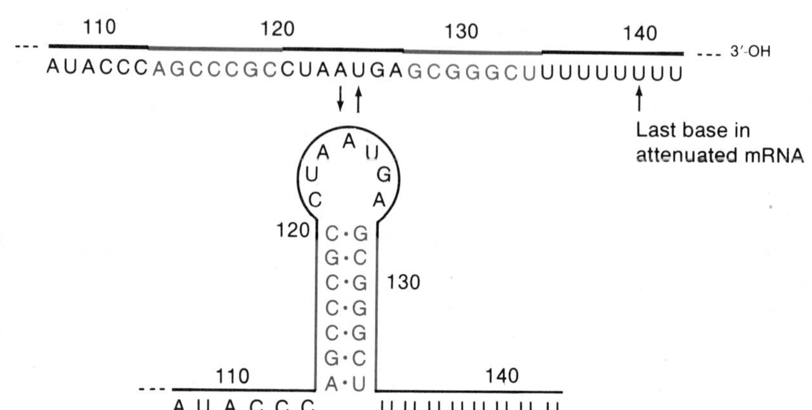

Figure 14-8
The terminal region (right end of *L* in Figure 14-19) of the *trp* leader mRNA. The base sequence given is extended past the termination site at position 140 to show the long stretch of U's. The red bases form an inverted repeat sequence that could lead to the stem-and-loop configuration shown (segment 3–4, Figure 14-10).

Thus, only when tryptophan is present does an active repressor molecule inhibit transcription. When the external supply of tryptophan is depleted (or reduced substantially), the equilibrium in the equation above shifts to the left, the operator is unoccupied, and transcription begins. This is the basic on-off regulatory mechanism.

In the on state a finer control, in which the enzyme concentration is varied by the amino acid concentration, is effected by (1) premature termination of transcription before the first structural gene is reached and (2) regulation of the frequency of this termination by the internal concentration of tryptophan. This modulation is accomplished in the following way.

A 162-base leader (noncoding) sequence is present at the 5′ end of the *trp* mRNA molecule. A mutant in which bases 123 through 150 are deleted synthesizes the *trp* enzymes in both derepressed cells and constitutive mutants at six times the normal rate, which indicates that bases 123–150 have regulatory activity. In nonmutant bacteria, after initiation of transcription most of the mRNA molecules terminate in this 28-base region, unless no tryptophan is present. The result of such termination is an RNA molecule that contains only 140 nucleotides and stops short of the genes encoding the *trp* enzymes. This 28-base region, in which termination occurs and is regulated, is called the **attenuator.** The base sequence (Figure 14-8) of the region in which termination occurs contains the usual features of a termination site—namely, a potential stem-and-loop configuration in the mRNA followed by a sequence of eight AT pairs (see Figure 9-8).

The leader sequence has several notable features:

1. An AUG codon and a later UGA stop codon in the same reading frame define a region encoding a polypeptide consisting of 14 amino acids—called the **leader polypeptide** (Figure 14-9).

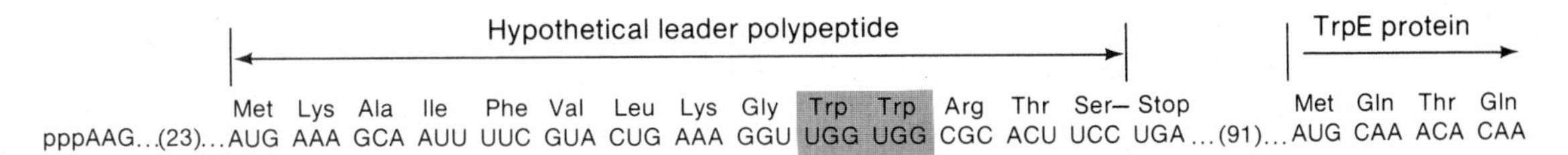

Figure 14-9
The sequence of the *trp* leader mRNA showing the hypothetical leader polypeptide, the two Trp codons (shaded red), and the beginning of the TrpE protein. The numbers (23 and 91) refer to the number of bases whose sequences are omitted for clarity.

2. Two adjacent tryptophan codons are located in the leader polypeptide at positions 10 and 11. We will see the significance of these repeated codons shortly.
3. Four segments of the leader RNA—denoted 1, 2, 3, and 4—are capable of base-pairing in two different ways—namely, forming either the base-paired regions 1–2 and 3–4 or just the region 2–3 (Figure 14-10). Two of these paired regions, 1–2 and 3–4, are also present in purified *trp* leader mRNA. The paired region 3–4 is in the terminator recognition region.

This arrangement enables premature termination to occur in the *trp* leader region by the following mechanism.

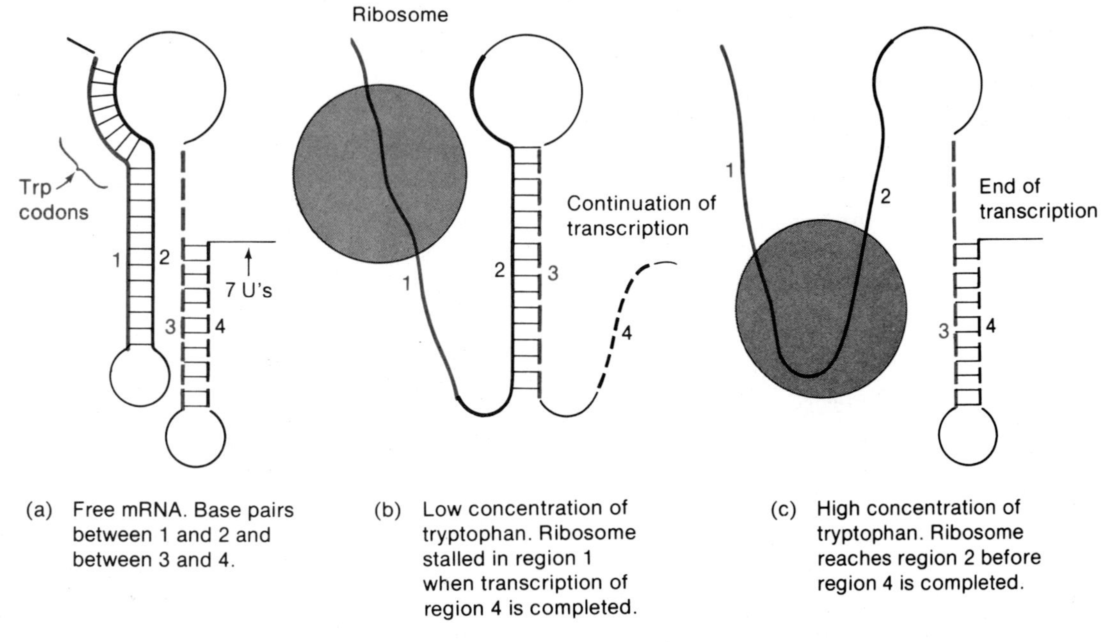

Figure 14-10
The accepted model for the mechanism of attenuation in the *E. coli trp* operon.

Termination of transcription is mediated through translation of the leader peptide region. Because there are two tryptophan codons in this sequence, the translation of the sequence is sensitive to the concentration of charged $tRNA^{Trp}$. That is, if the supply of tryptophan is inadequate, the amount of charged $tRNA^{Trp}$ will be insufficient and, hence, translation will be slowed at the tryptophan codons. Three points should be noted. (1) Transcription and translation are coupled, as is usually true in bacteria. (2) Since sequences 2 and 3 are paired in the duplex segments 1–2 and 3–4, then the region 2–3 cannot be present simultaneously with 1–2 and 3–4. (3) All base pairing is eliminated in the segment of the mRNA that is in contact with the ribosome.

Figure 14-10 shows that the end of the *trp* leader peptide is in segment 1. Usually a translating ribosome is in contact with about ten bases in the mRNA past the codons being translated. Thus, when the final codons of the leader are being translated, segments 1 and 2 are not paired. In a coupled transcription-translation system, the leading ribosome is not far behind the RNA polymerase. Thus, if the ribosome is in contact with segment 2 when synthesis of segment 4 is being completed, then segments 3 and 4 are free to form the duplex region 3–4 without segment 2 competing for segment 3. The presence of the 3–4 stem-and-loop configuration allows termination to occur when the terminating sequence of seven uridines is reached. If there is no added tryptophan, the concentration of charged $tRNA^{Trp}$ becomes inadequate and occasionally a translating ribosome is stalled for an instant at the tryptophan codons. These codons are located sixteen bases before the beginning of segment 2. Thus, segment 2 is free before segment 4 has been synthesized and the duplex 2–3 region can form, terminating translation. In the absence of the 3–4 stem and loop, termination does not occur and the complete mRNA molecule is made, including the coding sequences for the *trp* genes. Hence, if tryptophan is present in excess, termination occurs and little enzyme is synthesized; if tryptophan is absent, termination does not occur and the enzymes are made. At intermediate concentrations the fraction of initiation events that result in completion of *trp* mRNA will depend on how often translation is stalled, which in turn depends on the concentration of tryptophan.

Many operons responsible for amino acid biosynthesis (for example, the leucine, isoleucine, phenylalanine, and histidine operons) are regulated by attenuators equipped with the base-pairing mechanism for competition described for the *trp* operon. In the histidine operon, which also has an attenuator system (that is, prematurely terminated mRNA), a similar base sequence encodes a leader polypeptide having *seven* adjacent histidine codons (Figure 14-11(a)). In the phenylalanine operon seven phenylalanine codons are also present in the leader but they are divided into three groups (panel (b)).

(a)

	Met	Lys	His	Ile	Pro	Phe	Phe	Phe	Ala	Phe	Phe	Phe	Thr	Phe	Pro	Stop	
5′	AUG	AAA	CAC	AUA	CCG	UUU	UUC	UUC	GCA	UUC	UUU	UUU	ACC	UCC	CCC	UGA	3′

(b)

	Met	Thr	Arg	Val	Gln	Phe	Lys	His	His	His	His	His	His	His	Pro	Asp	
5′	AUG	ACA	CGC	GUU	CAA	UUU	AAA	CAC	CAC	CAU	CAU	CAC	CAU	CAU	CCU	GAC	3′

Figure 14-11
Amino acid sequence of the leader peptide and base sequence of the corresponding portion of mRNA from (a) the phenylalanine operon and (b) the histidine operon. The repeating amino acid is shaded in red.

A regulatory mechanism of this type cannot occur in eukaryotes because transcription and translation cannot be coupled; that is, transcription occurs in the nucleus and translation takes place in the cytoplasm.

AUTOREGULATION

Many proteins are made from transcripts that are initiated at a constant rate. However, with some gene products the requirements of a cell vary greatly and the rate of transcription of the corresponding gene matches the need. One mechanism for the regulation of synthesis of monocistronic mRNA is **autoregulation.** In the simplest autoregulated systems the gene product is also a repressor: it binds to an operator site adjacent to the promoter. When the concentration of the gene product exceeds what the cell can use, a product molecule occupies the operator and transcription will be inhibited. At a later time the need may be greater, molecules will be consumed, and the concentration of unbound molecules will decrease. In these conditions, the molecule bound to the operator will leave the site, the promoter will be free, and transcription will occur. The synthesis of most repressor proteins—for example, the *lacI* product and the phage λ immunity repressor—and many enzymes needed at all times are autoregulated.

FEEDBACK INHIBITION

If a culture of bacteria in which the *trp* operon has been derepressed is suddenly exposed to tryptophan, repression is rapidly established, and little further synthesis of the *trp* enzymes occurs. However, the *trp* enzymes persist and were it not for a second type of regulatory mechanism, wasteful synthesis of tryptophan and needless consumption of precursors and energy would occur. This state of affairs is avoided by

feedback inhibition, a mechanism by which the *activity* of the enzymes in a particular pathway is turned off by the product of the pathway.

The most common type of feedback inhibition is effected by an inhibition of the first enzymatic step of a biosynthetic pathway, caused by the product of the pathway. For example, in the pathway

$$A \xrightarrow{1} B \xrightarrow{2} C \xrightarrow{3} D$$

catalyzed by enzymes 1, 2, and 3, the product D would act on enzyme 1, which is clearly the most economical mode of inhibition.

Some biosynthetic pathways are responsible for the synthesis of two products from a common precursor. A hypothetical example of such a **branched pathway** is the following:

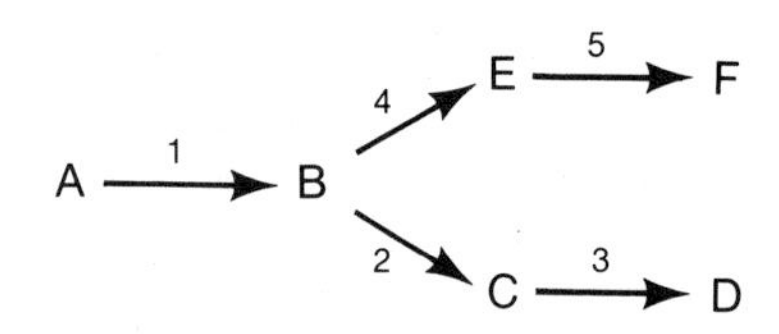

In this type of pathway it would be undesirable for a single product to inhibit enzyme 1 because both branches would be blocked. In general, the most economical kind of inhibition prevails—namely, D inhibits enzyme 2 and F inhibits enzyme 4. In this way, neither D nor F prevents the synthesis of the other. Conversion of A to B is wasteful if both D and F are present; therefore, in many branched pathways it is found that D and F together inhibit enzyme 1.

The elegance of feedback inhibition in a branched pathway is shown in Figure 14-12, which shows a well-studied multibranched pathway in which lysine, methionine, and threonine are synthesized from aspartate. Note how these amino acids inhibit enzymes 4, 6, and 8 immediately after the main branch in the pathway. Furthermore, three **isoenzymes** (different proteins that catalyze the same reaction)—1a, 1b, and 1c—are separately inhibited by lysine, homoserine, and threonine, respectively. Each of the three isoenzymes carries out the reaction at a different rate, each one synthesizing as much aspartyl phosphate as is needed to form the appropriate amount of the product that inhibits the isoenzyme. Thus, when an isoenzyme is inhibited, the amount of aspartyl phosphate required to synthesize the needed amino acids is still made.

Feedback inhibition occurs in both prokaryotes and eukaryotes but has been more fully described in the former. It can occur in a variety of

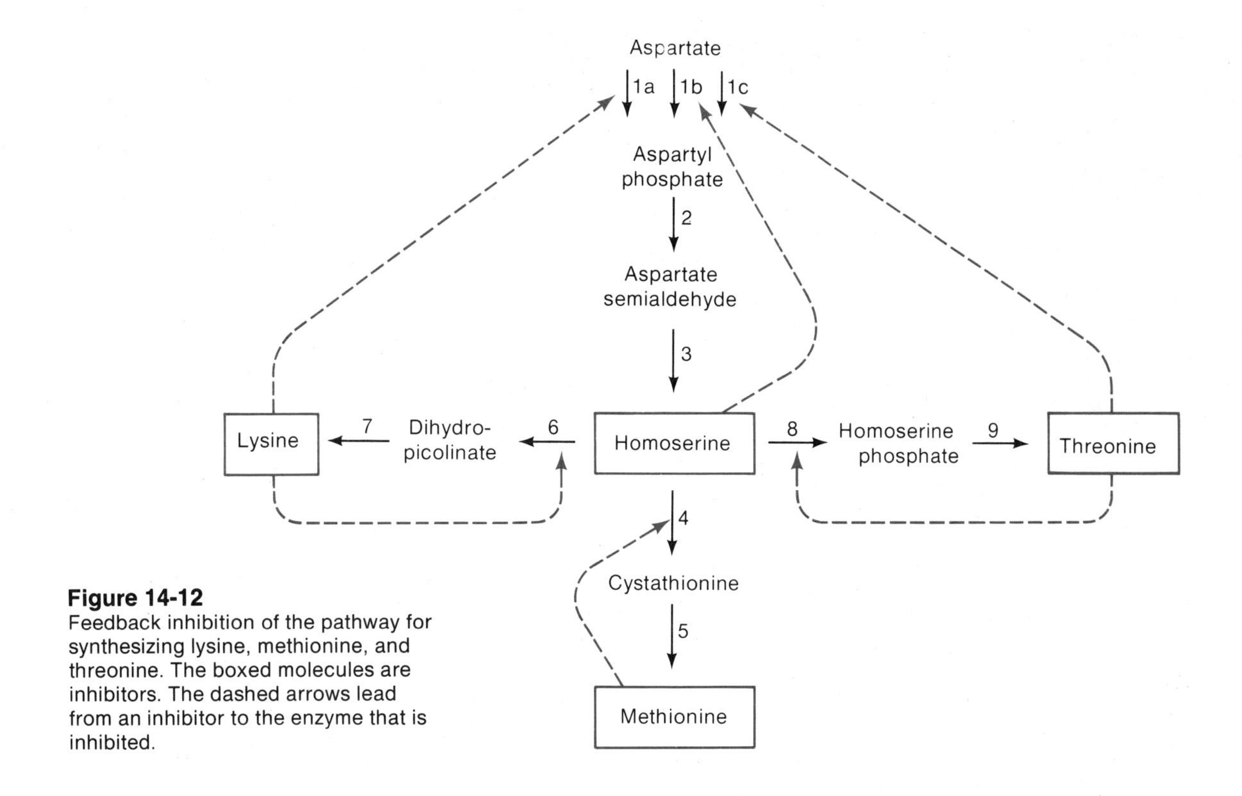

Figure 14-12
Feedback inhibition of the pathway for synthesizing lysine, methionine, and threonine. The boxed molecules are inhibitors. The dashed arrows lead from an inhibitor to the enzyme that is inhibited.

ways—for example, binding of the product, which is also the inhibitor, to the enzyme, causing a change in shape of the enzyme and thereby altering its catalytic properties.

15 Bacteriophages

Bacteriophages, or phages, have played an important role in the development of molecular biology. At the present time a few phages are the most completely understood of any organisms. Because of their lesser complexity than bacteria and higher cells and the availability of an enormous number of mutants, phages have been extraordinarily useful in the study of basic processes, such as replication, transcription, translation, and regulation. In a short book it is not possible to look into the molecular biology of phages in any depth. Instead, we shall examine some basic features of phage biology and provide a few examples of reproductive strategies by looking at particular stages of the life cycles of certain phages. For more information, consult *MB*, Chapter 15.

A bacteriophage is, first of all, a bacterial parasite. By itself, it can persist, but a phage can neither grow nor replicate except within a bacterial cell. Most phages possess genes encoding a variety of proteins. However, all known phages use the ribosomes, protein-synthesizing factors, amino acids, and energy-generating systems of the host cell.

Each phage must perform some minimal functions for continued survival. These are the following:

1. To protect its nucleic acid from environmental chemicals that could alter the molecule (for example, break the molecule or cause a mutation).

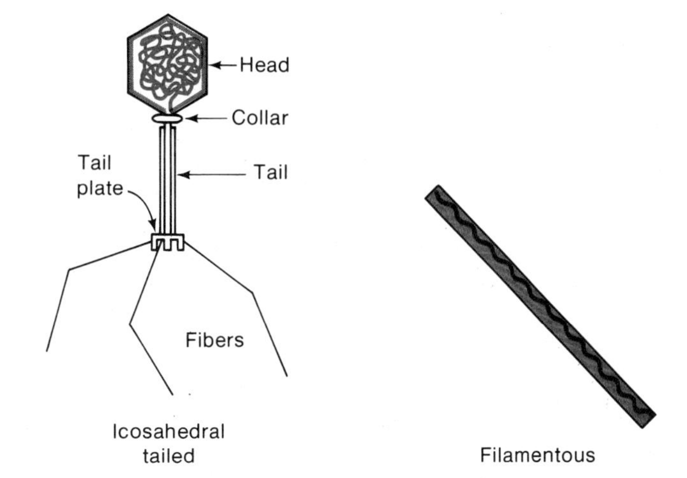

Figure 15-1
Two-dimensional view of three basic phage structures. The nucleic acid is shown in red.

2. To deliver its nucleic acid to the inside of a bacterium.
3. To convert an infected bacterium to a phage-producing system which yields a large number of progeny phage.
4. To release phage progeny from an infected bacterium.

These problems are solved in a variety of ways by different phage species.

Phage particles also differ in their physical structures from species to species and often certain features of their life cycles are correlated with their structure. The three basic structures are shown in Figure 15-1. The most common type of nucleic acid in phages is double-stranded linear DNA; however, double-stranded circular DNA, single-stranded linear and circular DNA, and single- and double-stranded linear RNA are also found (Table 15-1). The molecular weight of the nucleic acid also varies over a hundredfold range from one species to the next, which contrasts with bacteria, for which the variation is rarely more than about ten percent. Phages with larger nucleic acid molecules have more complex life cycles and are less dependent on bacterial enzymes for their reproduction. An unusual feature of the DNA of some phages is the presence of bases other than the standard A, T, G, C, as shown in the table. For example, T4 contains glucosylated 5-hydroxymethylcytosine instead of cytosine and SP01 has hydroxymethyluracil instead of thymine. The nucleic acid is always isolated from the environment, and thereby protected from harmful substances by an enclosing protein shell called either the **coat** or the **capsid.**

Table 15-1
Properties of the nucleic acid of several phages

Phage	*Host*	*DNA or RNA*	*Form*	*Molecular weight, $\times 10^6$*	*Unusual bases*
ϕX174	E	DNA	ss, circ	1.8	None
M13, fd, f1	E	DNA	ss, lin	2.1	None
PM2	PB	DNA	ds, circ	9	None
186	E	DNA	ds, lin	18	None
B3	PA	DNA	ds, lin	20	None
Mu	E	DNA	ds, lin	25	None
T7	E	DNA	ds, lin	26	None
λ	E	DNA	ds, lin	31	None
N4	E	DNA	ds, lin	40	None
P1	E	DNA	ds, lin	59	None
T5	E	DNA	ds, lin	75	None
SPO1	B	DNA	ds, lin	100	HMU for thymine
T2, T4, T6	E	DNA	ds, lin	108	Glucosylated HMC for cytosine
PBS1	B	DNA	ds, lin	200	Uracil for thymine
MS2, Qβ, f2	E	RNA	ss, lin	1.0	None
ϕ6	PP	RNA	ds, lin	2.3, 3.1, 5.0*	None

Note: Abbreviations used in this table are: E, *E. coli;* B, *Bacillus subtilis;* PA, *Pseudomonas aeruginosa;* PB, *Ps. aeruginosa BAL-31;* PP, *Ps. phaseolica;* ss, single-stranded; ds, double-stranded; circ, circular; lin, linear; HMU, hydroxymethyluracil; HMC, hydroxymethylcytosine.

*ϕ6 contains three molecules.

STAGES IN THE LYTIC LIFE CYCLE OF A TYPICAL PHAGE

Phage life cycles fit into two distinct categories—the **lytic** and the **lysogenic** cycles. A phage in the lytic cycle converts an infected cell to a phage factory, and many phage progeny are produced. A phage capable only of lytic growth is called **virulent.** The lysogenic cycle, which has been observed only with phages containing double-stranded DNA, is one in which no progeny particles are produced; the phage DNA usually becomes part of the bacterial chromosome. A phage capable of such a life cycle is called **temperate.*** In this section only the lytic cycle is outlined. The lysogenic cycle will be described later.

There are many variations in the details of the life cycles of different virulent phages. There is, however, what may be called a basic lytic cycle, which is the following (Figure 15-2):

1. *Adsorption of the phage to specific receptors on the bacterial surface.* These receptors are very varied and serve the bacteria for purposes other than phage adsorption.

*Most temperate phages also undergo lytic growth under certain circumstances.

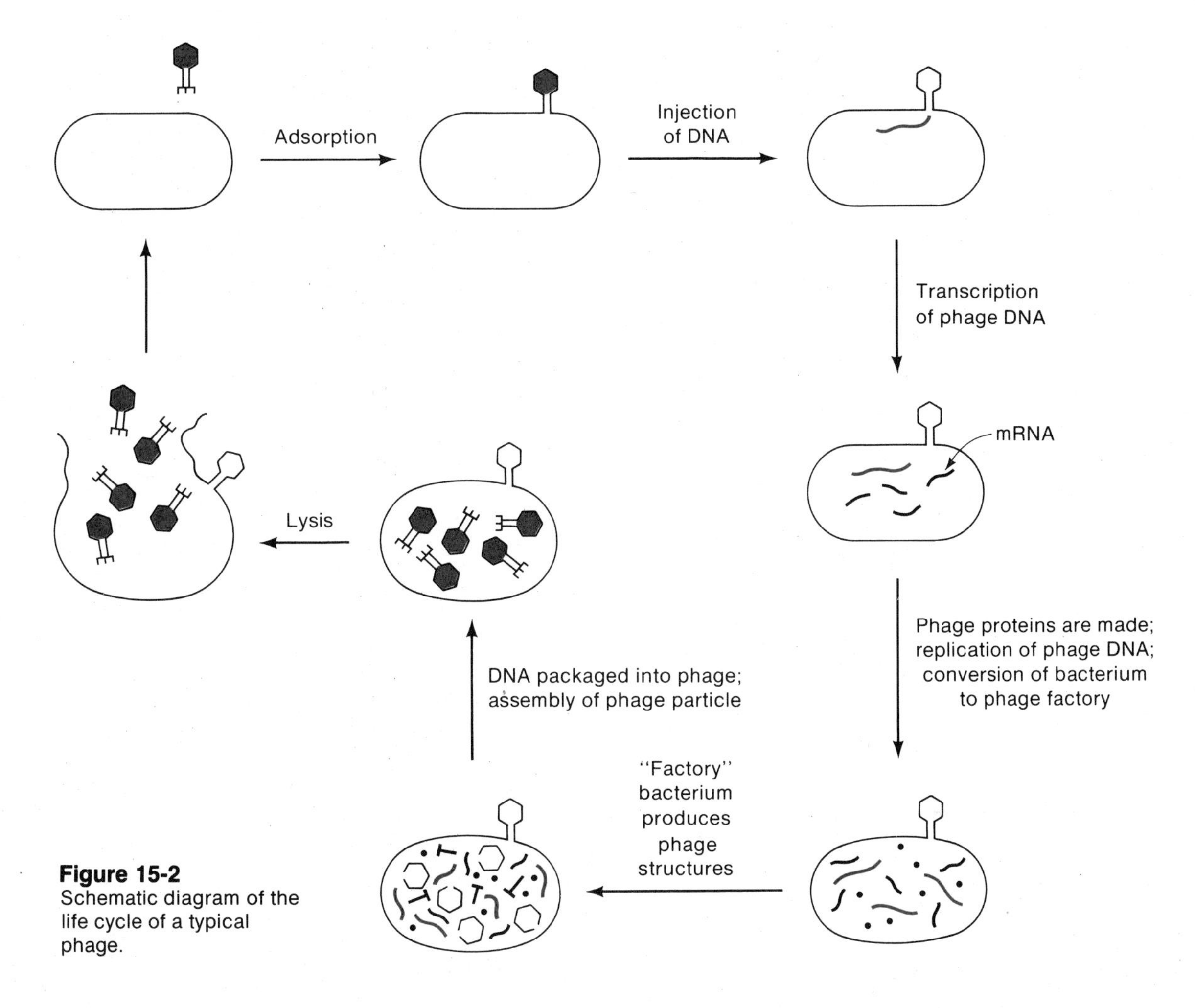

Figure 15-2
Schematic diagram of the life cycle of a typical phage.

2. *Passage of the DNA from the phage through the bacterial cell wall.* Some types of tailed phages use an injection sequence shown schematically in Figure 15-3.

3. *Conversion of the infected bacterium to a phage-producing cell.* Following infection by most phages a bacterium loses the ability either to replicate or to transcribe its DNA; sometimes it loses both. This shutdown of host DNA or RNA synthesis is accomplished in many different ways depending on the phage species.

4. *Production of phage nucleic acid and proteins.* By several mechanisms the phage directs the synthesis of a replicative system that specifically makes copies of phage nucleic acid. This programming is

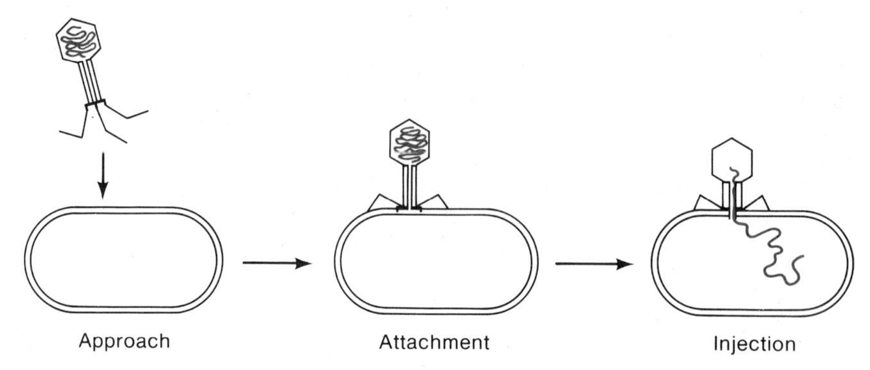

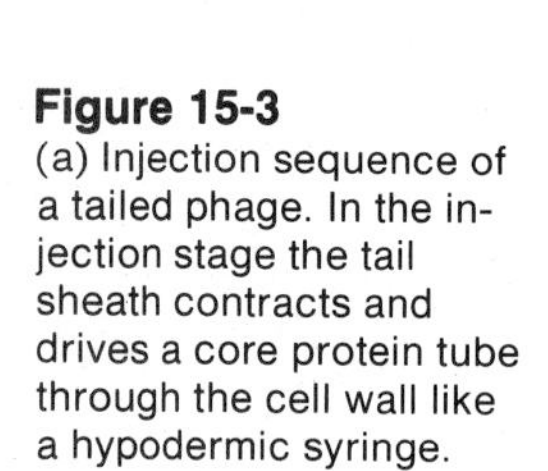

Figure 15-3
(a) Injection sequence of a tailed phage. In the injection stage the tail sheath contracts and drives a core protein tube through the cell wall like a hypodermic syringe.

accomplished either by synthesis of phage-specific polymerases or by addition of specificity elements to bacterial enzymes. Transcription is almost always initiated by the bacterial RNA polymerase but after the first transcription event either the bacterial polymerase is modified to recognize phage promoters or a phage-specific RNA polymerase is synthesized. Transcription is regulated and phage proteins are synthesized sequentially in time as they are needed. Distinct classes of mRNA are made at various times after infection; the major division is between "early" mRNA and "late" mRNA. Early mRNA usually encodes the enzymes required for takeover of the bacterium, for DNA replication, and for the synthesis of late mRNA. Late mRNA encodes the components of the phage particle, proteins needed for packaging of nucleic acid in the particles, and enzymes required to break open the bacterium. The more complex phages usually synthesize several classes of early mRNA, which are made in a particular time sequence.

5. *Assembly of phage particles.* This process is often called **morphogenesis.** Two types of proteins are needed for the assembly process: **structural proteins,** which are present (possibly in modified form) in the phage particle, and **catalytic proteins,** which participate in the assembly process but do not become part of the phage particle. A subset of the latter class consists of the maturation proteins, which convert intracellular phage DNA to a form appropriate for packaging in the phage particle. Usually 50–1000 particles are produced, the number depending on the phage species.

6. *Release of newly synthesized phage.* With most phages a phage protein called a **lysozyme** or an **endolysin** is synthesized late in the cycle of infection. This protein causes disruption of the cell wall. Other proteins called **membranases** dissolve the cell membrane and together with the lysozyme cause total destruction of the cell, so that phage are

released to the surrounding medium. This disruptive process is called **lysis.**

The events just described occur in an orderly sequence, as exemplified by the life cycle of *E. coli* phage T4 listed below (times in minutes at 37°C):

$t = 0$	Phage adsorbs to bacterial cell wall. Injection of phage DNA probably occurs within seconds of adsorption.
$t = 1$	Synthesis of host DNA, RNA, and protein is totally turned off.
$t = 2$	Synthesis of first mRNA begins.
$t = 3$	Degradation of bacterial DNA begins.
$t = 5$	Phage DNA synthesis is initiated.
$t = 9$	Synthesis of "late" mRNA begins.
$t = 12$	Completed heads and tails appear.
$t = 15$	First complete phage particle appears.
$t = 22$	Lysis of bacteria; release of about 300 progeny phage.

The total duration of the cycle is typical, most phages having a life cycle of 20–60 minutes, which is comparable to the generation times of most bacteria. The life cycles of animal viruses are considerably longer—24 to 48 hours—but this is again comparable to the life cycle of an animal cell.

SPECIFIC PHAGES

On a biological scale of complexity, a phage is a relatively simple form of life. However, phages are sufficiently complex that there is not a single phage type for which all of the molecular details of its life cycle are completely understood. The major progress has been with a small number of phage species that grow in *E. coli, Bacillus subtilis,* and *Salmonella typhimurium.* Special features of each phage have resulted in particular phages being more suited to the study of certain processes. For example, regulation of transcription is best understood in phages λ and T7, the study of morphogenesis has been more successful with phages T4 and λ than with other phages, phages P1 and P22 gave the first clues to understanding transduction, T5 has yielded the best information about DNA injection, T4 has been most profitable in the study of DNA synthesis, and the study of T4 and T7 has shown clearly how a phage takes over a bacterium.

In the following sections several well-understood features of a few phages will be described.

E. Coli Phage T4

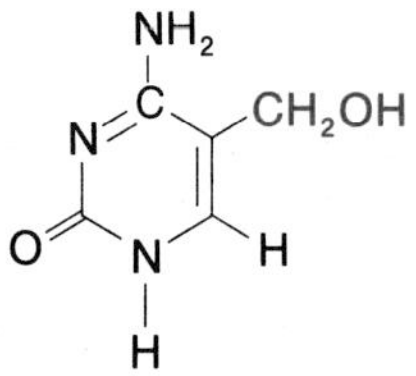

5-Hydroxymethylcytosine (HMC)

Figure 15-4
5-Hydroxymethylcytosine. If the CH_2OH (red) in HMC were replaced by hydrogen, the molecule would be cytosine.

E. coli phage T4 is a well-studied phage with an unusual DNA that lacks cytosine; instead, there is a modified form of cytosine that is called **5-hydroxymethylcytosine (HMC)**, which base-pairs with guanine (Figure 15-4). This base is further modified by glucosylation—that is, sugar is coupled to the OH group of HMC. This has the consequence that purified T4 DNA is somewhat resistant to a variety of DNases.

HMC has an important function in the T4 life cycle. T4 is a rapidly multiplying phage, producing about 500 progeny per infected cell in 22 minutes. The high rate of T4 DNA synthesis in infected *E. coli*, which is greater than that in an uninfected cell, is accomplished in two ways: (1) the number of enzymes needed to make the DNA precursor nucleoside triphosphates is increased by synthesizing phage-specified enzymes and (2) *E. coli* DNA is degraded by phage-encoded nucleases to mononucleotides, which can be converted to triphosphates. The phage-specified DNases could not distinguish T4 DNA from *E. coli* DNA were it not for the presence of the HMC in the phage DNA—HMC-containing DNA is resistant to the enzymes.

E. coli does not possess enzymes for forming HMC. This synthesis is instead accomplished by two phage enzymes, which convert dCMP to dHDP:

$$\text{dCMP} \xrightarrow{\text{T4 hydroxymethylase}} \text{dHMP}$$

$$\text{dHMP} \xrightarrow{\text{HM kinase}} \text{dHDP}$$

The *E. coli* enzyme, nucleoside phosphate kinase, which forms all triphosphates in *E. coli*, then converts dHDP to dHTP, the immediate precursor of the HMC in the DNA. These pathways are shown in Figure 15-5.

When *E. coli* DNA is degraded, dCMP is produced. Whereas most of this is converted to dHMP, as shown above, some will be converted to dCTP, which could be incorporated into progeny T4 DNA. This would yield a T4 DNA molecule that could be attacked by the phage-encoded nucleases that degrade *E. coli* DNA. Thus, a mechanism exists for preventing such incorporation: a phage-specified enzyme, dCTPase, degrades any dCDP and dCTP that is formed, forming dCMP again (Figure 15-5). Ultimately, all of the dCMP is in the form of dHMP.

HMC creates another problem for T4, since *E. coli* possesses an endonuclease that attacks certain sequences of nucleotides containing HMC. To avoid this damage the HMC residues in T4 DNA are glucosylated. This is accomplished by two phage enzymes, α-glucosyl transferase (αgt) and β-glucosyl transferase (βgt), each of which transfers a glucose from uridine diphosphoglucose (UDPG) to HMC that

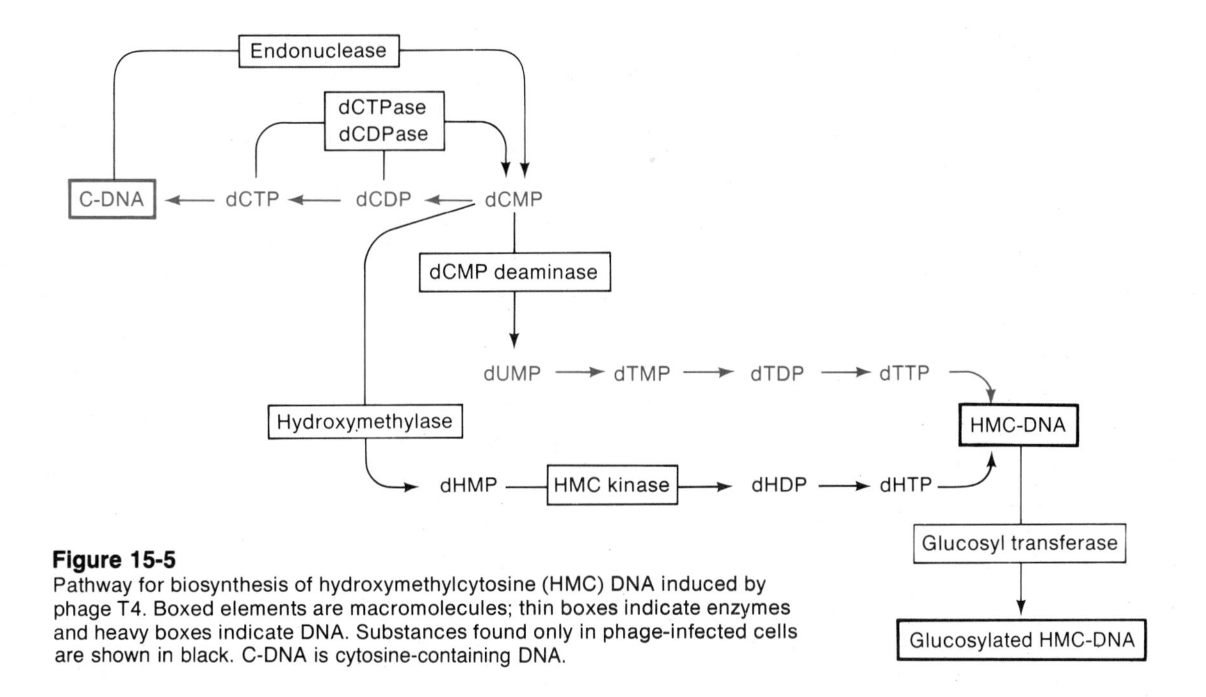

Figure 15-5
Pathway for biosynthesis of hydroxymethylcytosine (HMC) DNA induced by phage T4. Boxed elements are macromolecules; thin boxes indicate enzymes and heavy boxes indicate DNA. Substances found only in phage-infected cells are shown in black. C-DNA is cytosine-containing DNA.

is already in DNA. Thus glucosylation is a postreplicative modification. The *E. coli* endonuclease is inactive against glucosylated DNA; therefore, glucosylation is a protective device. A simple genetic experiment shows that this is the only essential function of glucosylation. A T4 αgt^- mutant cannot carry out glucosylation, so its newly synthesized DNA is destroyed by the *E. coli* HMC endonuclease. However, if an *E. coli* mutant *(rglB⁻)*, which lacks this nuclease, is used as a host bacterium, both T4 αgt^- and T4 βgt^- mutants grow normally even though nonglucosylated DNA is produced.

Another property of T4 DNA is that it is terminally redundant (Figure 15-6)—that is, a sequence of bases (about one percent of the total) is repeated at both ends of the molecule. Terminal redundancy is a feature of the DNA of many phage species and can be generated in several ways (see below).

Another feature of T4 DNA is that although each phage particle contains one DNA molecule, the molecules differ from phage to phage,

5′
A B C D E F G ... W X Y Z
A′B′C′D′E′F′G′ ... W′X′Y′Z′
3′

Figure 15-6
A terminally redundant molecule

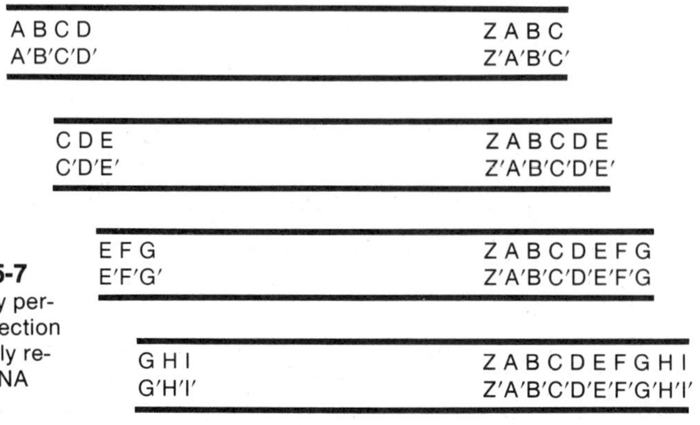

A B C D Z A B C
A′B′C′D′ Z′A′B′C′

C D E Z A B C D E
C′D′E′ Z′A′B′C′D′E′

E F G Z A B C D E F G
E′F′G′ Z′A′B′C′D′E′F′G

G H I Z A B C D E F G H I
G′H′I′ Z′A′B′C′D′E′F′G′H′I′

Figure 15-7
A cyclically permuted collection of terminally redundant DNA molecules.

even if the population was produced by replication of a single phage particle and its progeny. A sample of T4 DNA molecules is **cyclically permuted,** the meaning of which is shown in Figure 15-7. The figure indicates schematically that the termini of the DNA molecules in the population can be found at many different bases within an overall sequence—in reality, probably at any point in the base sequence. Note that cyclic permutation is a property of a phage population, whereas terminal redundancy is a property of an individual phage DNA molecule. With T4 both terminal redundancy and cyclic permutation are a consequence of the mechanism by which DNA is packaged into the phage head, as next described.

The mechanism of packaging of T4 DNA in a phage head is not yet elucidated, though the process can be carried out *in vitro*. The basic problem is how the long strand of DNA is tightly folded so that it will fit in the phage head. It is thought that packaging begins by the attachment of one end of a DNA molecule to a protein contained in the phage head. Then either a condensation protein or a small, very basic molecule (such as a polyamine) induces folding (Figure 15-8). The best-under-

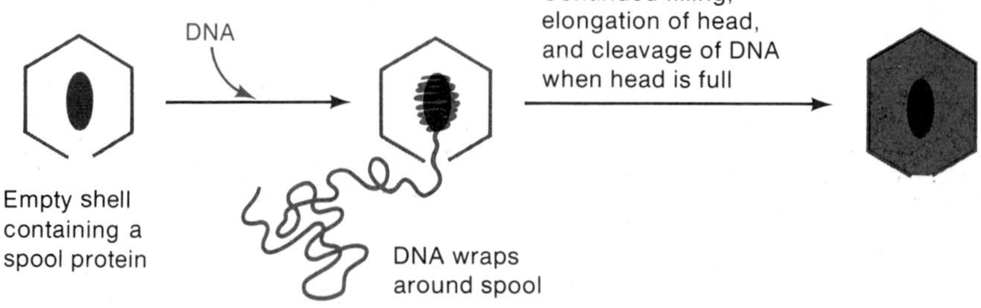

DNA
Empty shell containing a spool protein
DNA wraps around spool
Continued filling, elongation of head, and cleavage of DNA when head is full

Figure 15-8
Proposed model for filling a T4 head. Cleavage and rearrangement of head proteins is known to occur at several stages of the process.

stood feature of the process is that the molecule is cut from long concatemers. The cuts are not made in unique base sequences in the DNA because, if they were, T4 DNA could not be cyclically permuted. Instead, the cuts are made at positions that are determined by the amount of DNA that can fit in a head. Presumably a free end of the DNA molecule enters the head and this continues until there is no more room; then the concatemer is cut. This is known as the **headful mechanism** and it explains how both terminal redundancy and cyclic permutation arise (Figure 15-9). The essential point is that the DNA content of a T4 particle is greater than the length of DNA required to encode the T4 proteins. Thus, when cutting a headful from a concatemeric molecule, the final segment of DNA that is packaged is a duplicate of the DNA that is packaged first—that is, the packaged DNA is terminally redundant. The first segment of the second DNA molecule that is packaged is not the same as the first segment of the first phage. Furthermore, since the second phage must also be terminally redundant, a third phage-DNA molecule must begin with still another segment. Thus, the collection of DNA molecules in the phage produced by a single infected bacterium is a cyclically permuted set.

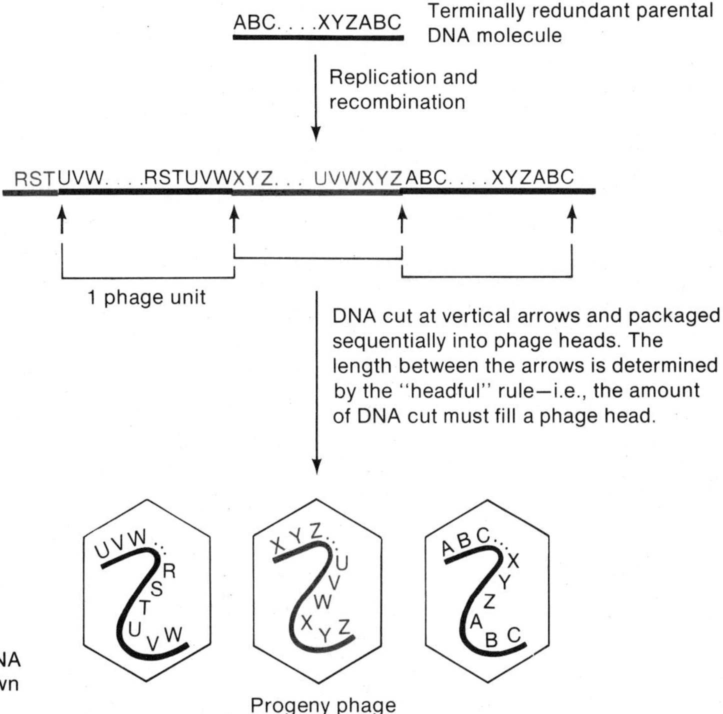

Figure 15-9
Origin of cyclically permuted T4 DNA molecules. Alternate units are shown in different colors for clarity only.

E. Coli Phage T7

Figure 15-10
Electron micrograph of phage T7. Note the very short tails. (Courtesy of Robley Williams.)

Phage T7 is a midsized phage (Table 15-1). Its DNA has a molecular weight of about 26×10^6 and contains equal amounts of adenine, thymine, guanine, and cytosine, and has no unusual bases. It has a terminal redundancy of 160 base pairs but it is not cyclically permuted. The DNA molecule is encased in a head that is attached to a very short tail (Figure 15-10).

T7 has been of interest for many years because it has a somewhat small number (49–50) of genes, which simplifies analysis of its life cycle. Furthermore, most of the gene products have been identified by gel electrophoresis, 42 have been purified (by early 1984), and many have been sequenced by chemical means. Of great importance is the recent elucidation of the complete sequence of 39,930 base pairs of T7 DNA; this information allows one to locate, by direct inspection, all of the promoters, termination sites, regulatory sites, spacers, leaders, initiation codons, termination codons, and genes.

A great deal is known about the molecular biology of T7. However, we shall only discuss how it regulates its transcription, and this only in barest essentials. The life cycle of T7 can be divided into three transcriptional stages, as shown in Figure 15-11. Transcription is temporally regulated, as shown in the figure, with transcript I made first. A basic principle of transcription of phage DNA is that *the first transcript must be made by the host RNA polymerase*, as no other enzyme is available; this is true of transcript I. One of the gene products in transcript I is a protein kinase that phosphorylates and thereby

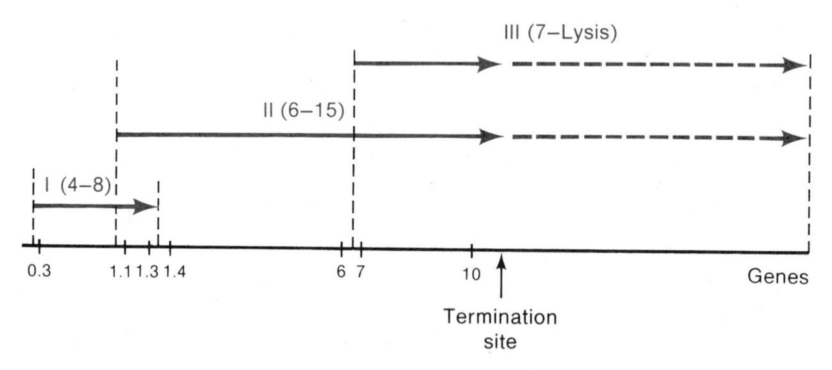

Figure 15-11
A simplified transcription map of T7. The DNA and gene numbers are in black; the three classes of transcripts are in red. The red numbers indicate the time interval (in minutes at 37°C) during which the transcript is made, the termination site indicated by the black arrow is about 90 percent efficient; thus 10 percent of the class II and III molecules are extended further, as indicated by the dashed lines.

inactivates *E. coli* RNA polymerase; this is the first stage in takeover of the bacterium by T7, though the inactivation is not complete. A second protein made from transcript I is T7 RNA polymerase; this enzyme is essential for further transcription since it alone (not *E. coli* RNA polymerase) recognizes the promoters for transcripts II and III. T7 RNA polymerase does not recognize the termination site for transcript I and hence allows polymerization to proceed past this site. Transcript II encodes the replication proteins, the major protein from which the phage head is constructed, and a second inhibitor of *E. coli* RNA polymerase. Synthesis of the latter protein completes the takeover of the bacterium and ensures that no energy is wasted making unnecessary bacterial RNA. A termination site for transcript II (black arrow in the figure) causes termination with 90 percent efficiency, meaning that 10 percent of the transcripts proceed on to the end of the DNA molecule. The extended region encodes the tail proteins. Since the tail is very short (Figure 15-10), less tail protein is needed than head protein; the extension of transcript II in 10 percent of the initiation events maintains the appropriate ratio of head and tail proteins. The lysis proteins are also included in the extended segment, but their concentration is not high enough to cause lysis during the period in which transcript II is being made. The DNA replication proteins are enzymes and hence are needed in very small amounts compared to the head and tail proteins. Therefore, late in the infection there is no need to make these enzymes; hence, at this time transcript III, which encodes all of the structural proteins and the lysis enzymes, but not the replication proteins, is made. Two properties of T7 are responsible for the reduction in the synthesis of transcript II and the initiation of transcript III: synthesis of a general inhibitor of transcription, and slow injection of T7 DNA. The promoter for transcript II (*pII*) is a very weak promoter compared to that for transcript III (*pIII*). One of the proteins encoded in transcript II causes an overall reduction in all transcription (the mechanism is unknown); when this protein is made, transcription is initiated only from the strong promoter. However, since T7 RNA polymerase can act at both *pII* and *pIII*, a mechanism is needed to prevent activation of *pIII* in the interval when the DNA replication enzymes must be synthesized. The mechanism is one that has not been observed very often with phages. Most phage species inject their DNA within about 30 seconds of adsorption, whereas with T7 it takes about 10 minutes. Thus, *pIII* is not injected until about seven minutes after infection, which delays synthesis of transcript III.

Let us now summarize how the three transcripts are made in an orderly way. At first, *E. coli* RNA polymerase makes transcript I. An inhibitor, translated from transcript I, prevents further synthesis of transcript I. T7 RNA polymerase, translated from transcript I, enables transcript II to be made. An inhibitor, translated from transcript II, prevents further synthesis of this transcript. Slow injection delays entry of *pIII* into the cell; however, once *pIII* is available, transcript III is made. The net result is early synthesis of the DNA replication proteins and late synthesis of the structural proteins of the phage particle.

E. Coli Phage ϕX174

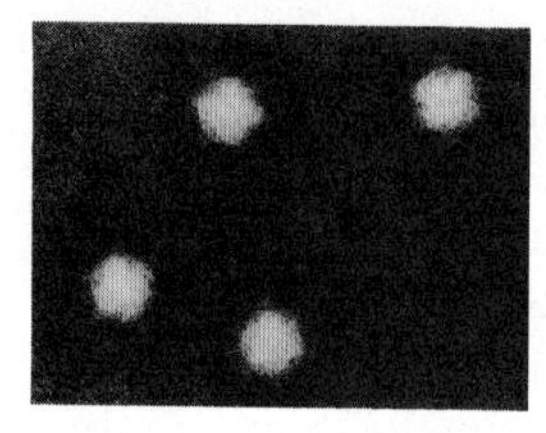

Study of *E. coli* phage ϕX174 has been important for many reasons: it was the first organism to be discovered containing single-stranded DNA as its genetic material and having DNA in a circular form. Its DNA was also the first to be replicated *in vitro* to yield biologically active DNA, and study of its mode of replication led to the hypothesis of rolling-circle replication. The phage is one of the smallest known (Table 15-1), contains 11 genes, and has a fairly simple life cycle. Each particle contains one circular, single-stranded DNA molecule consisting of 5386 nucleotides, whose base sequence is known. This strand is known as the (+) strand. The mode of DNA replication of ϕX174 DNA is clearly different from that of double-stranded DNA, and it is that which will concern us here.

The mechanism by which a base sequence is transmitted to daughter molecules is complementary base-pairing; thus, at some stage of the replication cycle of ϕX174 a DNA single strand (a (−) strand) must be synthesized whose sequence is complementary to the (+) strand. There are two ways that such a molecule could be produced: (1) a complementary strand could be synthesized and removed from the (+) strand as it is being made, like RNA transcribed from DNA, or (2) the circular single strand could be converted to a circular double-stranded molecule. Mechanism 2 is the one used by ϕX174, though some phages containing linear single-stranded DNA use the first mechanism.

Following infection of *E. coli* with ϕX174 and entry of the DNA in the cell, the first stage of replication occurs: conversion of the phage DNA molecule to a double-stranded supercoil by the *E. coli* replication system. Conversion occurs in four steps: synthesis of an RNA primer, extension of the primer by DNA polymerase III until replication of the circle is complete, removal of the RNA nucleotides and replacement with DNA nucleotides by polymerase I, and final sealing by DNA ligase. This supercoiled intermediate is known as RFI (replicating form I). In the second stage of replication RFI replicates many times to produce about 60 circular double-stranded molecules per cell. The second stage

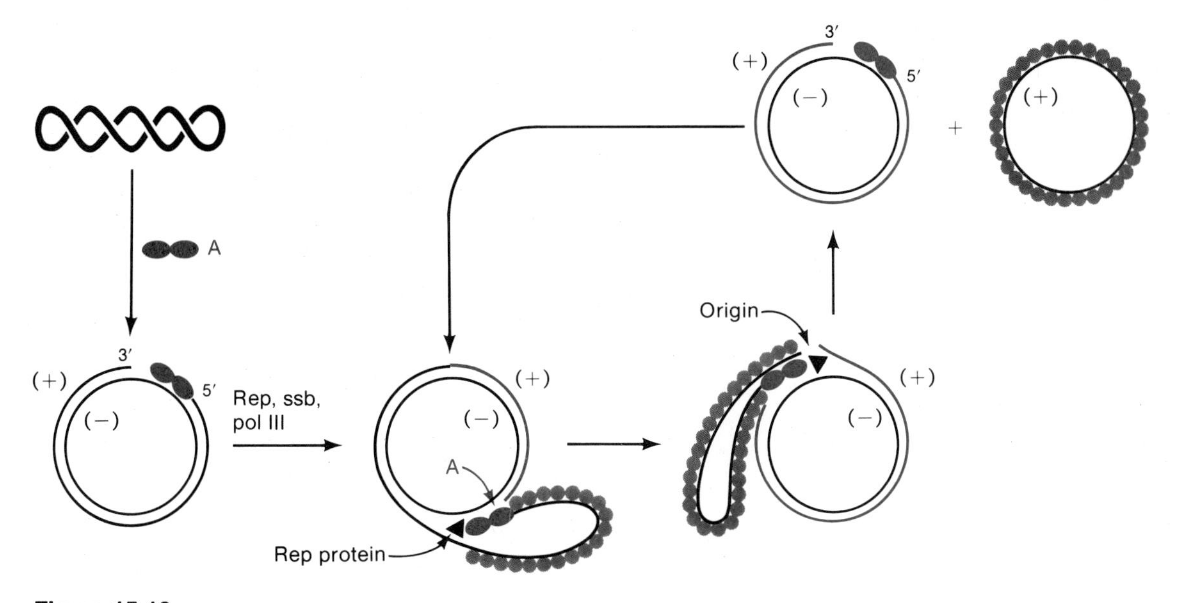

Figure 15-12
A diagram of looped rolling-circle replication of phage ϕX174. The gene-*A* protein nicks a supercoil and binds to the 5′ terminus of a strand (known as the (+) strand) whose base sequence is the same as that of the DNA in the phage particle. Rolling circle replication ensues to generate a daughter strand (red) and a displaced (+) single strand that is coated with ssb protein and still covalently linked to the A protein. When the entire (+) strand is displaced, it is cleaved from the daughter (+) strand and circularized by the joining activity of the A protein. The cycle is ready to begin anew. Note that the (−) strand is never cleaved.

of ϕX174 replication—synthesis of (+) strands for packaging in the phage coat—then occurs.

A variant of the rolling circle mode, called **looped rolling-circle replication,** is used to generate a progeny single-stranded circle from a double-stranded circular template. For phage ϕX174 this is accomplished in the following way (Figure 15-12). A phage protein called the gene-*A* protein makes a nick at the origin and in the process becomes covalently linked to the newly formed 5′-P terminus. Using the *E. coli* proteins–Rep, ssb, and polymerase III—chain growth occurs from the 3′-OH group, displacing the broken parental strand (which we call the (+) strand). This strand becomes coated with ssb protein and is not replicated. Leading strand synthesis continues until the origin is reached. At this point, the *A*-gene product binds to the 3′-OH group of the (+) strand and, using the energy obtained from the original nicking event, it joins the 3′-OH and 5′-P groups of the (+) strand, dissociates, and attaches to the newly synthesized (+) strand. This process can continue indefinitely, generating numerous circular (+) strands. Note that in looped rolling-circle replication, the displaced strand never

Table 15-2
Replication cycle of ϕX174

Stage	Time, minutes at 33°C	Event
ss → RF	0–1	Adsorption and penetration; viral single strand converted to parental RF; RF transcribed
RF → RF	1–20	Parental RF replicates and 60 progeny RF are formed
	25	Replication of RF molecules and host DNA stops
RF → ss	20–30	About 35 rolling circles form from progeny RF; from these, about 500 ss molecules are produced; DNA is inserted into phage particles
	40	Lysis occurs

Note: Abbreviations used are: ss, single-stranded phage DNA; RF, circular replication form.

exceeds the length of the circle, in contrast with ordinary σ replication. A summary of the replication cycle, with the timing of each stage, is given in Table 15-2.

All transcription of ϕX174 DNA is from the (−) strand and hence does not occur until the first RFI is made. Therefore, synthesis of RFI must utilize host enzymes only, as is the case. Details about transcription and about packaging of DNA into the phage head, both of which are well-understood phenomena, can be found in *MB*, pages 653–655.

E. Coli Phage λ: the Lytic Cycle

E. coli phage λ (Figure 15-13) may be the best understood of the double-stranded DNA phages, especially with respect to the elegant way in which transciption is regulated. The study of λ has made many important contributions to molecular biology, such as the discovery of bidirectional replication, transcription termination and the Rho protein, antitermination proteins, DNA ligase, DNA gyrase, site-specific recombination, circular DNA molecules, break-and-rejoin genetic recombination, and SOS repair, to name just a few.

λ is of particular interest because it can engage in both a lytic cycle, in which progeny phage are produced, and a lysogenic cycle, in which phage DNA becomes incorporated into the bacterial chromosome. Special features of the regulatory systems of λ, not found with all phages, determine the pathway that is used in any given infection. In

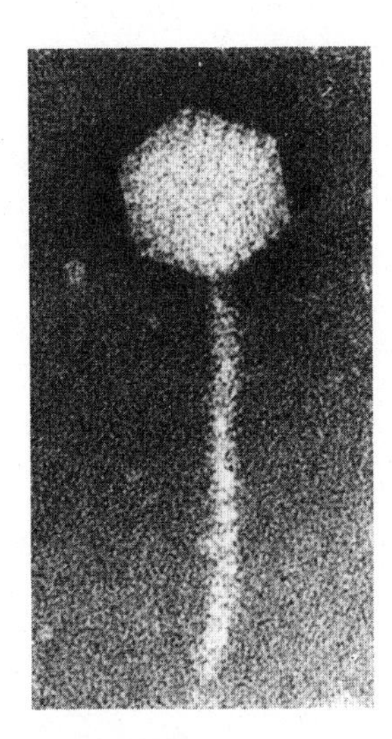

Figure 15-13
Electron micrograph of phage λ. (Courtesy of Robley Williams.)

this section, we will discuss significant features of the lytic cycle. The lysogenic cycle is described in the next section.

The DNA of λ has two components: a double-stranded segment containing 48,489 nucleotide pairs of known sequence, and two 12-nucleotide single-strand extensions at the 5′-P terminus of each strand. The two terminal single strands have complementary base sequences and are called **cohesive ends.** Immediately following injection of the DNA the cohesive ends base-pair, yielding a circle (Figure 15-14). The adjacent 5′-P and 3′-OH groups are quickly sealed by DNA ligase and then the circle is supercircled.

The arrangement of the genes of phage λ is shown in Figure 15-15. As is the case for many phages, the genes are clustered according to function. For example, the head, tail, replication, and recombination genes form four distinct clusters. Many λ proteins—for example, regulatory proteins and those responsible for DNA synthesis—act at particular sites in the DNA. In general these proteins are situated adjacent to their sites of action (when there is a single site)—for instance, the origin of DNA replication lies within the coding sequence for gene *O*, which encodes a DNA replication-initiation protein.

Transcription of λ occurs in several stages, with gene clustering allowing synthesis of related genes at the appropriate time. Similar to

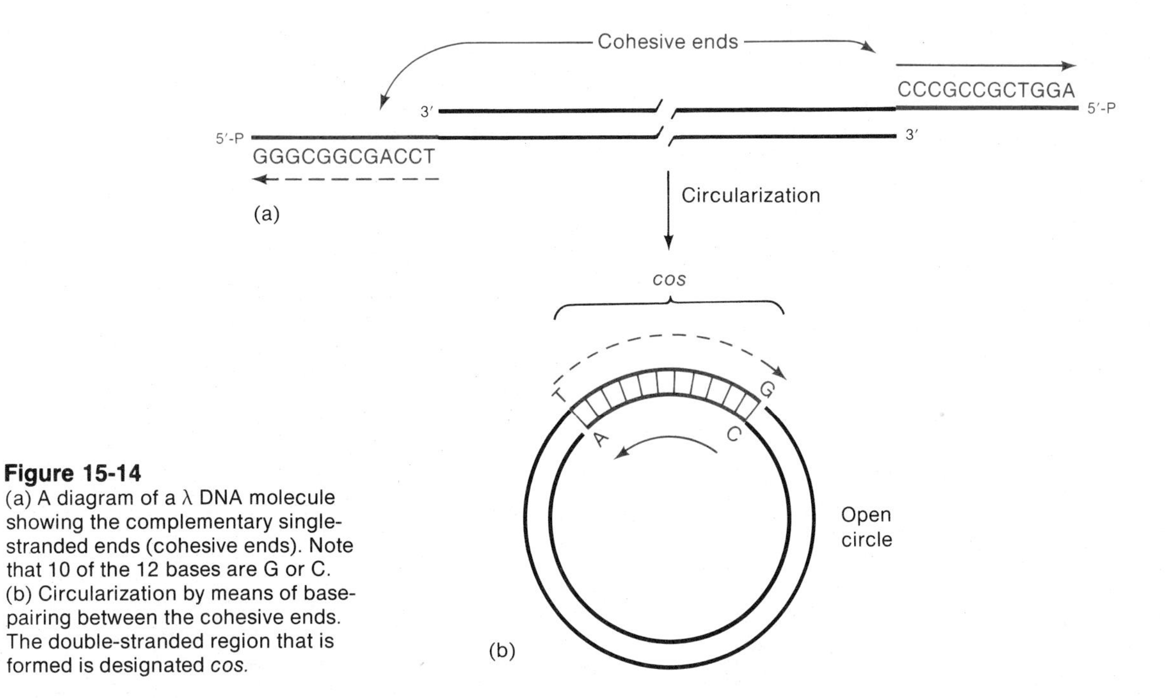

Figure 15-14
(a) A diagram of a λ DNA molecule showing the complementary single-stranded ends (cohesive ends). Note that 10 of the 12 bases are G or C. (b) Circularization by means of base-pairing between the cohesive ends. The double-stranded region that is formed is designated *cos*.

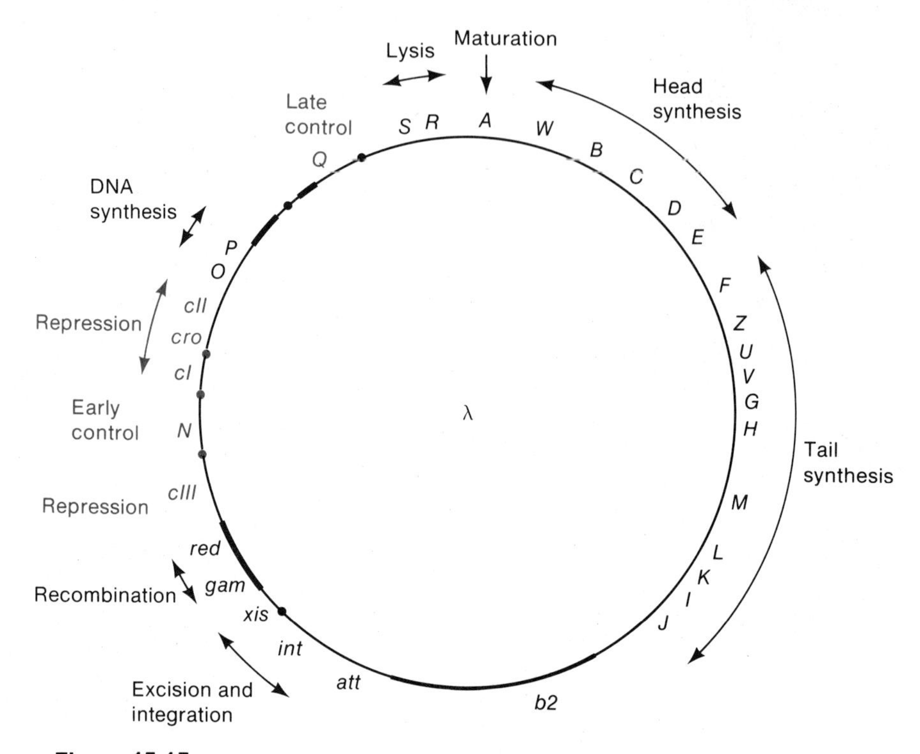

Figure 15-15
Genetic map of phage λ. Regulatory genes and functions are given in red. All genes are not shown. Major regulatory sites are indicated by black solid circles. Regions non-essential for both the lytic and lysogenic cycles are denoted by a heavy line.

T7, λ makes an early mRNA, which encodes regulatory proteins and DNA replication enzymes, and a late mRNA, which encodes the structural proteins of the phage particle. In contrast with T7, *E. coli* RNA polymerase is used for all transcription events. With most phages (and those discussed so far) transcription is regulated by controlling the availability of particular promoters. For example, T7 early mRNA is made by *E. coli* RNA polymerase, which does not recognize the late promoters, and T7 RNA polymerase functions at the late, but not the early, promoter. Other phages—for example, T4—use the host polymerase throughout the life cycle but modify the enzyme by the addition of accessory molecules that enable it to recognize other promoters. Control of promoter availability is used in the lysogenic cycle of λ, but in the lytic cycle a totally different system is used. With λ all promoters are recognized by the host enzyme, but modification of the enzyme determines whether particular transcription-*termination* sites are recognized. That is, unmodified RNA polymerase initiates at a promoter and stops at a particular termination site, but the modified enzyme passes the termination site and stops at a later site. This mechanism produces a temporal delay in the production of particular proteins in the following way. Consider a DNA segment containing the sequence *p–A–B–t1–C–t2*, in which *p* is a promoter, *ABC* are genes, and *t1* and *t2* are termination sites. Initially, RNA polymerase will attach to *p* and transcribe *A* and *B*. Let us assume that *B* encodes a regulatory protein that can bind to RNA polymerase so that *t1* is ignored. In that case, before the *C* product can be made, the *B* product must be made and joined with RNA polymerase; thus, the *C* product will be synthesized at a later time than the *A* product.

Figure 15-16 shows a portion of the genetic map of λ, with three regulatory genes (*cro, N, Q*), three promoters (*pL, pR, pR2*) and five termination sites (*tL1, tR1, tR2, tR3, tR4*). Seven mRNA molecules are also shown as red arrows; the L and R mRNA series are transcribed leftward and rightward, respectively, from complementary DNA strands. The black arrows indicate the site of action of the regulatory proteins. The critical genes whose products modify RNA polymerase are *N* and *Q*; their products are called **antiterminators**. Synthesis of each of the mRNAs—L1, R1, and R4—begins at the same time. The terminator tR1 is a weak one, so very little of the *O* and *P* products are made. Once the *N* product is translated from L1, it binds to the *nutL* and *nutR* sites on the DNA molecule. During the next passage of RNA polymerase along the DNA, the enzyme picks up the *N* protein from these sites and is thereby modified. The terminators *tL1*, *tR1*, and *tR2* are then bypassed, leading to production of L2 (which we will not

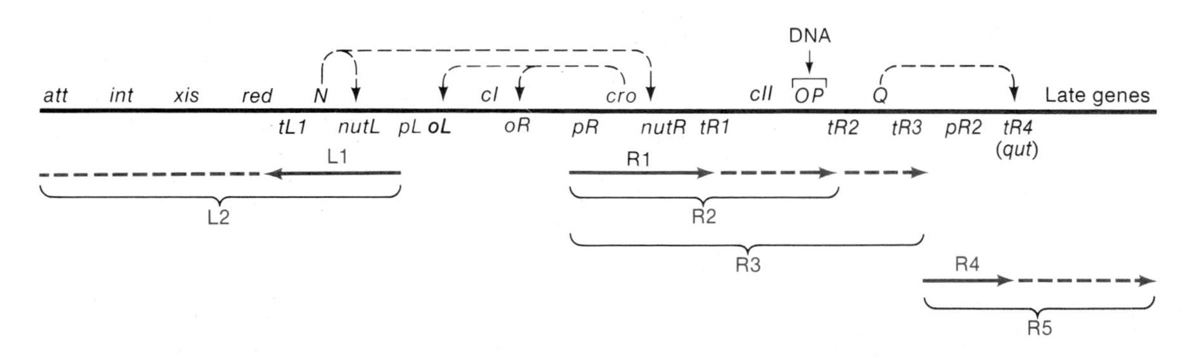

Figure 15-16
A genetic map of the regulatory genes of phage λ. Genes are listed above the line; sites are below the line. The mRNA molecules are red. The dashed black arrows indicate the sites of action of the N, Cro, and Q proteins.

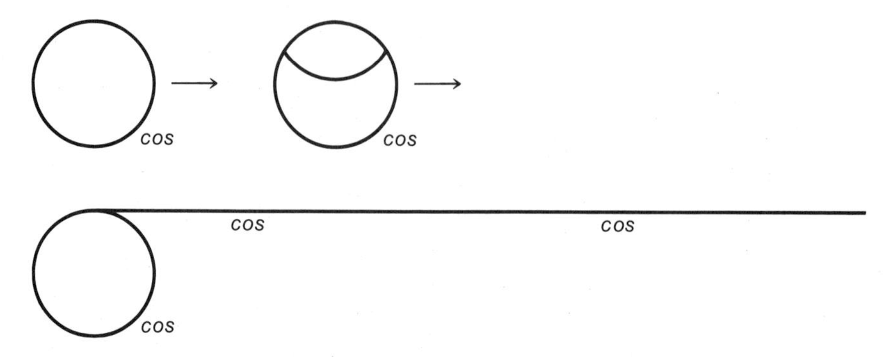

Figure 15-17
The replication sequence of phage λ.

discuss further) and R3. Translation of R3 yields the DNA replication proteins, allowing replication to begin, and the product of gene Q. The Q protein then binds to the site *qut*, where it is picked up by RNA polymerase, which is thereby modified so that it ignores *tR4*. Extension of R4 yields R5, the late mRNA, from which heads, tails, other structural proteins, and the lysis enzyme are made. Note that the delay of production of the late proteins compared to the early proteins is simply a result of the time required to transcribe and translate the N and Q genes. Details of the regulation of transcription and also of the roles of the genes *cro* and *cI* can be found in *MB, pages* 640–645 and 679–683.

It was mentioned earlier that λ DNA circularizes shortly after infection and that the cohesive ends base-pair to form the double-stranded *cos* region. Clearly, at some stage of the life cycle the single-stranded termini must be regenerated, since the DNA is linear within a phage head. This is accomplished by a cutting system, called **terminase** (or **Ter**), which cuts within the *cos* region of progeny DNA. However, the cuts are not made in progeny circles; instead, DNA replication and cutting are coordinated in a way that is found with many phage species. The earliest stage of replication of λ DNA is the bidirectional θ mode seen in Figures 7-4 and 7-24. Progeny circles do not continue to replicate in this way but switch to the rolling circle mode, yielding a long linear branch (Figure 15-17.) Note that the branch contains several *cos* sites compared to a circle, which contains only one. Progeny λ DNA molecules are cut from the branch of the rolling circles by terminase. The circular portion of the rolling circle is not cut (this would prevent further replication), because terminase requires two *cos* sites or one *cos* site plus one free single-stranded terminus. Thus, excision of λ units occurs by sequential cleavage from the free end of the branch, as shown in Figure 15-18.

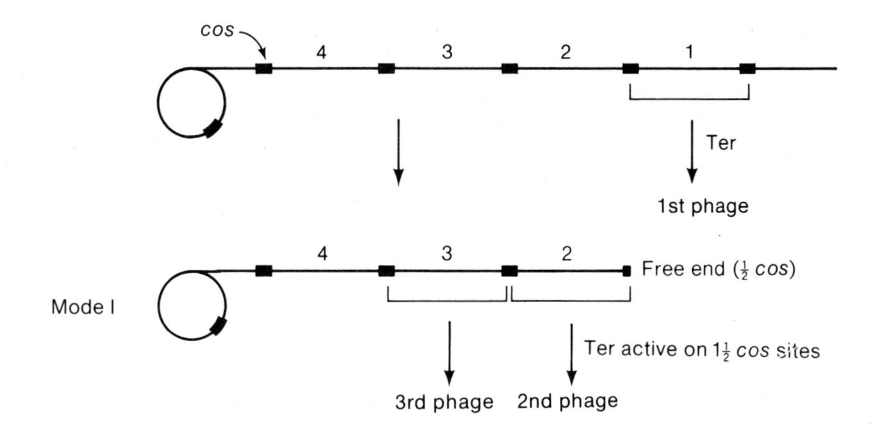

Figure 15-18
The mode of cutting λ units from a branch of a rolling circle. The units are cut sequentially from the free end by terminase.

E. Coli Phage λ: the Lysogenic Life Cycle

There are two types of lysogenic cycles. The most common one, for which *E. coli* phage λ is the prototype, follows (Figure 15-19):

1. A DNA molecule is injected into a bacterium.
2. After a brief period of transcription, needed to synthesize an integration enzyme, transcription is turned off by a repressor.
3. A phage DNA molecule is inserted into the DNA of the bacterium, forming a **prophage.**
4. The bacterium continues to grow and multiply and the phage genes replicate as part of the bacterial chromosome.

The second and less common type, for which *E. coli* phage P1 is the prototype, differs from the preceding one in that there is no integration system and the phage DNA becomes a plasmid (an independently replicating circular DNA molecule) rather than a segment of the host chromosome. In this chapter we will primarily consider the first type of life cycle.

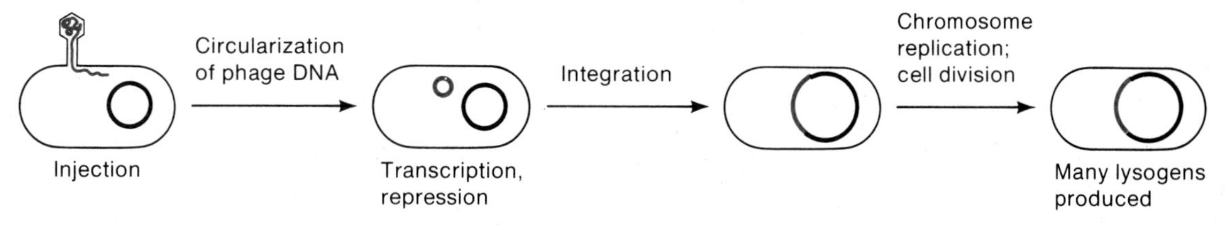

Figure 15-19
The general mode of lysogenization by insertion of phage DNA into a bacterial chromosome.

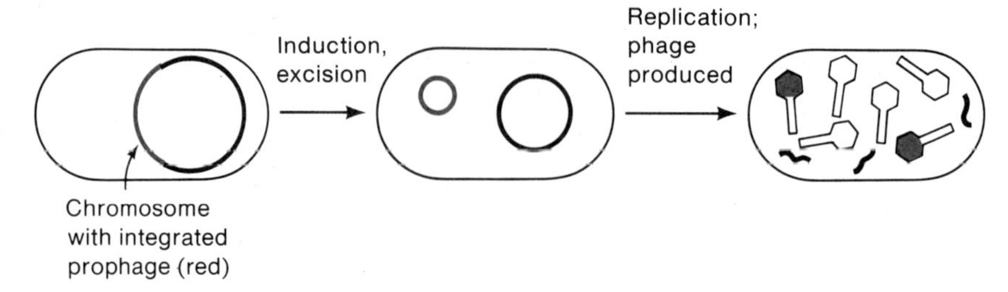

Figure 15-20
An outline of the events in prophage induction. The prophage DNA is in red. The bacterial DNA is omitted from the third panel for clarity.

Two important properties of lysogens are the following: (1) A lysogen cannot be reinfected by a phage of the type that first lysogenized the cell. (2) Even after many cell generations, a lysogen can initiate a lytic cycle; in this process, which is called **induction**, the phage genes are excised as a single segment of DNA (Figure 15-20).

The molecular mechanism for immunity and the circumstances that give rise to induction will be described shortly.

The resistance of a λ lysogen to infection by λ is called **immunity.** The cause of the phenomenon is the following. Phage λ contains a repressor-operator system (Figure 15-21). The repressor gene is called *cI*; the repressor protein binds to two operators, *oL* and *oR*, which are adjacent to two promoters *pL* and *pR*. The letters L and R mean leftward and rightward and refer to the direction of synthesis of two early mRNA molecules, when the genetic map is drawn in a standard orientation (see Figure 15-15).

In a lysogen the cI repressor is synthesized continuously and in slight excess with respect to the operators. Both *oL* and *oR* sites contain bound repressor molecules so that *pL* and *pR* are unavailable to RNA polymerase. Thus, in a lysogen, transcription from the two early promoters is prevented. It will be seen later that this is sufficient to keep the prophage in an "off" state; thus the lysogen grows indefinitely. If a normal cell is infected by λ, the two operators of the incoming λ DNA molecule will be unoccupied, since phage repressor has not yet been made and transcription will occur. However, if a phage tries to infect a lysogen, the excess repressor molecules already present in the lysogen

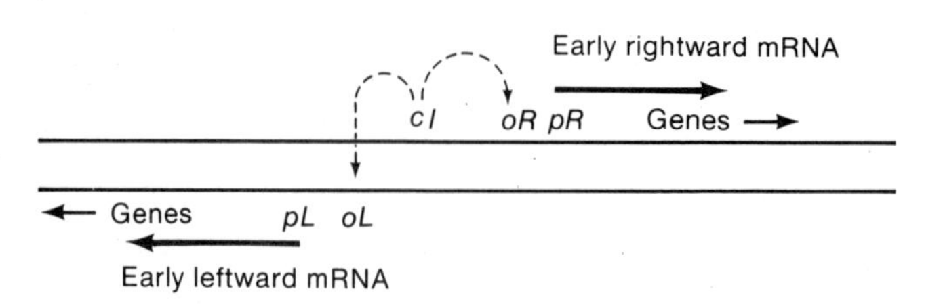

Figure 15-21
The repressor-operator system of λ, showing the two early mRNA molecules. Symbols: *cI*, repressor gene; *p*, promoter; *o*, operator; *L*, left; *R*, right.

bind to the two operators on the infecting DNA molecule before an RNA polymerase can bind to the *pL* and *pR* sites. This operator-binding prevents the phage from proceeding into lytic development. This inhibition is referred to as **resistance to homoimmune superinfection.**

What happens to superinfecting DNA? It is able to form a supercoil—phage gene products are not needed for supercoiling—but it cannot replicate. The bacterium is unaffected by the presence of this DNA molecule and grows and divides normally, so that the superinfecting DNA is progressively diluted out.

There are many temperate phages of *E. coli* other than λ. Two related to λ are phages 21 and 434. Each of these phages has its own immune system—that is, its own repressor and repressor-specific operators. Thus, the 434 repressor cannot bind to a λ operator and a λ repressor cannot bind to a 434 operator. Such a pair of phages is said to be **heteroimmune** with respect to one another. A temperate phage can form a plaque on a heteroimmune lysogen because the repressor made in the lysogen does not bind to the operator of the superinfecting phage. This is summarized in Table 15-3. The immunity region of the DNA, which includes the *cI* gene, operators, and promoters, is denoted by *imm*—specifically, *imm*λ, *imm21* and *imm434*. Interesting hybrid phages, which have been very useful in the laboratory, have been created by crossing two heteroimmune phages and selecting a recombinant containing the immunity region of one phage and the remaining genes of the other heteroimmune phage. A prominent example is a λ-434 hybrid, which is genotypically designated λ*imm434*.

Induction is also based on the repressor-operator interaction. If, for any reason, the repressor in a lysogenic cell is inactivated, the λ operators will be free and transcription will begin. In addition to all of the gene products that are made in the lytic cycle and are necessary for phage production, an enzyme called **excisionase** is made. This enzyme

Table 15-3
Ability of different phages to form plaques on homoimmune and heteroimmune lysogens

Superinfecting phage	*Lysogen*		
	B(λ)	B(*434*)	B(*21*)
λ	−	+	+
434	+	−	+
21	+	+	−

Note: The notation B(P) denotes a bacterium B lysogenic for phage P. + = forms a plaque; − = does not form a plaque.

cuts the prophage DNA from the bacterial chromosome and forms a circle; thus, the λ DNA is in the same state as in the beginning of a lytic cycle. The molecular mechanism of induction is not well understood. It is known that the repressor is cleaved by a protease, but the events leading to induction are unclear. The initial signal seems to be damage to DNA, in particular, single-stranded DNA produced by the repair of damage. If DNA is damaged, cell death might occur, so it is to the advantage of the prophage to get out of the bacterium. Precisely how the damaged DNA activates the protease that cleaves the repressor is not known, but is an active field of study. Ultraviolet radiation is a powerful inducing agent because of its ability to damage DNA.

In the lysogenic cycle of λ and other phages, a phage DNA molecule is inserted (or integrated, to use an equivalent term) into the bacterial chromosome to form a prophage. In order to form a lysogen, (1) repression must be established and (2) a prophage must be formed by insertion of a λ DNA molecule.

When λ DNA integrates, it is inserted at a preferred position in the *E. coli* chromosome; this site is between the *gal* operon and the *bio* (biotin) operon and is called the λ attachment site or, genotypically, *att*. For most temperate phages integration occurs at one preferred site, though there are a few phages that either have several preferred sites or appear to be able to insert their DNA anywhere in the *E. coli* chromosome. An important feature of the insertion mechanism was derived from the genetic observation that the gene order in the λ prophage is a permutation of the gene order of the DNA in the phage (see Figure 15-22). This permutation was explained by a model (due to Allen Campbell and widely known as the **Campbell model**) for the mechanism of integration (Figure 15-23). In this correct model, λ DNA is circularized first; and then prophage integration occurs as a physical breakage and rejoining of phage and host DNA—precisely, between the bacterial DNA attachment site and another attachment site in the phage DNA that is located near the center of the phage DNA molecule. A phage protein, **integrase** (its gene designation is *int*), recognizes the phage DNA and bacterial DNA attachment sites and catalyzes the physical exchange. This results in integration of the λ DNA molecule into the bacterial DNA. As a consequence of the circularization and

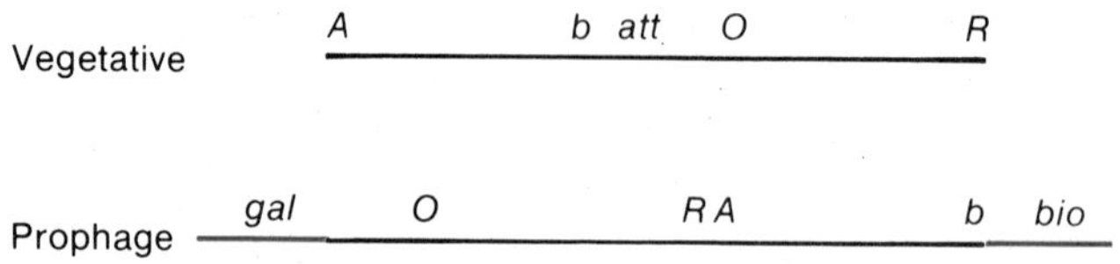

Figure 15-22
The order of the genes on the DNA molecule in the phage head (vegetative order) and in the prophage (prophage order). The genes have been selected arbitrarily to provide reference points.

Vegetative: *A* — *b att O* — *R*

Prophage: *gal* — *O* — *R A* — *b* — *bio*

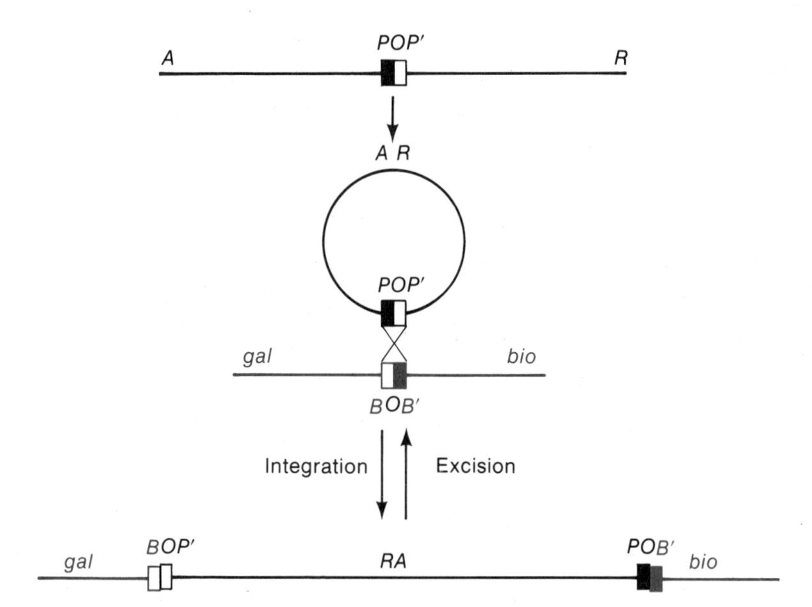

Figure 15-23
The Campbell model for the mechanism of prophage integration and excision of phage λ. The phage attachment site has been denoted *POP′* in accord with subsequent findings. The bacterial attachment site is *BOB′*. The prophage is flanked by two new attachment sites denoted *BOP′* and *POB′*.

integration, the linear order of the genes in the infecting λ DNA molecule is permuted, as shown in the figure. A question that arose after this model was proposed was: why doesn't the integrase excise the prophage shortly after integration occurs, since the prophage is flanked by two attachment sites? Furthermore, excision would seem to be the preferred reaction since, kinetically, two attachment sites in the same DNA molecule (that is, the host chromosome) ought to interact with each other more rapidly than two attachment sites in different DNA molecules (phage and bacterial DNA). Why this does not happen became clear when it was discovered that, inasmuch as the DNA attachment sites in the phage and the bacterium are not the same, they recombine to form prophage attachment sites that, as a consequence, also differ from the sites joining the bacterial and phage DNA.

All of the λ attachment sites have three different components. One of these is common to all sites and is denoted by the letter *O*.* The phage attachment site is written *POP′* (P for phage) and the bacterial attachment site is written *BOB′* (B for bacteria). Thus, in the integration reaction two new attachment sites, *BOP′* and *POB′*, are generated (Figure 15-23). **These are often written *attL* and *attR* to designate the left** and right prophage-attachment sites, respectively. Integrase cannot

*Another common notation is a raised dot, •, so that the bacterial attachment site, which we write *BOB′*, may be written *B • B′*. In this book we prefer the *O*.

catalyze a reaction between *BOP′* and *POB′*, so that the reaction

$$BOB' + POP' \xrightarrow{\text{Integrase}} BOP' + POB'$$

Bacterium Phage **Prophage**

is irreversible if integrase is the only enzyme present.

The result of the integration reaction is that the λ DNA is linearly inserted between the *gal* and *bio* loci and henceforth is replicated simply as a segment of *E. coli* DNA. The first evidence for insertion, gained from genetic experiments, showed (1) a permuted gene order in the prophage, (2) genetic linkage between *gal* and the genes to the right of *POP′*, (3) genetic linkage between *bio* and the genes to the left of *POP′*, and (4) a greater distance between the *gal* and *bio* genes in a lysogen than in a nonlysogen. Final proof for insertion came from the following physical experiment. A prophage was formed in a circular plasmid whose molecular weight was 105×10^6 and which contained *BOB′*. The plasmid remained circular but its molecular weight increased by the amount of one λ molecule (31×10^6) to 136×10^6.

Integrase has been purified, the base sequences of the *BOB′* and *POP′* sites are known, and the integration reaction can be carried out *in vitro*. Its mechanism is fairly well understood and is presented in *MB*, pages 673–675.

The synthesis of integrase is coupled to the synthesis of the cI repressor. This is efficient because integrase and the repressor are both needed in the lysogenic cycle and neither is needed in the lytic cycle. An outline of the regulatory pathway responsible for this coupling is shown in Figure 15-24. In the absence of a positive regulatory element (the product of the λ gene *cII*) the promoters for both the *int* and *cI* genes are unavailable to RNA polymerase. Shortly after infection, the *cII* gene product is translated from R3 mRNA. If the concentration of the protein is high enough, the *cII* product binds to sites near the promoters for the *cI* and *int* genes (designated *pre* and *pI*, respectively) and

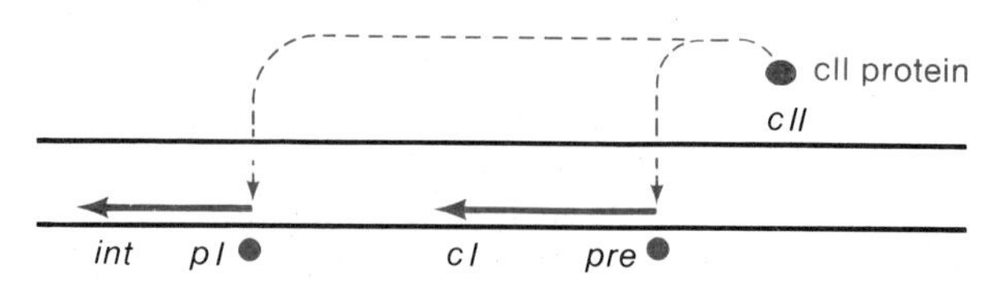

Figure 15-24
The role of the cII protein in stimulating synthesis of *cI* and *int* mRNA from *pre* and *pI* respectively. The mRNA molecules are drawn as red arrows which indicate the direction of RNA synthesis. The arrows are located nearer the DNA strand that is copied. Dashed arrows point to the sites of binding of the cII protein (red dots).

thereby renders them accessible to RNA polymerase. (In this respect the cII protein is similar to the cAMP-CAP complex in the *lac* operon.) RNA polymerase then transcribes both genes and the gene products are made. Synthesis of the cII protein is also the key step in the decision between a lytic and lysogenic pathway in a particular infection; however, details of its role are not yet understood.

TRANSDUCING PHAGES

Some phage species, known as **transducing phages**, are able to package bacterial DNA as well as phage DNA. Particles containing bacterial DNA, called **transducing particles**, are not always formed and usually constitute only a small fraction of a phage population. Such particles can arise by two mechanisms, aberrant excision of a prophage and packaging of bacterial DNA fragments. We begin with a discussion of the first mechanism.

One mechanism by which transducing particles form is shown in Figure 15-25, which depicts the formation of the galactose-and biotin-transducing forms of phage λ—namely, λ*gal* and λ*bio*.

When λ prophage is induced, an orderly sequence of events ensues in which the prophage DNA is precisely excised from the host DNA. In the case of phage λ this is accomplished by the combined efforts of the *int* and *xis* genes acting on the left and right prophage attachment sites. At a very low frequency—namely, in about one cell per 10^6–10^7

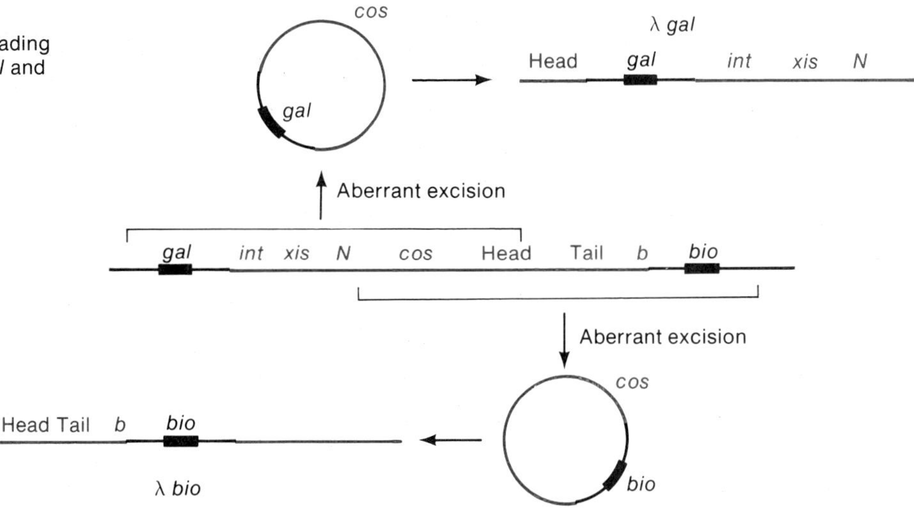

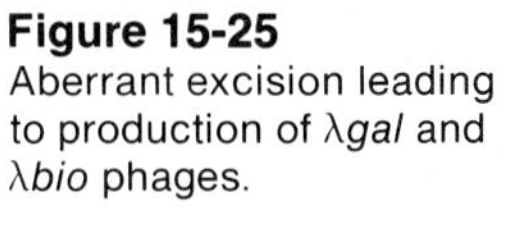
Figure 15-25
Aberrant excision leading to production of λ*gal* and λ*bio* phages.

cells—an excision error is made; two incorrect cuts are made—one within the prophage and the other cut in the bacterial DNA. The pair of abnormal cuts will not always yield a length of DNA that can fit in a λ phage head—it may be too large or too small. However, if the spacing between the cuts produces a molecule between 79 percent and 106 percent of the length of a normal λ phage-DNA molecule, packaging can occur. Since the prophage is located between the *E. coli gal* and *bio* genes and because the cut in the host DNA can be either to the right or the left of the prophage, transducing particles can arise carrying the *bio* genes (cut to the right) or the *gal* genes (cut to the left). Formation of the λ*gal*- and λ*bio*-transducing particles entails loss of λ genes. The λ*gal* particle lacks the tail genes, which are located at the right end of the prophage; the λ*bio* particle lacks the *int, xis,* . . . genes from the left end of the prophage. The number of missing phage genes of course depends on the position of the cuts that generated the particle and thus correlates with the amount of bacterial DNA in the particle. The missing phage genes come from the prophage ends, but because of the permutation of the gene order in the prophage and the phage particle, the deleted phage genes are always from the central region of the phage DNA, as shown in the figure.

Since these transducing particles lack phage genes, they might be expected to be nonviable. This is indeed true of the λ*gal* type, which lack genes essential for synthesis of the phage tail and, in the case of the larger deletion-substitutions, the genes for the phage head also. Thus, these particles are incapable of producing progeny; they are defective and this is denoted by the symbol d—thus, a *gal*-transducing particle is written λd*gal* (and sometimes λdg). The λd*gal* particle contains the cohesive ends and all of the information for DNA replication and transcription; it therefore goes through a normal life cycle, including bacterial lysis. In fact, if the head genes are not deleted, the concatemeric branch produced by rolling circle replication is cleaved by the Ter system. However, tails are not added to the filled head, no viable particles are produced in the lysate, and plaques are not formed on a host lawn. Note that there is no discrepancy between the formation of a λd*gal* transducing particle and its ability to reproduce itself. A λd*gal* particle fails to reproduce only because it lacks the tail genes, but such a particle arises from a normal prophage having a full set of genes.

The situation is quite different with λ*bio* particles for these usually lack only nonessential genes—*int, xis,* etc. These genes are needed for the lysogenic cycle but not for the lytic cycle, so these particles are able to replicate and to form plaques. To denote this, the letter p, for plaque-forming, is added; that is, the particle is called λp*bio*.

Transducing phages, like λ, able to form transducing particles containing bacterial genes from selected regions of the bacterial chromosome, are called **specialized transducing phages.** Other phage species are able to package fragments of bacterial DNA from all parts of the chromosome. These are called **generalized transducing phages.** The specialized transducing phages producing transducing particles only come from a lysogen. In contrast, the generalized transducing phages form transducing particles from infected cells. The mechanism by which these particles are produced is straightforward. Some phage species, of which T4 is an example, degrade bacterial DNA early in infection. With T4 the degradation is complete, but in other species, of which *E. coli* phage P1 and *Salmonella typhimurium* phage P22 are examples, fragments of chromosomal DNA comparable in size to phage DNA are present at the time phage DNA is packaged. The packaging system of λ is a site-specific one, requiring *cos* sites. However, some phages, like T4, either merely fill a head, or like P1 and P22, are able to package fragments having a range of sizes. The latter phages cannot distinguish bacterial DNA from phage DNA and occasionally package a piece of bacterial DNA. In a P1 population about one percent of the particles contain *only* a fragment of bacterial DNA.

Transducing particles are quite useful in genetic experiments. For example, they can be used to transfer a bacterial gene from one cell to another. They have also been used as means of isolating particular segments of a bacterial chromosome.

16 Regulation of Gene Activity in Eukaryotes

Eukaryotic cells are of two general types—free-living unicells, such as yeast and algae, and those resident in organized tissue. The needs of the latter class differ from the needs of both the former and of prokaryotes in that the environment of cells in tissue does not usually change drastically in time. During the growth phase of an organism cells differentiate in response to various signals, mostly unknown; however, once differentiated, the cells remain stable, producing particular substances either at a constant rate or in response to external signals such as hormones and temperature changes. The free-living unicellular eukaryotes and the prokaryotes share certain regulatory features, though the gene organization of the former is definitely that of eukaryotic cells. A great deal is known about regulation in yeast, but at a less profound level of understanding than that of *E. coli* operons. Whereas details of the mechanisms in yeast differ, often significantly, from those observed in bacteria, the overall regulatory strategy of yeast remains that of responding to large fluctuations in the availability of nutrients in the environment. For this reason, the emphasis of this section is on eukaryotic cells that form organized tissue. Most of the information comes from detailed studies of the DNA of mammals, amphibians (toads of the genus *Xenopus*), insects (*Drosophila*), birds (usually the chicken), and echinoderms (the sea urchin).

SOME IMPORTANT DIFFERENCES IN THE GENETIC ORGANIZATION OF PROKARYOTES AND EUKARYOTES

Numerous differences exist between prokaryotes and eukaryotes with regard to transcription and translation, and in the spatial organization of DNA, as described in Chapters 5 and 9. Seven of those most relevant to regulation are the following:

1. In a eukaryote usually only a single type of polypeptide chain can be translated from a completed mRNA molecule; thus, operons of the type seen in prokaryotes are not found in eukaryotes.
2. The DNA of eukaryotes is bound to histones, forming chromatin, and to numerous nonhistone proteins. Only a small fraction of the DNA is bare. In bacteria some proteins are present in the folded chromosome, but most of the DNA is free.
3. A significant fraction of the DNA of eukaryotes consists of a few nucleotide sequences that are repeated hundreds to millions of times. Some sequences are repeated in tandem but most repetitive sequences are not. Other than duplicated rRNA and tRNA genes and a few specific short sequences such as certain parts of promoters, bacteria contain few repeated sequences.
4. A large fraction of the base sequences in eukaryotic DNA is untranslated.
5. Eukaryotes possess mechanisms for rearranging certain DNA segments in a controlled way and for increasing the number of specific genes when needed. This seems to be rare in bacteria.
6. The bases of a gene and the amino acids of the gene product are usually not colinear in eukaryotes; introns are present in most eukaryotic genes and RNA must be processed before translation begins.
7. In eukaryotes RNA is synthesized in the nucleus and must be transported through the nuclear membrane to the cytoplasm where it is utilized. Such extreme compartmentalization probably does not occur in bacteria.

We shall see in this section how some of these features are incorporated into particular modes of regulation.

GENE FAMILIES

In prokaryotes genes having closely related functions are often organized in operons and are transcribed as part of a polycistronic mRNA molecule. Thus, the entire system is under control of one promoter region, and the system can be turned on and off by controlling the availability of the promoter. Without modification this method cannot be used in eukaryotes inasmuch as eukaryotic mRNA is usually monocistronic.

Many related eukaryotic genes can be functionally grouped into a set called a **gene family.** These genes are rarely as near to one another as the genes are in a bacterial operon, though they may be clustered (not adjacent, but within hundreds to thousands of nucleotide pairs of one another); however, frequently they are widely scattered and even located on different chromosomes. Gene families are currently classified as simple multigene families, complex multigene families, and developmentally controlled multigene families. An example of each type is shown in Figure 16-1.

A **simple multigene family** is one in which one or a few genes are repeated in a tandem array. The simplest example is the set of genes for

(a) Simple multigene family

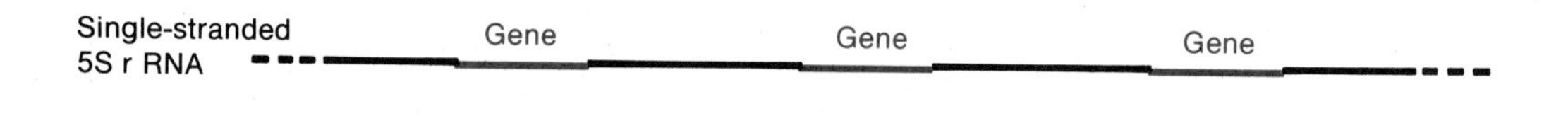

(b) Complex multigene families

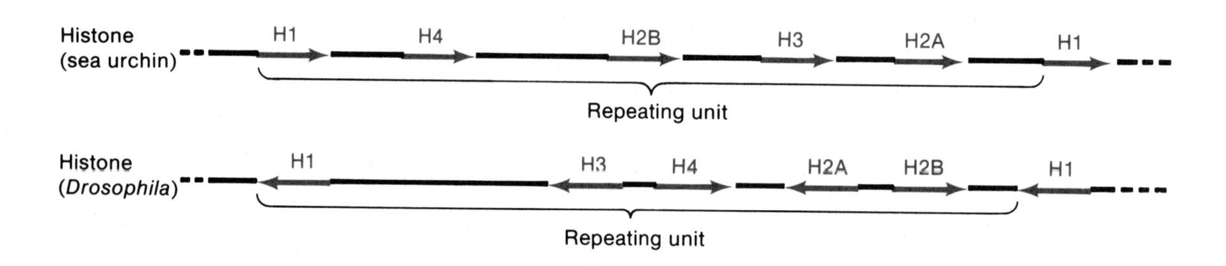

(c) Developmentally controlled complex multigene family

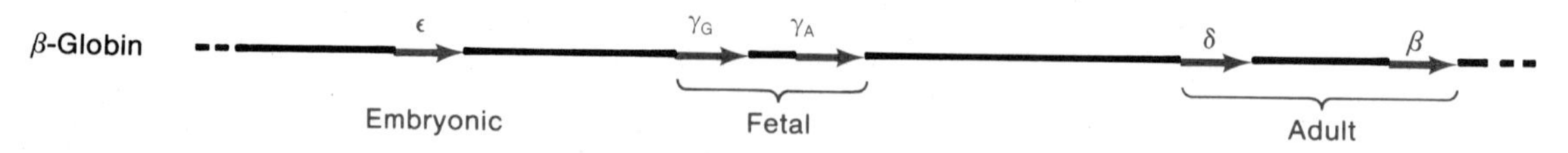

Figure 16-1
Four examples of gene families. The arrows indicate the direction of transcription. Genes are shown in red; spacers are black.

5S ribosomal RNA. In the toad *Xenopus* this set of rRNA genes forms a gigantic array in which each 5S rRNA gene sequence is separated from an adjacent gene sequence by a spacer (not an intron, because it is not within a gene) to form a gene cluster. The sizes of the spacers vary from two to six times the length of the 5S rRNA gene. Each 5S rRNA gene is transcribed as a separate RNA molecule that is cut and spliced to generate the finished 5S rRNA molecule. Several clusters of 5S RNA genes are present in *Xenopus*, each cluster containing hundreds to thousands of 5S rRNA genes. Presumably, the individual genes (or at least the cluster) are transcribed in response to a single signal.

The 5.8S, 18S, and 28S rRNA genes of *Xenopus* comprise another simple multigene family, in which the primary transcript contains all three rRNA sequences in the order 18S–5.8S–28S, each separated by a small spacer. The rRNA molecules are cleaved from the primary transcript by processing enzymes. In this case, the 1:1:1 ratio of the three species (the ratio present in a ribosome) is achieved by there being one copy of each per transcript, as is also the case for *E. coli* rRNA. Note that the 5S rRNA is transcribed separately from the 18S–5.8S–28S unit. How a ratio of one copy of 5S rRNA for each 18S–5.8S–28S unit is obtained is unknown. A simple explanation is that synthesis of the 5S rRNA transcript and the three-species transcript is initiated in response to the same signal.

The **complex multigene families** generally consist of a cluster of several functionally related genes, each transcribed independently (in contrast with the multigene transcript just described for the rRNA unit) and separated by a spacer. Two examples are shown in Figure 16-1 (b); each has features that indicate that there are several types of complex multigene families. For instance, in the sea urchin each of the five histone genes is separately transcribed as a single monocistronic RNA and each gene is transcribed in the same direction—that is, from the same DNA strand. In contrast, in *Drosophila* the five genes are transcribed in both directions, as shown in the figure. In yeast the organization of the histone genes differs even more; there are two quite distant gene clusters, each containing one *H2A* gene and one *H2B* gene, whereas the other histone genes are separate and scattered throughout the chromosome. In all cases, the genes are considered to constitute a family, because the gene products are related and are synthesized in a fairly fixed ratio.

The **developmentally controlled gene families** are regulated such that different genes are expressed at different stages of development of the organism—that is, they are expressed in a definite sequence in time. The best-characterized one is the globin family. Hemoglobin is a tetrameric protein containing two α subunits and two β subunits. However, there are several different forms of both α and β subunits differing by only one or a few amino acids, and the forms that are present depend on the stage of development of an organism. For example, the following developmental sequence shows the subunit types present in humans at various times after conception:

	Embryonic (*<8 weeks*)	*Fetal* (*8–41 weeks*)	*Adult* (*birth → henceforth*)
α-like	$\zeta_2 \rightarrow \zeta_1$	α_1 and α_2	α_1 and α_2
β-like	ϵ	γ_G and γ_A	β and δ

During the embryonic period the ζ_2 α-like chain, which appears first, is gradually replaced by the ζ_1 form. In the fetal and adult stages, the α_1 and α_2 types are both present but α_2 predominates; the two β-like types, γ_G and γ_A (called so because one has a glycine at a site at which the other has an alanine), are roughly equal in amount. In the adult stage, fifty times more β type is present than the δ form. The result of these changes is that before birth all the hemoglobin contains two ζ_2 chains and two ϵ chains ($2\zeta_2$,2ϵ), whereas after birth 98 percent contains two α_2 and two β subunits (customarily called hemoglobin A) and 2 percent contains two α_2 and two δ units (hemoglobin A_2).

The α- and β-like genes form separate clusters. The β cluster is shown in Figure 16-1(c). A remarkable property of each cluster is that the order of the genes is basically the order in which they are expressed in development. (This is not true of all developmentally controlled gene families.) Neither the significance of the different forms nor the way in which synthesis is programmed is known.

Examination of the globin clusters in many different organisms gives some indication of how clusters arose and of the evolutionary factors that may have provided for their maintenance. Primitive fish, marine worms, and insects have a single globin gene, whereas in amphibians α and β genes are closely linked on a single chromosome. The lower mammals and birds have different forms of both α and β genes, and the clusters are on different chromosomes. Thus, it has been postulated that ancestral globin (which may have appeared about 800 million years ago) was the product of a single gene and that the globin families evident today arose by a series of gene duplications, mutations, and transpositions from an ancestral gene. The following hypothetical (but likely) sequence of events has been suggested. The small animals were able to function with the limited O_2-carrying capacity of a single hemoglobin molecule. Spontaneous mutation gave rise to an altered globin that in a heterozygote was able to form a tetrameric molecule capable of carrying more O_2. For reasons beyond the scope of this book, the tetrameric molecule can deliver a larger fraction of its bound O_2 to tissue than the monomers can, which enabled animals of larger size to evolve. Later, during the evolution of mammals one of the two β-chain genes underwent mutation and duplication, again giving rise to the γ form of globin found in the fetus. Fetal hemoglobin has an even higher affinity for O_2 than adult hemoglobin and thus is advantageous for the rapidly developing fetus, perhaps enabling a more complex organism to arise. Further mutation and duplication occurred during primate evolution, giving rise to additional forms of hemoglobin.

The scheme just suggested is supported by the existence of remnants of past events of the type described. Study of various segments of the globin gene clusters in several organisms has indicated the presence of nonfunctional sequences that are nearly homologous to normal globin genes. The sequences are called **pseudogenes,** and are considered to be relics of evolution, derivatives of once-functional genes that were duplicated. Presumably a mutation occurred at one time that inactivated the gene product. Since at least one functional copy of the gene remained, there would have been no strong selective pressure to eliminate the gene, and in time additional mutations would have accumulated. Pseudogenes are widespread in gene clusters and are frequently defective in transcription, intron excision, and translation.

GENE DOSAGE AND GENE AMPLIFICATION

Some gene products are required in much larger quantities than others. A common means of maintaining particular ratios of certain gene products (other than by differences in transcription and translation efficiency, as discussed earlier) is by **gene dosage.** For example, if two genes *A* and *B* are transcribed at the same rate and the translation efficiencies are the same, 20 times as much of product A can be made as product B if there are 20 copies of gene *A* per copy of gene *B*. The histone gene family exemplifies a gene dosage effect: in order to synthesize the huge amount of histone required to form chromatin, most cells contain hundreds of times as many copies of histone genes as of genes required for DNA replication.

A special case of a gene dosage effect is **gene amplification,** in which the number of genes increases in response to some signal. The best-understood example of gene amplification is found in the development of the oocytes (eggs) of the toad *Xenopus laevis,* in which the number of rRNA genes increases by about 4000 times. This increase exists only during and for the purpose of development of an egg. The precursor to the oocyte, like all somatic cells of the toad, contains about 600 rRNA-gene (rDNA) units; after amplification about 2×10^6 copies of each unit are present. This large amount enables the oocyte to synthesize 10^{12} ribosomes, which are required for the protein synthesis that occurs during early development of the embryo.

Prior to amplification, the 600 rDNA units are arranged in tandem and each unit consists of the 18S–5.8S–28S sequence described in the preceding section—that is, one gene of each type, separated by a spacer. During amplification, over a three-week period during which the oocyte develops from a precursor cell, the rDNA no longer consists of a single contiguous DNA segment containing 600 three-gene units but instead is present as a large number of small circles and replicating rolling circles, each containing one or two three-gene units. The rolling circle replication accounts for the increase in the number of copies of the genes. The precise mechanism of excision of the circles and formation of the rolling circles is not known.

Once the oocyte is mature, no more rRNA needs to be synthesized until well after fertilization and into early development, at which time 600 copies is sufficient. Thus, the excess rDNA serves no purpose and it is slowly degraded by intracellular enzymes. Following fertilization the chromosomal DNA replicates and mitosis ensues, occurring repeatedly as the embryo develops. During this period the extra chromosomal rDNA does not replicate; degradation continues and by the time several hundred cells have formed, none of this DNA remains.

Amplification of rRNA genes during oogenesis occurs in many organisms including insects, amphibians, and fish. Amplification of a gene that encodes a protein has been observed in *Drosophila;* the genes that produce chorion proteins (a component of the sac that encloses the egg) are amplified in ovarian follicle cells just before maturation of an egg. In this case, as with the rRNA genes, amplification enables the cells to produce a large amount of protein in a short time. We will see later that if a large amount of protein is to be synthesized and a *long* time is available for the synthesis, gene amplification is unnecessary, for this can be accomplished by increasing the lifetime of the RNA.

REGULATION OF TRANSCRIPTION

The mRNA of eukaryotes is conveniently classified into groups based on the number of copies of a particular mRNA molecule present in each cell—single-copy (a term used both for a single copy and a few copies per haploid complement), moderately prevalent (a few to several hundred copies per cell), and superprevalent (hundreds to thousands of copies per cell). Single-copy and moderately prevalent mRNA primarily encode enzymes and structural proteins, respectively. Superprevalent mRNA molecules are transcribed from only a small fraction of eukaryotic genes, and their production is usually associated with a change in the stage of development—for example, an adult erythroblast cell in the bone marrow produces a huge amount of mRNA from which adult globin may be translated, whereas little or no globin is produced by precursor cells that have not yet become erythroblasts. A great deal of what is known about the regulation of transcription in higher eukaryotes concerns superprevalent mRNA.

Of the known regulators of transcription in higher eukaryotes the **hormones**—small molecules, polypeptides, or small proteins that are carried from hormone-producing cells to target cells—constitute a class that has been studied in some detail. Many of the sex hormones act by turning on transcription. If a hormone regulates transcription, it must somehow signal the DNA. Penetration of a target cell by a hormone and its transport to the nucleus is a much more complex process than entry of lactose into *E. coli* and is understood only in outline, as shown in Figure 16-2. Steroid hormones (H) are hydrophobic (nonpolar) molecules and pass freely through the cell membrane. A target cell contains a specific cytoplasmic receptor (R) that forms a complex (H-R) with the hormone. The receptor R usually undergoes some modification (in shape or in chemical structure) after the H-R complex has formed; the modified form of the receptor is denoted R′. The H-R′ complex then passes through the nuclear membrane and enters the nucleus. From this point on, little is known about most systems. It is likely that in the nucleus either the H-R′ complex or possibly the hormone alone engages

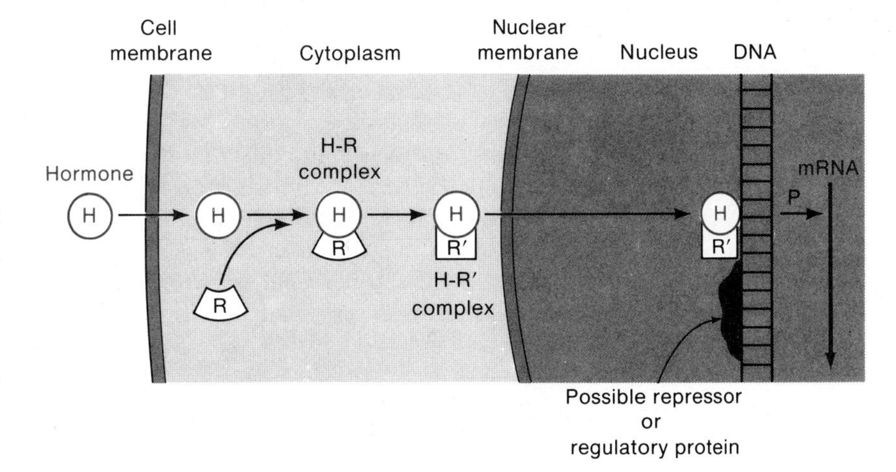

Figure 16-2
Schematic diagram showing how a hormone H reaches the DNA and triggers transcription by binding to a cytoplasmic receptor. A regulating protein may prevent the H-R′ complex from reaching a promoter (P) or may stimulate binding to the promoter.

in one of the following processes: (1) direct binding to DNA, (2) binding to an effector protein, (3) activation of a DNA-bound protein, (4) inactivation of a repressor, and (5) a change in the structure of chromatin to make the DNA available to RNA polymerase. In most hormone-activated systems studied to date it has not been possible to determine which process is involved. One well-understood example is the action of glucocorticoid, a hormone that inhibits glucose degradation and stimulates glucose synthesis in the liver. The activity of glucocorticoid in initiating transcription of the relevant genes involves the direct binding of a positively regulatory protein to a DNA molecule.

A well-studied example of induction of transcription by a hormone is the stimulation of the synthesis of ovalbumin in the chicken oviduct by the sex hormone **estrogen.** When chickens are injected with estrogen, oviduct tissue responds by synthesizing ovalbumin mRNA. This synthesis continues as long as estrogen is administered. Once the hormone is withdrawn, the rate of synthesis decreases. Both before giving the hormone and sixty hours after withdrawal, no ovalbumin mRNA is detectable.

When estrogen is given to chickens, only the oviduct synthesizes mRNA because other tissues lack the cytoplasmic hormone receptor. (This type of deficiency is the usual cause of insensitivity to a particular hormone.) The mechanism by which the receptor is synthesized in some but not all cells is not known.

REGULATION OF PROCESSING

Often two cells make the same protein but in different amounts, even though in both cell types the same gene is transcribed. This phenomenon is frequently associated with the presence of different mRNA molecules, which are not translated with the same efficiency. In the synthesis of α-amylase in the rat, different mRNA from the same gene result from the use of different processing sequences. The rat salivary gland produces more of the enzyme than the liver, though the same coding sequence is transcribed. In each cell type the same primary transcript is synthesized, but two different splicing mechanisms are used. The initial part of the primary transcript is shown in Figure 16-3. The coding sequence begins 50 base pairs within exon 2 and is formed by joining exon 3 and subsequent exons. In the salivary gland the primary transcript is processed such that exon S is joined to exon 2 (that is, exon L is removed as part of introns 1 and 2). In the liver exon L is joined to exon 2, because exon S is removed along with intron 1 and with the leader L. The exons S and L become alternate leaders of amylase mRNA, which somehow results in translation at different rates.

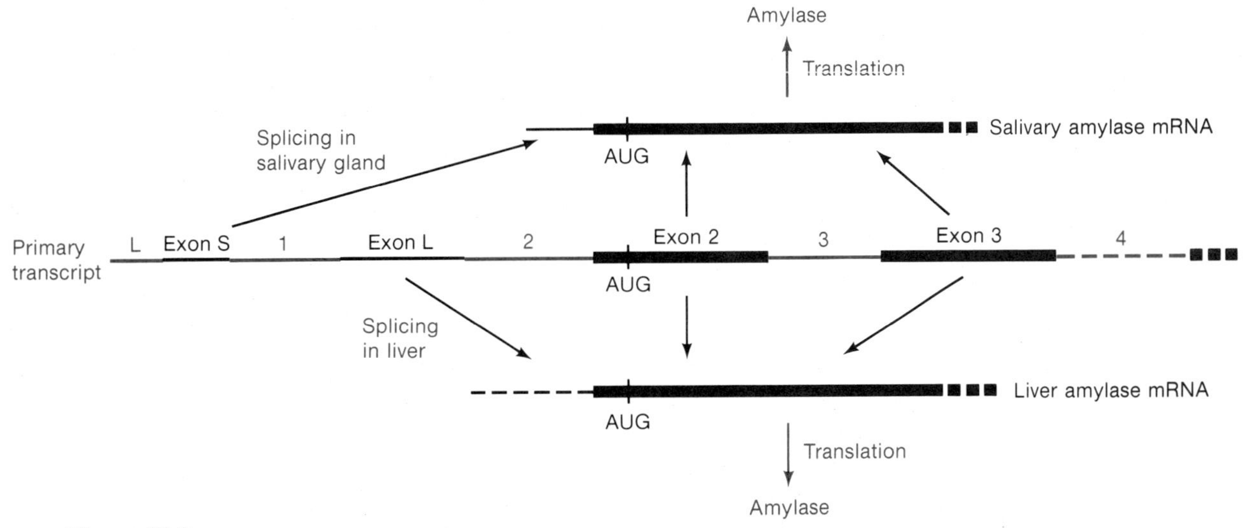

Figure 16-3
Production of distinct amylase mRNA molecules by different splicing events in cells of the salivary gland and liver of the mouse. The leader L and the introns are in black. The exons are red. The coding sequence begins at the AUG in exon 2.

HYPERSENSITIVE SITES AND UPSTREAM REGULATORY SITES

In regulation of transcription in prokaryotes, availability of the promoter for binding of RNA polymerase is critical, and in negatively regulated operons the binding of a protein (a repressor) to an operator is often sufficient to prevent transcription. In eukaryotes DNA does not exist as a free molecule but is tightly bound to histones, so the question of how RNA polymerase comes into contact with promoters is an important one. The answer is not yet clear, but several observations suggest that chromatin may have a special structure in promoter regions. If free DNA molecules are exposed to most DNases, the DNA is rapidly broken. However, the rate of breakage of chromatin is much less than that of free DNA, because the histones reduce access of the DNA to the enzymes. When chromatin isolated from a particular cell type is treated with certain DNases, specific regions of the DNA are cleaved at a higher rate than the bulk of the DNA; the positions of these regions vary from one cell type to another. These regions are called **hypersensitive sites;** their importance is that their existence and positions correlate with transcriptional activity of particular genes. For example, in chromatin isolated from chicken reticulocytes (which synthesize globin), sites in the globin cluster are preferentially broken, whereas chicken oviduct chromatin contains hypersensitive sites in the ovalbumin region. In contrast, there are no such sites in the ovalbumin region of reticulocyte chromatin nor in the globin region of oviduct chromatin. Evidence from study of the *Sgs4* gene of *D. melanogaster* suggests that the hypersensitive sites may actually be essential for gene activity. A mutant has been found in which the hypersensitive region is deleted but in which the coding region of the Sgs4 protein is intact. This mutant fails to make Sgs4 protein, suggesting that deletion of the hypersensitive region prevents activity of the gene. The mechanistic relation between a hypersensitive site and transcription is unknown. In one case the DNA of the hypersensitive site in a viral DNA molecule is bare, lacking nucleosome structure, but such an observation has not yet been made for cellular DNA. Hypersensitive sites are usually several hundred base pairs *upstream* from the start point for transcription, which seems quite far compared to the size of the binding site for RNA polymerase (Figure 16-4). (By "upstream" is meant "against the direction of transcription.")

Upstream elements of importance have also been observed in other experiments. For example, in several yeast genes, mutations have been isolated that eliminate or reduce transcription. These mutations define regions called **upstream activation sites,** whose properties are quite different from the transcription-regulating promoter-operator regions in prokaryotes. In a bacterial operon the operator and promoter are adjacent, and binding a regulatory protein to an operator either blocks (a repressor) or stimulates (a molecule of the cAMP-CAP type) the binding of RNA polymerase to the promoter. If the promoter and operator are physically separated (by experimental manipulation), the regulatory region becomes ineffective—that is, a repressor cannot prevent initiation of transcription nor can a positive regulator stimulate transcription. However, upstream activation sites are not only usually 50–300 nucleotide pairs from the promoter but their locations can be experimentally changed by as many as a few hundred nucleotide pairs without substantially diminishing their activity. Upstream activation sites have been discovered so recently that their role in regulation is unknown, but they seem to be a common feature of eukaryotic systems.

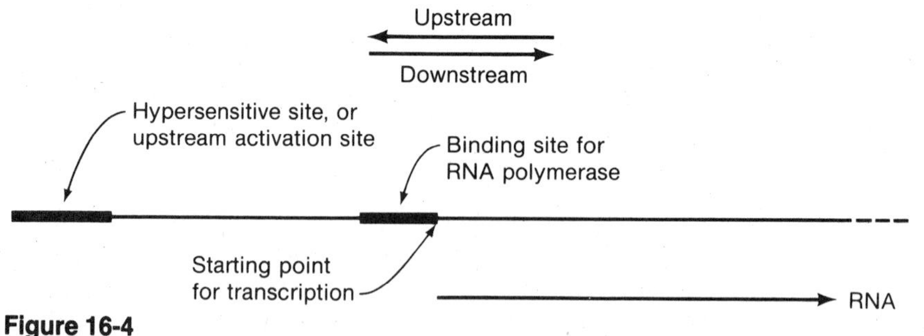

Figure 16-4
The relative sizes and locations (drawn approximately to scale) of a hypersensitive site or upstream activation site (at roughly the same position) and the binding site for RNA polymerase in a typical eukaryotic gene. The transcribed region would be about ten times longer than shown. The meaning of the terms "upstream" and "downstream" is also shown.

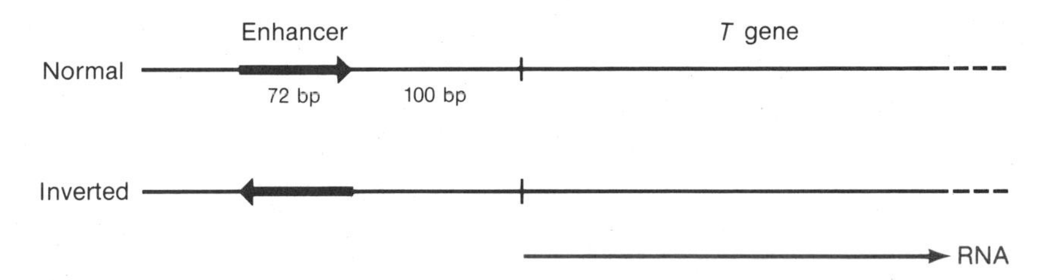

Figure 16-5
Position of the enhancer sequence near the virus SV40 *T* gene with respect to the transcription start site. If the normal sequence is experimentally inverted, RNA is still made at a high rate. Experimental deletion of the enhancer reduces transcription more than 100-fold.

Another type of region whose location can also be changed without detracting from its effectiveness is the **enhancer element.** The prototype enhancer was discovered in the animal virus SV40. It is a sequence of 72 nucleotide pairs located more than 100 nucleotide pairs *upstream* from the gene *T*. Three properties of this sequence are general features of enhancers. (1) Experimental deletion of the sequence reduces transcription of this gene by more than a hundredfold. (2) By experimental manipulation the SV40 enhancer can be moved further upstream, downstream, and even to the other side of the promoter without significant loss of activity. (3) The sequence can also be experimentally reversed in direction and retain enhancing activity (**Figure 16-5**). Furthermore, by the genetic engineering techniques described in Chapter 13, the sequence has been isolated and moved next to genes in other organisms and the enhancing effect remains. In several viral systems enhancers have been found in some hypersensitive sites, but the relation is unclear. Little is known about the molecular basis of enhancement; the most popular hypothesis is that enhancers alter chromatin structure and provide an entry site for RNA polymerase.

TRANSLATIONAL CONTROL

In bacteria most mRNA molecules are translated about the same number of times, with only fairly small variation from gene to gene. In eukaryotes translational regulation occurs in which an mRNA molecule is not translated at all until a signal is received. Other types of control include (1) regulation of the lifetime of a particular mRNA molecule and (2) regulation of the rate of overall protein synthesis. In this section examples of each of these modes of regulation will be presented.

An important example of translational regulation is that of **masked mRNA.** Unfertilized eggs are biologically static, but shortly after fertilization many new proteins must be synthesized—for example, the proteins of the mitotic apparatus, the cell membranes, as well as others. Unfertilized sea urchin eggs store large quantities of mRNA for many months in the form of mRNA-protein particles made during formation of the egg. This mRNA is translationally inactive, but within minutes after fertilization, translation of these molecules begins. Here the timing of translation is regulated; the mechanisms for stabilizing the mRNA, for protecting it against RNases, and for activation are unknown.

Translational regulation of a second type also occurs in mature unfertilized eggs. These cells need to maintain themselves but do not have to grow or undergo a change of state. Thus, the rate of protein synthesis in eggs is generally low. This is not a consequence of an inadequate supply of mRNA but of a limitation of an as-yet-unidentified element, called the **recruitment factor,** which apparently interferes with formation of the ribosome-mRNA complex.

A dramatic example of translational control is the extension of the lifetime of silk fibroin mRNA in the silkworm *Bombyx mori*. During cocoon formation the silk gland of the silkworm predominantly synthesizes a single type of protein, silk fibroin. Since the worm takes several days to construct its cocoon, it is the *total amount* and not the rate of fibroin synthesis that must be great; the silkworm achieves this by synthesizing a fibroin mRNA molecule that is very long-lived.

Transcription of the fibroin gene is initiated at a strong promoter by an unknown signal and about 10^4 fibroin mRNA molecules are made in a period of several days. (This synthesis is an example of transcriptional regulation.) A typical eukaryotic mRNA molecule has a lifetime of about three hours before it is degraded. However, fibroin mRNA survives for several days during which each mRNA molecule is translated repeatedly to yield 10^5 fibroin molecules. Thus, each gene is responsible for the synthesis of 10^9 protein molecules in four days. Altogether the silk gland makes 300 μg or 10^{15} molecules of fibroin during this period. If the lifetime of the mRNA were not extended, either 25 times as many genes would be needed or synthesis of the required fibroin would take about 100 days; therefore, 10^6 genes or 5×10^5 diploid cells would be needed.

Another example of an mRNA molecule with an extended lifetime is the mRNA encoding casein, the major protein of milk, in mammary glands. When the hormone prolactin is provided to the gland, the lifetime of casein mRNA increases. Synthesis of the mRNA also continues, so the overall rate of production of casein is markedly increased by the hormone. When the prolactin is withdrawn, the concentration of casein mRNA decreases because the RNA is degraded more rapidly.

MULTIPLE PROTEINS FROM A SINGLE SEGMENT OF DNA

In prokaryotes coordinate regulation of the synthesis of several gene products is accomplished by regulating the synthesis of a single polycistronic mRNA molecule encoding all of the products. The analogue to this arrangement in eukaryotes is the synthesis of a **polyprotein,** a large polypeptide that is cleaved after translation to yield individual proteins. Each protein can be thought of as the product of a single gene. In such a system the coding sequences of each gene in the polyprotein unit are not separated by stop and start codons but instead by specific *amino acid sequences* that are recognized as cleavage sites by particular protein-cutting enzymes. Polyproteins have been observed with up to ten cleavage sites; the cleavage sites are not cut simultaneously, but are cut in a specific order. Use of a polyprotein serves to maintain an equal molar ratio of the constituent proteins, as was seen earlier in the production of *Xenopus* rRNA; moreover, delay in cutting at certain sites introduces a temporal sequence of production of individual proteins, a mechanism frequently used by animal viruses.

Many examples of polyproteins are known. The synthesis of the RNA precursors uridine triphosphate and cytidine triphosphate proceeds by a biosynthetic pathway that at one stage has the reaction sequence

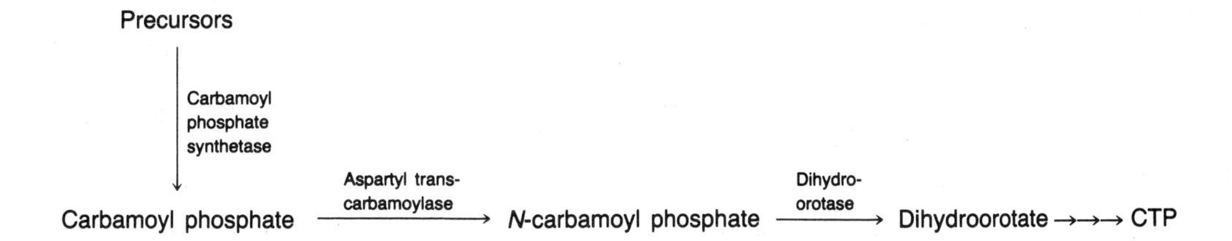

In bacteria each of the three enzymes is made separately. In yeast and *Neurospora* only two proteins are made; one is dihydroorotase and the other is a large protein that is cleaved to form carbamoyl phosphate synthetase and aspartyl transcarbamoylase. Mammals synthesize a single tripartite protein, which is cleaved to form the three enzymes.

Earlier it was noted that two different amounts of amylase are made in the salivary gland and in the liver by differential splicing of a single RNA molecule. The *same* protein was made in both tissues, though in different amounts. In chicken skeletal muscle two different forms of the muscle protein myosin are made from the same gene; these are called alkali light chains LC1 and LC3. This gene has two different transcription initiation sites (TATA sequences), which yield two different primary transcripts. These two transcripts are processed differently to form mRNA molecules encoding distinct forms of the protein (**Figure 16-6**). In *Drosophila* the mRNA is processed in four different ways, the precise mode depending on the stage of development of the fly. One class of myosin is found in pupae and another in the later embryo and larval stages. How the mode of processing is varied is not known.

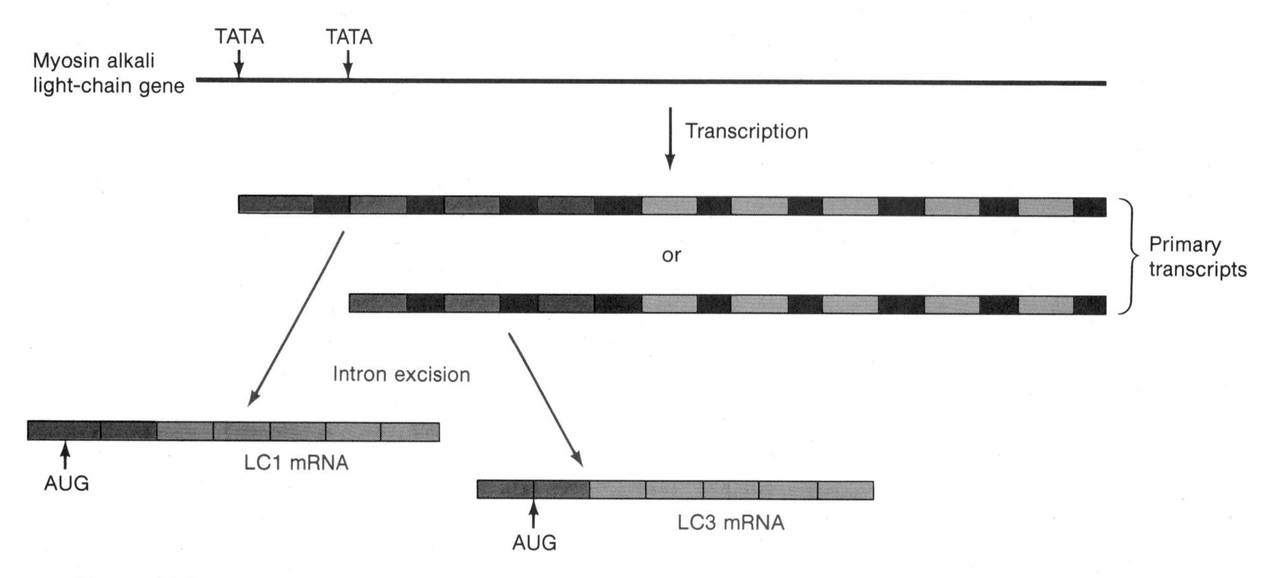

Figure 16-6
The chicken LC1/LC3 gene. Two distinct TATA sequences lead to the production of different primary transcripts, which contain the same coding sequences. Two modes of intron excision lead to the formation of distinct mRNA molecules, which encode proteins having different amino-terminal regions and the same carboxyl-terminal region.

GENE REARRANGEMENT: JOINING CODING SEQUENCES IN THE IMMUNE SYSTEM

In eukaryotes activation of gene expression is often accomplished by the juxtaposition of different segments of DNA. One example is the joining of a gene with its promoter, which has been observed in yeast and Drosophila. Another example is the joining of different structural genes to form a functional gene product, as in the production of **immunoglobulins (antibodies)**; this is the topic of this section.

Humans are able to synthesize more than 10^6 different antibody molecules—"different" in the sense of being an antibody to a different target molecule or **antigen.** If each antibody molecule were encoded in a distinct gene, a significant fraction of the DNA of a single cell would have to be used for antibody synthesis. The amount of DNA required could be reduced substantially if an antibody contained three subunits A, B, and C, and there were 100 different *A* genes, 100 different *B* genes, etc. In this way, $100 \times 100 \times 100 = 10^6$ different proteins would be generated by 300 genes. If each of the genes were active in every cell—that is, if 300 different proteins were synthesized in every antibody-producing cell—then each cell would produce 10^6 different antibodies. However, it is known that a mature antibody-producing cell makes only one type of antibody, so there must be some mechanism that is responsible for programming each cell. In this section it will be seen that such programming does occur.

Immunoglobulin G (IgG) is one of several classes of immunoglobulins. It is a tetramer containing two L chains and two H chains, as described in Chapter 4. Both the L and H chains consist of two segments, the constant (C) and the variable (V) region. The V region is responsible for recognition and binding a specific antigen. In this section, we consider how the enormous number of amino acid sequences of the V regions are formed from a comparatively small number of genes. There are two types of L chains, called κ and λ. We will only describe how the κ type is produced; the basic mechanism for the λ type is similar, differing only in detail.

Many genes can be used to form a κ-type L chain; they are of three types—*V*, *J*, and *C*. There are roughly 300 different *V* genes (which are responsible for the synthesis of the first 95 amino acids of the variable region), 4 different *J* genes (which encode the final 12 amino acids of the variable region and join the V and C regions), and 1 copy of the *C* gene (which encodes the constant region). In an embryonic cell the *V* genes form a tight cluster, the *J* genes form a second tight cluster quite far from the *V*-gene cluster, and the *C* gene follows not far after the *J*-gene cluster. All of the genes are on the same chromosome, as shown in Figure 16-7. Note that each *V* gene is preceded by leader regions where transcription can be initiated; the *J* and *C* genes are not preceded by leaders.

Regions encoding particular IgG molecules have been cloned (by recombinant DNA techniques) from various mouse cell lines, each producing a particular IgG molecule. The *V*-*J*-*C* region has also been cloned from mouse embryo cells, which have not yet been committed to

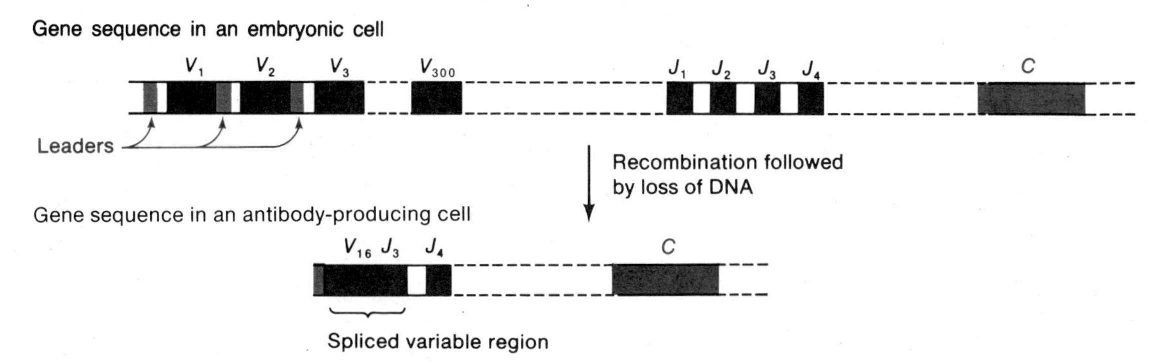

Figure 16-7
DNA splicing in the formation of the variable region of an L chain. The V_{16}–J_3 joint has been removed, presumably by a site-specific recombination event. The *V* genes to the left of V_{16} have been removed either by homologous recombination or by site-specific recombination within the leader sequences.

antibody synthesis and thus presumably contain an unaltered master set of genes for antibody synthesis. For each clone obtained from an antibody-producing cell line it has been found that a large segment of the embryonic DNA sequence is not present and that the missing segment is always a sequence between the particular *V* gene that encodes the first 95 amino acids of the V region and the *J* gene that encodes the last 12 amino acids of the V region. This is explained by a gene rearrangement in which DNA between the particular *V* and *J* genes is deleted. An example of one such rearrangement is shown in Figure 16-7. Many different gene sequences encoding particular IgG proteins have been cloned and it has been found that in each clone a different segment of DNA is not present. For example, in the figure the DNA between V_{16} and J_3 is absent; in another clone there might be a V_{210}-J_1 junction instead. In both cases a *V* gene and a *J* gene have become *spliced* together to form a complete gene for the variable region. Note that this is DNA splicing and not the RNA splicing that has been discussed in Chapter 9.

Studies of the base sequence of cloned IgG DNA sequences show that the junction between a particular *V* gene and a particular *J* gene is not always the same. That is, the two terminal triplets of the juxtaposing *V* and *J* genes can exchange at any one of four sites that yields a triplet. The meaning of this statement is shown in Figure 16-8. In the example shown there are three possible amino acids at the joint, which adds diversity to the number of possible variable regions.

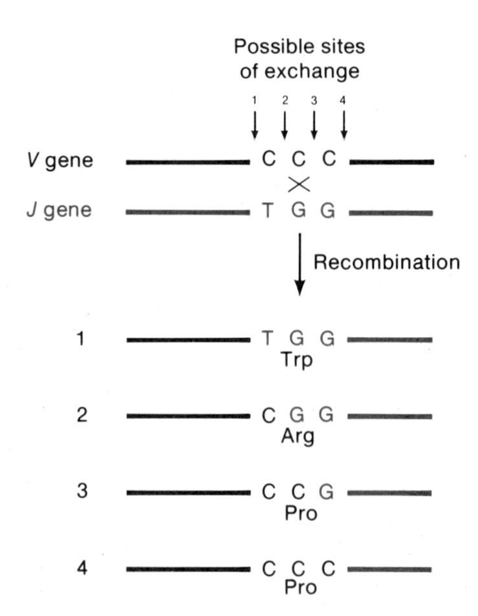

Figure 16-8
Four possible junctions at a *V-J* joint giving rise to three different amino acids.

Since there are 300 different *V* genes, 4 different *J* genes, and (on the average) 2.5 different amino acids at the junction, there are, then, $300 \times 4 \times 2.5 = 3000$ different variable regions. The H chain genes are organized in a similar but not identical way (there are four different types of genes) and there are about 5000 variable regions in the H chains. Thus, since each IgG molecule contains two identical H chains and two identical L chains, there are about $3000 \times 5000 = 1.5 \times 10^7$ different IgG molecules can be formed from the V_L, J_L, V_H, and J_H genes. This is ample to account for the diversity of antibody molecules.

The splicing of the *J* genes just described is thought to occur by means of a site-specific recombination mechanism. The signal for the *V-J* joining is probably a pair of specific base sequences to the right and left of the *V* and *J* genes, respectively. A base sequence

V gene–CACAGTG–11 bases–ACATAAAC

consisting of a 7-base-pair palindrome, an 11-base-pair spacer, and an 8-base-pair (A+T)-rich sequence is immediately to the right of each *V* gene. To the left of each *J* gene is the sequence

GTTTTTGT–22 bases–CACTGTG–*J* gene

consisting of (reading right to left) of a 7-base-pair palindrome identical (except for the fourth base) to the one that is adjacent to the *V* gene, a 22-base-pair spacer, and an 8-base-pair (A+T)-rich sequence almost complementary to the one near the *V* gene. Presumably, those sequences are recognized by the splicing system. Neither the enzymes required for splicing nor the nature of the chemical exchange nor the chemical events that trigger splicing are known.

Figure 16-9 shows that DNA splicing does not fully generate an L chain sequence since (1) the spacer between the *J* and *C* gene remains and (2) the actual L chain has amino acids derived from only one *V* gene and one *J* gene and the spliced DNA usually contains many *V* and *J* genes. The correct amino acid sequence is obtained by a final RNA-splicing event, as shown in the figure. Note how the particular *V-J* joint determines the RNA splicing pattern. For example, the RNA removed is always a segment between the leader and the *V* gene and between the *C* gene and the right end of the *J* gene in the *V-J* joint.

A major question is whether DNA splicing occurs in response to a particular antigen. The best evidence suggests that the events occur at random in the course of development and that when differentiation is complete, there are about 10^6 different antibody-producing cell types, each able to make a single type of antibody. How, then, does an

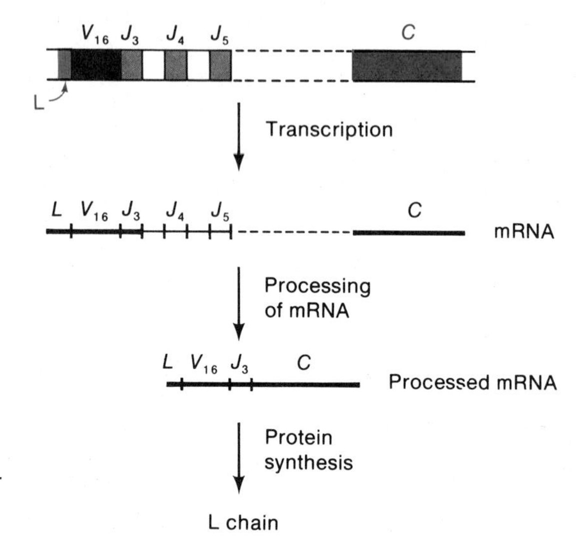

Figure 16-9
Production of the L chain of a particular IgG molecule. Solid regions indicate coding sequences used to generate the final L chain.

organism know when to make a huge amount of a particular antibody in response to a particular antigen? The answer to this question is not known with certainty but the following mechanism, called **clonal selection,** is probably correct—at least in outline. Each antibody-producing cell type makes a small amount of specific antibody. Some of this antibody becomes bound to the cell surface. When such a cell is exposed to the specific antigen that can react with the specific antibody, a complex antigen-antibody network forms, because each antibody molecule can interact with two antigen molecules and each antigen can interact with two antibody molecules (Chapter 4). By continual joining of antigens and antibodies, a single mass of material (called a cell cap) forms on the cell surface. The cell responds to this by an infolding of the cell membrane around the material. Once the cap has been taken into the cell, in a way that is totally obscure, cell division is stimulated and extensive production of antibody ensues. Thus, a clone of cells is generated which synthesizes a large amount of the specific antibody.

An example of the synthesis of a family of proteins—the amphibian and avian egg yolk proteins forming vitellogenin—which utilizes many of the processes described in this chapter, can be found in *MB*, pages 960–966.

Problems

The problems that follow are quite simple. For more challenging ones the student should consult *Problems for Molecular Biology*.

CHAPTER 1

1. (a) Is a cell in which the DNA is enclosed in a nuclear membrane a prokaryote or a eukaryote?
   (b) Classify the following cells as either prokaryotes or eukaryotes: bacteria, fungi, algae, yeasts, amoebae, wheat cells, human liver cells.
2. A bacterium divides every 35 minutes. If a culture containing 10^5 cells per ml is grown for 175 minutes, what will be the cell concentration?
3. Match the terms with the definitions:
   Terms
   1. Minimal medium. 2. Prototroph. 3. Plating. 4. Auxotroph.
   Definitions
   A. A cell that can grow in a minimal medium
   B. Depositing bacteria or phage on an agar surface for growth.
   C. A cell that requires for growth organic substances other than an organic carbon source.
   D. A growth medium containing salts and only one organic compound, which is used as a carbon source.
4. Define haploid and diploid.
5. Mutants that fail to synthesize a substance X have been found in four complementation groups, none of which are

cis-acting. How many proteins are required to synthesize X?

6. Four genes *kyuA*, *kyuB*, *kyuC*, and *kyuQ* are known to be required to synthesize substance Q and each biochemical reaction can be detected. The reaction sequence is $P \rightarrow B \rightarrow C \rightarrow A \rightarrow Q$ in which the product of a gene *kyuX* is needed to synthesize substance X. Addition of ^{14}C-P yields ^{14}C-Q.
   (a) A mutant is found for which addition of ^{14}C-P yields ^{14}C-A but no ^{14}C-Q. In what gene is this mutation?
   (b) Another mutant is found for which there is no conversion of ^{14}C-P to any other substance. Furthermore, addition of ^{14}C-A fails to yield ^{14}C-Q. What kind of mutant is this one?
7. In *E. coli* phage β some mutations in the phage gene *G* are compensated for by an additional mutation in gene *H* and some mutations in gene *H* are compensated for by mutations in gene *G*. What do these facts indicate?
8. (a) The following recombination have been observed: $a \times c$, 2%; $b \times c$, 13%; $b \times d$, 4%, $a \times b$, 15%; $c \times d$, 17%; $a \times d$, 19%. What is the gene order?
   (b) In the cross $aBd \times AbD$ what is the frequency of production of *ABD* progeny?
9. You are interested in the biosynthetic pathway of compound X and isolate 10 different (independently isolated) mutants of *E. coli* which require compound X for growth. The mutations are mapped and their approximate positions are given in Figure 1-9.
   (a) What is the minimum number of genes involved in the synthesis of compound X?
   (b) Why must your answer be the minimum estimate?
10. The complementation data shown in Table 1-10 are observed. The numbers refer to particular mutations. The symbols + and − indicate that the two mutations do and do not complement respectively. How many genes are represented? Assign the mutations to the genes.

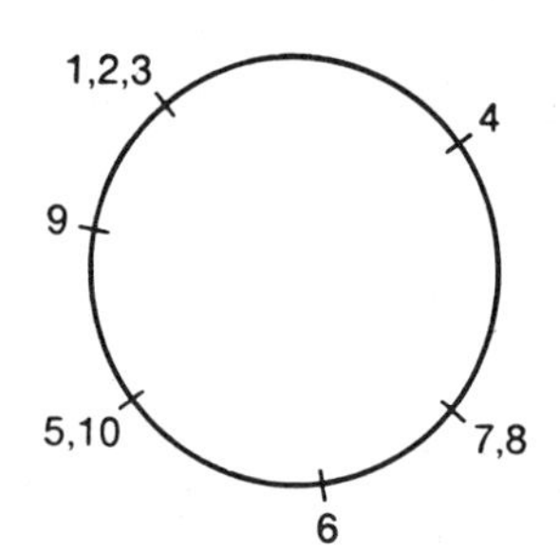

Figure 1-9

Table 1-10

	Mutants						
	1	*2*	*3*	*4*	*5*	*6*	*7*
1	−	+	+	+	+	+	−
2		−	+	−	+	+	+
3			−	+	+	−	+
4				−	+	+	+
5					−	+	+
6						−	+
7							−

CHAPTER 2

1. Which amino acid is unable to form a proper peptide bond?
2. What is a common state of cysteine in a protein?
3. Which amino acids can engage in hydrogen bonding?
4. What chemical groups are at the ends of a protein molecule?

5. In a nucleic acid molecule, which carbon atoms of the deoxyribose bear a phosphate, a hydroxyl group, and the base?
6. Which nucleic acid base is unique to DNA? to RNA?
7. What is the difference between a nucleoside and a nucleotide?
8. In a nucleic acid which carbon atoms are connected by a phosphodiester group?
9. (a) What chemical groups are at the end of a single DNA strand?
(b) What two chemical groups are at the end of a double-stranded DNA molecule?
10. What term describes the tendency of a macromolecule to fold so nonpolar groups cluster?
11. A mixture of different proteins is subjected to electrophoresis in three polyacrylamide gels, each having a different pH value. In each gel five bands are seen.
(a) Can one reasonably conclude that there are only five proteins in the mixture? Explain.
(b) Would the conclusion be different if a mixture of DNA fragments was being studied?

CHAPTER 3

1. What is the base sequence of the DNA strand complementary to each of the base sequences that follow?
(a) T G A T C A G G T C G A C
(b) A T A T A T A T A T A T A T
2. Which base pair, adenine-thymine or guanine-cytosine, contains a greater number of hydrogen bonds?
3. For double-stranded DNA
(a) What is the relation between the value of ([A] + [G]) and ([C] + [T]), in which the brackets indicate molar concentration?
(b) Of ([A] + [T]) and ([G] + [C])?
4. Order the DNA molecules shown below from lowest to highest melting temperature:

(I) A A G T T C T C T G A A
T T C A A G A G A C T T

(II) A G T C G T C A A T G C A G
T C A G C A G T T A C G T C

(III) G G A C C T C T C A G G
C C T G G A G A G T C C

5. Consider a long linear DNA molecule, one end of which is rotated four times with respect to the other end, in the unwinding direction. The two ends are then joined. If the molecule is to remain in the underwound state, how many base pairs will be broken? If the molecule is allowed to form a supercoil, how many nodes will be present?
6. State which of the following operations will induce formation of a supercoil: joining the ends of a linear DNA and doing nothing else; twisting the two ends of a linear DNA and then joining the ends together; joining the ends of a linear DNA and then twisting the circle.
7. Describe several ways in which a linear DNA molecule can form a circle.
8. Suppose ten protein molecules are bound to a circular DNA molecule having a nick. Each bound protein molecule breaks one base pair. The nick is then sealed with DNA ligase (an enzyme that seals 3′-OH–5′-P nicks) and the protein molecules are removed by treatment with a protease. What will be the shape of the DNA molecule after removal of the protein?
9. An exonuclease is added to a solution of

single-stranded DNA. Only a small fraction of the nucleotides is solubilized. Give several possible explanations for the resistance to the enzyme.

10. A single-stranded DNA molecule is being sequenced by the Maxam-Gilbert procedure. It is sufficiently long that all bands are not resolved in the gel at one time. Therefore, the sample is divided into two parts and each is electrophoresed for a different length of time (short run and long run), as shown below. What is the base sequence of the DNA?

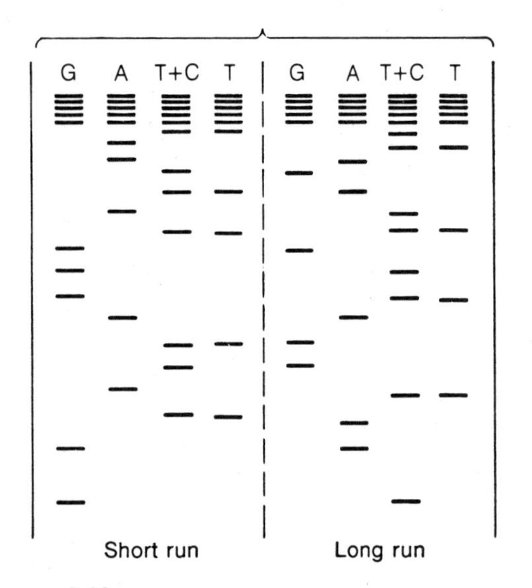

Figure 4-10

CHAPTER 4

1. (a) Which amino acids are polar and which are nonpolar?
   (b) Is isoleucine more nonpolar than alanine? Why?
2. About which bond in the polypeptide backbone is there no free rotation?
3. Name the kinds of bonds in proteins in which the side chains of each of the following amino acids might participate: cysteine; arginine; valine; aspartic acid.
4. What would you guess to be the environment of a glutamine that is internal? What is the environment of an internal lysine?
5. Which of the following sets of three amino acids are probably clustered within a protein? (1) Asn, Gly, Lys; (2) Met, Asp, His; (3) Phe, Val, Ile; (4) Tyr, Ser, Lys; (5) Aly, Arg, Pro.
6. State whether the following polypeptides would aggregate to form a multisubunit protein consisting of identical subunits and give the reason for your conclusion.
   (a) The folded molecule contains two distinct surface regions whose shapes are complementary and no polar amino acids are nearby.
   (b) There is a large hydrophobic cluster that is in a crevice just below the surface
   (c) A large hydrophobic patch is flanked by two lysines.
   (d) The surface has a region in which positively and negatively charged amino acids alternate and are in a linear array.
7. Is an enzyme likely to have a very rigid configuration? Explain.
8. What is meant by induced fit?
9. In general, a diploid cell containing the + and − alleles encoding an enzyme E has an E^+ phenotype. You have isolated a particular mutant that yields an E^- phenotype even when the + allele is present. Suggest a possible explanation for the observed phenotype.
10. Many proteins consist of several identical subunits. Some of these proteins have a single binding site, whereas others have several identical binding sites. What might you predict about the locations of the binding sites in these two classes of multisubunit proteins?

CHAPTER 5

1. Define the terms chromatin, histones, nucleosome, core particle, and linker DNA.
2. Suggest a function for the chromosome scaffolding protein in *E. coli*.
3. A DNA-protein complex has been isolated and you wish to determine whether the DNA and protein are covalently linked.
   (a) Which of the following experimental results would give information about covalent linkage? (1) Treatment of the complex with 2 M NaCl does not dissociate the protein from the DNA; (2) treatment of the complex with a reagent that breaks down hydrophobic interactions does not dissociate the complex; (3) digestion with a protease removes all detectable protein from the DNA.
4. A DNA-binding protein binds tightly to double-stranded DNA and very poorly to single-stranded DNA. It can bind to all base sequences with equal affinity. In 1 M NaCl binding is poor. What is the probable binding site?
5. Acridine molecules bind to DNA by intercalating between the base pairs. If the Cro protein and λ DNA are mixed, the Cro-DNA complex forms. What would happen if acridine were added to the DNA before the Cro protein was added?

CHAPTER 6

1. An early structural model for DNA was the so-called tetranucleotide hypothesis: four nucleotides (one each of adenine, cytosine, guanine, and thymine) were thought to be covalently linked to form a planar unit. It was suggested that units are linked together to yield a repeating polymer of the tetranucleotide. Explain why, if DNA had such a structure, it would probably not have become the genetic material of cells?
2. A criticism of the transformation experiments was that DNA might somehow be involved in the biosynthesis of the polysaccharide coat of the virulent *Pneumococcus*. Thus, the nonvirulent mutant would be lacking some step required for polysaccharide synthesis and this step could be bypassed by the addition of DNA. By this argument, DNA would not be a genetic substance. What kinds of experiments could be done to eliminate this argument?
3. Suppose that when the Hershey-Chase experiment was done, the following results were obtained. After blending and centrifugation, the ^{35}S sedimented into the pellet and the ^{32}P remained in the supernatant. Microscopic observation showed that after blending, all of the bacteria had been broken into small fragments. Electron microscopic examination of the pellet showed that there were no whole cells but just phage shells attached to small fragments of something (you do not know what). What else would you have to know, measure, or determine to enable you to draw the conclusion that DNA is the genetic material of the phage? Tell what reservations, if any, you might have about your conclusion.

CHAPTER 7

1. State whether each of the following is true or false.
   (a) In the synthesis of DNA the covalent

bond which forms is between a 3′-OH and a 5′-P group.

(b) In general, the DNA replicating enzyme in *E. coli* is DNA polymerase I.

(c) A single strand of DNA can be copied if the four nucleoside triphosphates and polymerase I are provided.

(d) If polymerase I is added to the four nucleoside triphosphates without a DNA template, DNA is synthesized but with a random base sequence.

(e) An RNA primer must be complementary in base sequence to some region of the DNA to initiate DNA synthesis.

2. From what substrates is DNA polymerized? What properties do all known DNA polymerases share?

3. What is meant by the terms primer and template?

4. What is the role of RNA in DNA replication?

5. What is the chemical group (3′-P, 5′-P, etc.) at the sites indicated by the dots labeled a, b, and c?

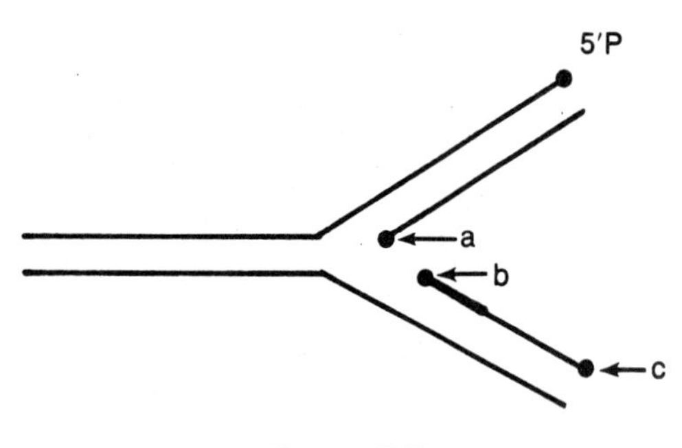

Figure 7-5

6. If one of the following enzymes is absent, not even one nucleotide can be added at the replication fork. Which enzyme is it? (1) Polymerase I (polymerizing activity); (2) polymerase I (5′ → 3′ exonuclease activity); (3) polymerase III, (4) DNA ligase.

7. To join together two precursor fragments, which of the following sequences of enzymes is probably used? Assume both fragments are already made.

(1) Polymerase I (5′ → 3′ exonuclease), polymerase I (polymerase), ligase.

(2) Polymerase I (5′ → 3′ exonuclease), polymerase III, ligase.

(3) Ribonuclease, polymerase III, ligase.

(4) Primase, polymerase I, ligase.

8. Distinguish the roles of helicases and ssb proteins in DNA replication.

9. It has been stated that helicases are needed because no DNA polymerase can unwind a double helix. However, before helicases were discovered, enzymologists were able to replicate DNA using reaction mixtures that probably contained no helicases.

(a) What property of DNA allows advance of the replication fork without a helicase?

(b) What effect would you expect a helicase to have on either the rate or fidelity of replication?

10. Suppose you have a DNA molecule with a gap in one strand 5000 nucleotides long and terminated with a 3′-OH group and a 5′-P group. If a DNA polymerase is added to this molecule *in vitro* (with everything else needed to make DNA), will the DNA filling the gap be a single piece or consist of short fragments? Would you expect fragments of any kind if the gap was filled *in vivo*?

11. What is the chemical group (for example, 3′-P, 5′-P, and so on) which is at the indicated terminus of the daughter strand of the extended branch of the rolling circle shown in Figure 7-11?

12. Answer the following questions about rolling circle replication.

(a) How many *de novo* initiation events

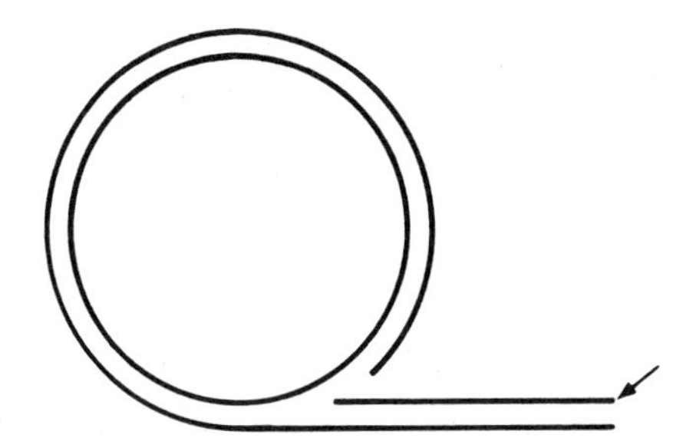

Figure 7-11

are needed in a complete replication cycle?

(b) If the circle has unit length, what is the maximum length of the linear branch that can be formed?

(c) What is the primer?

CHAPTER 8

1. Name the two most common changes in DNA that occur spontaneously without attack by external agents.
2. Name the repair system that
   (a) Cleaves thymine dimers.
   (b) Cleaves N-glycosylic bonds.
   (c) Cleaves phosphodiester bonds.
3. Uvr^+ bacteria possess the excision-repair system. The ability of ultraviolet-irradiated T4 phage to form plaques is the same on both Uvr^+ and Uvr^- bacteria. How might you explain this fact?
4. Name two features of SOS repair that distinguish it from all other repair systems.
5. An unexcised thymine dimer produces a partial block to DNA replication. It is observed that if there are unexcised thymine dimers in parental strand, the daughter strands contain large gaps which are frequently several thousand nucleotides long. Would you expect to find gaps in both daughter strands if there were only a single thymine dimer in the parent molecule? If not, in which strand? Explain.
6. A bacterial repair system called X removes thymine dimers. You have in your bacterial collection the wild-type (X^+) and an X^- mutant. Phage λ, when ultraviolet-irradiated and then plated, gives a larger number of plaques on X^+ than on X^- bacteria. It has been proposed on the basis of survival curve analysis that the X enzyme is inducible. To test this proposal, ultraviolet-irradiated λ phage are adsorbed to both X^+ and X^- bacteria in the presence of the antibiotic chloramphenicol (which inhibits protein synthesis). No thymine dimers are removed in the X^+ cell and 50 percent are removed in the X^- cell. In the absence of chloramphenicol, the same results were obtained.
   (a) Is X an inducible system?
   (b) Suppose 5 percent of the thymine dimers are removed in the presence of chloramphenicol and 50 percent in its absence; how would your conclusion be changed?

CHAPTER 9

1. (a) From what substrates is RNA made?
   (b) On what template?
   (c) With what enzyme?
   (d) Is a primer required?
2. Describe the differences, if any, between the chemical reactions catalyzed by DNA polymerase and RNA polymerase.
3. What chemical groups are present at the origin and terminus of a molecule of mRNA that has just been synthesized?
4. What are the functions of the core enzyme and the holoenzyme *in vivo*?
5. (a) What is mRNA?

(b) How does mRNA sometimes differ from a primary transcript?
(c) Define coding strand and antisense strand.
(d) Define cistron and polycistronic mRNA.
(e) What parts of a mRNA molecule are not translated?

6. What is a transcription unit and is it the same thing as a gene?
7. What is meant by the terms "upstream" and "downstream"?
8. The Pribnow box sequence is an example of what is termed a "consensus sequence"—namely, a base sequence from which other sequences having similar functions can be obtained by changing only one or two bases.
(a) What is the evolutionary significance of such a sequence?
(b) What is the biochemical significance of a conserved base, such as the conserved T in the Pribnow box?
(c) What biochemical differences might you expect between sequences that differ only slightly in the nonconserved bases?
9. An RNA molecule is isolated having a 3′-OH terminus and a 5′-P terminus. What information does this fact provide?
10. These questions refer to eukaryotic RNA.
(a) What is a cap?
(b) At which end of mRNA is the poly(A)?
(c) Are there eukaryotic mRNA molecules that do not contain either feature?
11. (a) In what way, relating to polycistronic mRNA, do eukaryotic and prokaryotic protein synthesis differ?
(b) What is currently thought to be the mechanistic basis of that difference?
12. (a) What are intervening sequences or introns?
(b) What is mRNA "splicing"?

CHAPTER 10

1. A DNA molecule has the structure

TACGGGAATTAGAGTCGCAGGATC
ATGCCCTTAATCTCAGCGTCCTAG

The upper strand is the coding strand and is transcribed from left to right. What is the amino acid sequence of the protein encoded in this DNA molecule?
2. Which amino acids can replace arginine by a single base-pair change?
3. The amber codon UAG does not correspond to any amino acid. Some strains carry suppressors which are tRNA molecules mutated in the anticodon, enabling an amino acid to be placed in at a UAG site. Assuming that the anticodon of the suppressor differs by only one base from the original anticodon, which amino acids could be inserted at a UAG site? At a UAA site?
4. There are several arginine codons. Suppose you had a protein that contained only three arginines (Arg-1, Arg-2, Arg-3). In a particular mutant, Arg-1 is replaced by glycine. In another mutant, Arg-2 is replaced by methionine. In still another mutant, Arg-3 is replaced by isoleucine. Suppose several hundred other mutants at various sites are isolated. Which other amino acids would you expect to find replacing Arg-1, Arg-2, and Arg-3, assuming only single-base changes?
5. Suppose that you are making use of the alternating copolymer GUGUGUGUGU ... as an mRNA in an *in vitro* protein-synthesizing system. Assuming

that an AUG start codon is not needed in the *in vitro* system, what peptides are made by this mRNA?

6. Which of the following properties are essential for the functioning of an aminoacyl synthetase? (1) Recognition of a codon; (2) recognition of the anticodon of a tRNA molecule; (3) recognition of the amino-acid recognition loop of a tRNA molecule; (4) ability to distinguish one amino acid from another.

7. What are the *s* values of prokaryotic and eukaryotic ribosomes, their subunits, and their RNA molecules?

8. Which of the following are steps in protein synthesis in prokaryotes? (1) Binding of tRNA to a 30S particle; (2) binding of tRNA to a 70S ribosome; (3) coupling of an amino acid to ribosome by an aminoacyl synthetase; (4) separation of the 70S ribosome to form 30S and 50S particles.

9. Which of the following statements are true of tRNA molecules?

(1) They are needed because amino acids cannot stick to mRNA.

(2) They are much smaller than mRNA molecules.

(3) They are synthesized without the need for intermediary mRNA.

(4) They bind amino acids without the need of any enzyme.

(5) They occasionally recognize a stop codon if the Rho factor is present.

10. Translation has evolved in a particular polarity with respect to the mRNA molecule. What would be the disadvantages of having the reverse polarity?

11. Which of the following is the normal cause of chain termination?

(1) The tRNA corresponding to a chain-termination triplet cannot bind an amino acid.

(2) There is no tRNA with an anticodon corresponding to a chain-termination triplet.

(3) Messenger RNA synthesis stops at a chain termination triplet.

12. Which processes in protein synthesis require hydrolysis of GTP?

CHAPTER 11

1. Distinguish the terms nonsense mutation and missense mutation.

2. One variety of a temperature-sensitive mutation is the cold-sensitive (Cs) mutation, which has a mutant phenotype below a particular temperature. Table 11-2 describes several mutations

Table 11-2

	32°C	*37°C*	*42°C*
ess2(Ts)	+	−	−
ess5(Ts)	−	−	+

in an essential gene ess, in which + and − refer to colony formation and the lack thereof, respectively. What would be the phenotype of an *ess*(Ts)*ess5*(Cs) double mutant?

3. Several hundred independent missense mutants, altered in the A protein of tryptophan synthetase, have been collected. Originally, it was hoped that at least one mutant for each of the 186 amino-acid positions in the protein would be found. However, fewer than 30 of the positions were represented with one or more mutants. Suggest some possibilities to explain why this set of missense mutants was so limited.

4. *E. coli* polymerase I possesses several enzymatic activities. Two important activities are the polymerizing function and the $3' \rightarrow 5'$ exonuclease. Mutant polymerases have been found which either increase or decrease mutation rates in an organism containing the mutant enzyme. A mutant that increases the mutation rate is called a mutator; a mutant that decreases the mutation rate is called an antimutator. The mutator and antimutator activities are usually a result of changes in the ratio of the two enzymatic activities described above. How do you think the ratios change in a mutator and in an antimutator?
5. Which of the following amino acid substitutions would surely yield a mutant phenotype? (1) Pro to His; (2) Lys to Arg; (3) Ile to Thr; (4) Ile to Val; (5) Ala to Gly; (6) Phe to Leu; (7) Tyr to His; (8) Arg to Ser.
6. Define the terms transition and transversion.
7. Consider a bacterial gene containing 1000 base pairs. As a result of treatment of a bacterial culture with a mutagen, mutations in this gene are recovered at a frequency of one mutant per 10^5 cells. One of these mutants is taken and grown and a pure culture of this mutant is obtained. This culture is then treated with the same mutagen and revertants are found at a frequency of one per 10^5 cells. Would you expect the gene product obtained from the revertant to have the same amino acid sequence as the wild-type cell? Explain.
8. Since nonsense suppressors are mutant tRNA molecules, how does the cell survive loss of a needed tRNA by such a mutation?
9. An enzyme contains 156 amino acids. (This number has no significance for the problem.) Suppose amino acid 28, which is glutamic acid, is replaced by asparagine in a mutant, and, as a result, all activity is lost. In this mutant protein, amino acid 76, which is asparagine, is replaced by glutamic acid and full activity of the enzyme is restored. What can you say about amino acids 28 and 76 in the normal protein?

CHAPTER 12

1. Could you have a plasmid with no genes whatsoever?
2. (a) What mode of DNA replication is used in transfer?
   (b) What early enzymatic step is needed in transfer replication, but not in normal replication?
3. In plasmid transfer, what are the roles of DNA synthesis in the donor and the recipient?
4. What procedure can be used to transfer a nontransmissible plasmid from one strain to another?
5. An $F'(\text{Ts})lac^+$ plasmid has a temperature-sensitive mutation in its replication system.
   (a) What is the phenotype of an $F'(\text{Ts})$ lac^+/lac^- cell at 42°C?
   (b) An $F'(\text{Ts})lac^+/lac^-gal^+$ strain is grown for many generations and then plated at 42°C. Some Lac^+ colonies form at 42°C. How have these formed?
   (c) Some of the Lac^+ colonies in (b) are Gal^-. How have these formed?
6. When transposition occurs, two base sequences are duplicated. What are they?
7. What is meant by the terms "direct repeat" and "inverted repeat"? Use the sequence ABCD as an example.
8. Since there is at present (1985) no detailed understanding of the mechanism of transposition, what is the evidence that

compels one to conclude that DNA replication is an essential step?

9. An Amp-r plasmid whose replication is temperature-sensitive is introduced into an Amp-s cell by the $CaCl_2$-transformation method. After growth for several generations 10^7 cells are plated on agar containing ampicillin at 42° C. Fifty colonies form.
(a) Give three mechanisms which could explain the presence of these colonies.
(b) Which mechanisms would not occur in a $RecA^-$ cell?

10. A particular IS element has inserted into different positions in two λ phages. In one case the insertion is at position 55; in the other, at position 75 (both measured on a scale of 100 from the left end of the phage. The IS element has a length equal to 1 percent of the size of λ DNA. One of these phage variants shows polar effects on downstream genes; the other does not.
(a) Explain the difference in the polar effects.
(b) How could your explanation best be tested by electron microscopy?

11. Sometimes in a bacterial strain carrying two different plasmids, one of which contains a transposon, a single plasmid arises that has the genetic information of both plasmids. If the cell is $RecA^-$, this fusion can occur in two possible ways—(1) illegitimate recombination and transposon-mediated formation of a cointegrate or (2) some other form of nonhomologous recombination. What information would be necessary to distinguish these mechanisms unambiguously?

12. Two bacterial genes (*chk* and *lat*) are very near one another. When chk^- mutants are isolated (they arise spontaneously at a frequency of about 10^{-6} mutants per cell per generation), about 1 percent of these mutants are also lat^-. This is a surprisingly high frequency for spontaneous formation of double mutants. Could you guess what might be the cause of this phenomenon?

CHAPTER 13

1. These questions concern restriction endonucleases.
(a) What is meant by a restriction enzyme?
(b) What do you think is the biological role of bacterial restriction enzymes?
(c) Why do bacteria not destroy their own DNA by their restriction enzymes?
(d) What is a type I and a type II enzyme?
(e) What common feature is present in every base sequence recognized by a restriction enzyme?

2. Two types of cuts are made by restriction enzymes. What are they and what terms are used to describe the termini formed?

3. What property must two different restriction enzymes possess if they yield identical patterns of breakage?

4. Restriction enzymes A and B are used to cut the DNA of phages T5 and T7, respectively. A particular T5 fragment is mixed with a particular T7 fragment and a circle forms that can be sealed by DNA ligase. What can you conclude about these enzymes?

5. A linear DNA molecule is digested with the EcoRI enzyme. Can all fragments circularize?

6. The enzyme terminal transferase is a DNA polymerase in the sense that it adds a nucleotide 5′-P to a free 3′-OH group. What property does it have that the usual polymerases do not have? Explain how terminal transferase can be used to join together two DNA molecules.

7. A Kan-r Amp-r plasmid is treated with the BglI enzyme, which cleaves the *amp* gene. The DNA is annealed with a BglI digest of *Drosophila* DNA and then used to transform *E. coli*.
(a) What antibiotic would you put in the agar to insure that a colony has the plasmid?
(b) What antibiotic-resistance phenotypes will be found on the plate?
(c) Which phenotype will have the *Drosophila* DNA?
8. Plasmid pBR607 DNA is circular and double-stranded and has a molecular weight of 2.6×10^6. This plasmid carries two genes whose protein products confer resistance to tetracycline (Tet-r) and ampicillin (Amp-r) in host bacteria. The DNA has a single site for each of the following restriction enzymes: EcoRI, BamHI, HindIII, PstI, and SalI. Cloning DNA into the EcoRI site does not affect resistance to either drug. Cloning DNA into the BamHI, HindIII and SalI sites abolishes tetracycline-resistance. Cloning into the PstI site abolishes ampicillin-resistance. Digestion with the following mixtures of restriction enzymes yields fragments with the sizes listed in Table 13-8. Position the PstI, BamHI. HindIII, and SalI cleavage sites on a restriction map, relative to the EcoRI cleavage site.

Table 13-8

Enzymes in mixture	*Molecular weights of fragments (millions)*
EcoRI, PstI	0.46, 2.14
EcoRI, BamHI	0.2, 2.4
EcoRI, HindIII	0.05, 2.55
EcoRI, SalI	0.55, 2.05
EcoRI, BamHI, PstI	0.2, 0.46, 1.94

CHAPTER 14

1. Distinguish negative and positive regulation.
2. What is the meaning of the following terms: repressed, induced, constitutive, coordinate?
3. Is the regulation of gene activity in prokaryotes mainly transcriptional or post-transcriptional? At what steps of RNA synthesis is transcriptional regulation exerted in different systems?
4. Suppose *E. coli* is growing in a growth medium containing lactose as the sole source of carbon. The genotype is $i^-z^+y^+$. Glucose is then added. Which one of the following will happen?
(1) Nothing
(2) Lactose will no longer be utilized by the cell.
(3) *Lac* mRNA will no longer be made.
(4) The repressor will bind to the operator.
5. A cell with genotype $i^+z^+y^+$ is in a growth medium containing neither glucose nor lactose (that is, it uses some other carbon source). How many proteins are bound to the DNA comprising the *lac* operon? How many if glucose is present?
6. A Lac$^+$ Hfr that is (its genotype is $i^+o^+z^+y^+$) is mated with a female that is $i^-o^+z^-y^-$. In the absence of inducer, α-galactosidase is made for a short time after the Hfr and female cells have been mixed. Explain why it is made and why only for a short time. What would happen if the female were $i^+o^cz^-y^-$?
7. Explain how lactose molecules first enter an uninduced $i^+z^+y^+$ cell to induce synthesis of β-galactosidase.
8. For each of the *E. coli* diploids that follow, indicate whether the strain is inducible or constitutive, or negative for

β-galactosidase and permease, respectively.

(a) $i^+o^+z^-y^+/i^+o^cz^+y^+$.

(b) $i^+o^+z^+y^+/i^+o^cz^+y^-$.

(c) $i^-o^+z^-y^+/i^-o^cz^+y^+$.

(d) $i^-o^+z^-y^+/i^+o^cz^+y^-$.

The symbols *i*, *o*, *z*, *y*, are abbreviations for *lacI*, *lacO*, *lacZ*, and *lacY*.

9. An *E. coli* mutant is isolated that renders the cell simultaneously unable to utilize a large number of sugars, including lactose, xylose, and sorbitol. However, genetic analysis shows that each of the operons responsible for utilization of these sugars is free of mutation. What are the possible genotypes of this mutant?

10. What is meant by an attenuator?

11. How many proteins are bound to the *trp* operon

(a) When tryptophan and glucose are present?

(b) When tryptophan and glucose are absent?

(c) When tryptophan is present and glucose is absent?

12. Consider a branched pathway, shown in Figure 14-12, which is regulated by feedback inhibition.

(a) Indicate the enzymes which are subject to feedback inhibition and for each, identify the inhibitor.

(b) Which steps are likely to be catalyzed by a set of isozymes? How many isozymes are probably in these steps?

CHAPTER 15

1. Bacteria are allowed to grow on an agar surface until a confluent turbid layer appears. Then, 10^3 T4 phage are spread on the surface. Six hours later (a time sufficient for plaque formation, if the phage had been added at the time the bacteria were placed on the agar) no plaques are evident. Explain.

2. Name the minimal (essential) structural components of the simplest phage particle.

3. Give several examples of mechanisms by which a phage converts a bacterium to a phage-producing machine.

4. What would you expect to happen if, after virulent phage infection (such as with T4), all the phage genes were transcribed and translated at once following phage DNA infection?

5. If an *E. coli* culture is simultaneously infected by phages T4 and T7, each at a multiplicity of infection of 5, only T4 phage will be produced. From what you

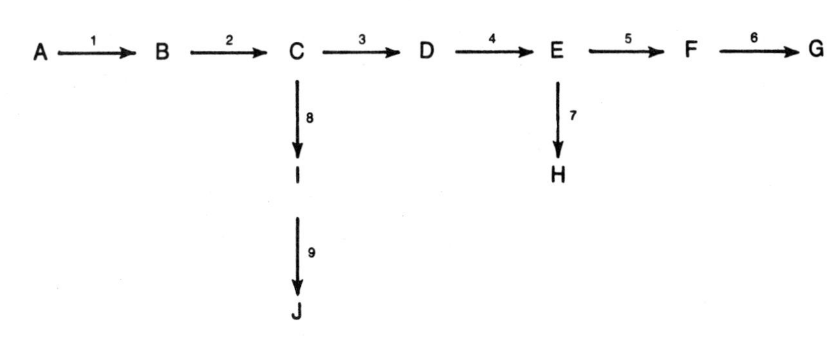

Figure 14-12

know about T4 biology, propose a simple explanation. What would you expect to happen if the T4 was added 10 minutes after addition of T7?

6. Describe the course of T4 phage DNA synthesis following infection of *E. coli* with a T4 mutant which cannot synthesize (a) cytidine hydroxymethylase, or (b) α-glycosyl-transferase.
7. If a T4 phage has a large deletion which of the following will be true?
   (1) The activity of one or more proteins will be greatly altered or altogether missing.
   (2) The phage DNA will be smaller.
   (3) The terminally redundant region will be larger.
   (4) The phage DNA will be the same size.
   (5) Cyclic permutation will be eliminated.
8. If an *E. coli* culture is heavily irradiated with ultraviolet light and then infected with T4 phage, the burst size of T4 is nearly normal. If it is instead infected with λ phage, very few phage are produced. Explain this difference.
9. How many λ particles can be packaged from a dimer circle? A trimer circle?
10. The *cI* gene of λ codes for the immunity repressor. If the λ is cI^-, the plaque is uniformly transparent. If it is ci^+, the plaque has a turbid center. Explain the plaque morphology. Would you expect T2 plaques ever to have turbid centers?
11. Suppose you have a λ mutant that makes a clear plaque on *E coli* strain A and a turbid plaque on strain B. How can this difference be explained?
12. A temperate phage has the gene order *A B C D E F G H* and a prophage gene order *G H A B C D E F*. What information does this give you about the phage?
13. Phage variants have been prepared containing many different attachment sites.
    (a) What attachment sites are in λ *bio* and λ *gal* phage?
    (b) What attachment sites are generated by Int-mediated recombination between λ *bio* and λ *gal*?
14. When Hfr males conjugate with F^- cells lysogenic for λ, zygotes normally survive. However, when Hfr males lysogenic for λ conjugate with F^- nonlysogens, zygotes produced from matings that have lasted for almost two hours lyse, owing to the zygotic induction of λ.
    (a) How can you explain zygotic induction?
    (b) How can you determine the locus of the integrated λ prophage?
15. Infection of a λ lysogen with ten λ phage particles does not result in phage development. This is the immune response. If the multiplicity of infection is 30, phage development occurs. In a dilysogen 60 phage particles per cell are needed to initiate a successful infection. Explain.

CHAPTER 16

1. (a) Why are eukaryotic genes not organized in operons?
   (b) How does a gene family differ from an operon?
2. In discussing gene families one frequently mentions embryonic, fetal, and adult forms of a protein. To what phenomenon do these terms refer? What term is used to describe a gene family of this sort?
3. (a) Are all members of a gene family transcribed in the same direction?
   (b) Are all members of a gene family actively transcribed in all tissues of a single organism?
   (c) Are all members of a gene family on the same chromosome?
   (d) What is a fairly common location for

repetitive sequences in clustered gene families?

4. Differentiated cells are often called upon to produce an enormous amount of a particular protein. What methods are used when only a limited time is available for this synthesis and when a long time is available?
5. Describe in outline the mechanism for generating antibody diversity.
6. (a) If you wanted to redesign the mouse genome to increase antibody diversity, but by adding the least amount of DNA, which component would you increase? (b) If an organism has 150 different *V* genes, 12 *J* genes, and 3 possible *V-J* joints for L chains, and can make 5000 different H chains, how many different antibodies can be made?
7. A particular hormone typically acts on only one or a small number of cell types. What is the usual property of a cell for determining whether a cell does or does not respond to a hormone?
8. A particular cell makes substance X in response to a hormone H. Prior to exposure of a culture of the cells to H, a large RNA molecule can be found in the nucleus. It is capped and has a poly(A) tail. After addition of H, this RNA is not found but a second RNA, which in an *in vitro* protein-synthesizing system makes X, is found in cytoplasmic polysomes. Both RNA molecules hybridize to the same cloned DNA molecule. What simple mechanism might be used for hormonal control of synthesis of X?
9. Several cases are known in which a single effector molecule regulates the synthesis of different proteins encoded in distinct mRNA molecules 1 and 2. Give a brief possible molecular explanation for each of the following observations made when the effector is absent.
(a) Neither nuclear nor cytoplasmic RNA can be found that hybridizes to either of the genes encoding molecules 1 and 2.
(b) Nuclear but not cytoplasmic RNA can be found that hybridizes to the genes encoding molecules 1 and 2.
(c) Both nuclear and cytoplasmic RNA but not polysome-associated RNA can be found that hybridizes to the genes encoding molecules 1 and 2.
10. A gene family containing three genes is being studied. Messenger RNA molecules for the three genes appear in the cytoplasm at nearly the same time following turn-on of transcription. It has been hypothesized that a single transcript is made and that it is processed in three different ways (perhaps at random to yeild the three mRNA molecules). The principal evidence is that the same sequence of four nucleotides is adjacent to the cap in all three mRNA molecules. The entire gene family has been cloned as a single unit in the plasmid pBR322. The

Figure 16-10

plasmid DNA is partly denatured and all three mRNA molecules are added using conditions in which DNA-RNA hybrids form. Thus, some of the plasmid molecules have R loops, that is, bubbles in which one of the DNA strands is joined to mRNA rather than to its complement. The molecules in Fig. 16-10 are observed in electron micrographs of these mRNA-containing plasmids (other forms are also seen). Could the hypothesis be correct?

Answers

CHAPTER 1

1. (a) Eukaryote.
(b) Only bacteria are prokaryotes.

2. Growth is for five generations. Thus, the initial number increases by a factor of 25 or 32 and the final concentration is 32 $\times 10^5 = 3.2 \times 10^6$ cells/ml.

3. 1, D; 2, A; 3, B; 4, C.

4. Haploid—having only one copy of each chromosome; diploid—having two copies of each chromosome.

5. In general, at least four. However, intragenic complementation may be possible. That is, if the protein molecule is folded in such a way that it consists of two interacting regions, a deleterious change in one region (which destroys the interaction) could be compensated for by a change (which, by itself, is also deleterious) in the other region. In other words, the two incorrectly folded regions could interact correctly to yield a functional protein.

6. (a) *kyuQ*.
(b) A regulatory mutant that prevents synthesis of each gene product.

7. That the *G*-gene and *H*-gene products probably join together to form an active macromolecular complex. The regions in which mutations occur (that can be compensated or are compensatory) are probably in the binding sites.

8. (a) The gene order is *a c b d*.
(b) $0.15 \times 0.04 = 0.006$.

9. (a) Six genes.
(b) Some of the closely linked mutations might lie in closely linked genes, that is, they might complement one another. Moreover, other genes in the pathway for which no mutants have yet been isolated might exist. Ten mutations distributed over seven or eight genes according to the Poisson distribution would stand a good chance of leaving one gene untouched.

10. Three genes. The groups are: (1,7), (2,4), and (3,6).

CHAPTER 2

1. Proline.
2. Two cysteines cross-linked by a disulfide bond.
3. All, using the CO and NH groups in the peptide bond. In addition, all polar amino acids can form hydrogen bonds via their side chains.
4. A carboxyl and an amino group.
5. 1′, base; 3′, OH; 5′, phosphate.
6. Thymine, DNA; uracil, RNA.
7. A nucleoTide is a nucleoSide phosphate.
8. The 3′ and 5′ carbons.
9. (a) 3′-OH and 5′-P.
   (b) One 3′-OH and one 5′-P group.
10. A hydrophobic interaction.
11. (a) No, since differences in shape can affect mobility—that is, two proteins differing in both molecular weight and shape might have the same electrophoretic mobility.
    (b) Yes, for all fragments have the same basic shape and same charge-to-mass ration.

CHAPTER 3

1. (a) ACTAGTCCAGCTG.
   (b) TATATATATATAT.
2. GC has 3.
3. (a) [purines]/[pyrimidines] = ([A] + [G]) = ([C] + [T]) = 1.
   (b) This ratio depends on the particular DNA molecule.
4. The order should be I, II, III, since melting temperature increases with increasing (G+C) content.
5. $4 \times 10 = 40$ base pairs. Four nodes.
6. Only twisting followed by joining.
7. A linear DNA molecule can form a circle by cohesion of single-stranded termini; by recombination between redundant termini; and by exonucleolytic conversion of double-stranded termini to single stranded termini, followed by hydrogen bonding between single strands.
8. Ten base pairs are contained in one turn of a DNA helix. Thus, the molecule will be a supercoil having one node—that is, a figure-8.
9. Possibilities include a circular DNA molecule, altered termini (for example, altered 3′-P or 5′-OH groups), a blocked terminus (for example, blocked by an ester), and a hairpin producing a double-stranded terminus.
10. The short run gives the sequence GGTACTAGGGTATCAAT. The long run gives the sequence CAATGGATCGTCAGATC. Presumably the CAAT is common to both so the sequence is GGTACTAGGGTATCAATGGATCGTCAGATC.

CHAPTER 4

1. (a) The polar amino acids are arginine, asparagine, aspartic acid, cysteine, glutamic acid, glutamine, histidine, lysine, serine, threonine, and tyrosine. The nonpolar amino acids are alanine, glycine, isoleucine, leucine, methionine, phenylalanine, proline, tryptophan, and valine.
   (b) Isoleucine is more nonpolar than alanine, because it has a long, nonpolar side chain. Proline is unable to form a proper peptide because it lacks a free amino group.
2. The peptide bond.
3. Cysteine—disulfide; arginine—ionic; valine—hydrophobic; aspartic acid—ionic and hydrogen bonds. All of course could be in van der Waals bonds.
4. Both will tend to be near an amino acid of opposite charge.

5. Set 3, in a hydrophobic cluster.
6. (a) Yes. A van der Waals attraction would favor aggregation. This would be aided by a hydrophobic interaction if the side chains on the surface were very nonpolar.
(b) No. For geometric reasons the hydrophobic regions probably cannot come into contact.
(c) No. Charge repulsion will effectively counteract any tendency to form a hydrophobic cluster.
(d) Yes. If the geometry is appropriate, the alternation of unlike charges will allow unlike charges in the two molecules to attract.
7. No. In general, enzymes must be slightly flexible in order to adapt their shape to the substrate and to carry out the required chemical reaction.
8. The flexing of an enzyme about a substrate.
9. The enzyme could be a multisubunit protein in which one defective subunit is sufficient to eliminate enzymatic activity.
10. When there is a single binding site, it is probably located at or near the junction of all of the subunits. This location is certainly unlikely if there are several identical binding sites. If the number of binding sites equals the number of subunits, one might guess that the binding sites are far from all regions of contact between the subunits. If the number of binding sites is half the number of subunits, each site probably includes the contact region of two subunits.

CHAPTER 5

1. Chromatin—the DNA-histone complex present in eukaryotic chromosomes; histones—a class of five proteins bound to DNA in chromatin; nucleosome—a DNA-histone complex consisting of a DNA-histone core particle, linker DNA, and histone H1; core particle—a DNA-histone complex consisting of an octamer (two copies each of H2A, H2B, H3, and H4) and a 140-base-pair DNA molecule; linker DNA—the segment of DNA that is not bound to a histone octamer.
2. It probably enables the DNA molecule, whose length is about 400 times that of a bacterium, to fit neatly inside the cell. It may also be important in guiding daughter DNA molecules to daughter cells.
3. (a) None. Results (1) and (2) test only one type of linkage. Result (3) is independent of the nature of the linkage.
(b) Complete digestion with a protease and a nuclease and identification of a component which is neither a free amino acid or a free nucleotide but consists of one amino acid and one nucleotide. One could show that the amino acid and nucleotide are joined by observing comigration during electrophoresis or chromatography; since amino acids and nucleotides do not bind noncovalently, one can assume that the amino acid and the nucleotide are covalently joined in this complex.
4. The lack of sequence specificity suggests that either the phosphates or deoxyribose are involved in binding. The effect of 1 M NaCl indicates ionic binding. Thus, the phosphates are likely to be very important components of the binding site.
5. These molecules are intercalating agents and would move the DNA bases apart. Since the Cro protein binds stereospecifically, it is likely that the Cro-DNA complex would not form. Also, the intercalating agent might block specific contact points.

CHAPTER 6

1. As a simple repeating polymer consisting of a single unit, it cannot carry information for an amino acid sequence.
2. Repeat the transformation with other genetic markers.
3. Blending has apparently broken open the cells and the fragments of the cell wall plus any material that has adhered to them are in the pellet. Thus you could show that protein is in the pellet (plus some uninjected DNA) and that injected DNA is in the supernatant.

CHAPTER 7

1. (a) and (e) are true.
2. Deoxynucleoside triphosphates. Addition to 3′-OH group; requirement for a template.
3. Primer—a nucleotide bound to DNA and having a 3′-OH group; template—a polynucleotide strand whose base sequence can be copied.
4. RNA serves as a primer.
5. (a) 5′-P. (b) 3′-OH. (c) 5′-P.
6. (3).
7. Sequence 1.
8. A helicase unwinds a helix. The ssb proteins prevent the helix from rewinding (forming intermolecular hydrogen bonds) and prevent intramolecular base pairing from occurring.
9. (a) Breathing.
   (b) Without a helicase, replication depends on breathing as the sole cause of unwinding. Thus, replication is much slower when there is no helicase. There is no effect on fidelity.
10. The gap will be filled by a single piece. No fragments would be formed as long as one end of the gap had a 3′-OH group.
11. 3′-OH.
12. (a) One.
    (b) There is no maximum.
    (c) One of the parental single strands.

CHAPTER 8

1. Spontaneous deamination of cytosine and depurination.
2. (a) Photoreactivation.
   (b) A base-N-glycosylase.
   (c) Excision repair.
3. Either the damage to T4 DNA is, for some reason, not repairable by the Uvr system or T4 possesses its own repair system. The latter is correct.
4. It is inducible and it is error-prone.
5. If the dimer is in the strand being copied by the leading strand, only the leading strand would have a gap because the open helix stabilized by ssb protein allows synthesis of the lagging strand. The important question is whether a dimer is a block to helicase. If so, a dimer in the template for the lagging strand would block advance of the leading strand and produce gaps in both leading and lagging strands. There is no information at present on this point.
6. (a) X is not inducible, because the enzymes were present before protein synthesis was blocked by chloramphenicol.
   (b) X would probably be considered inducible. The residual 5 percent could be due to either a second noninducible system, which can excise thymine dimers, or to a small amount of synthesis of X proteins when chloramphenicol is present.

CHAPTER 9

1. (a) Ribonucleoside 5′-triphosphates.
   (b) Double-stranded DNA.

(c) RNA polymerase
(d) No.

2. The reactions are identical—namely, reaction of a nucleoside 5′-triphosphate with a 3′-OH terminus of a nucleotide to form a 5′-3′-dinucleotide. The substrates differ in that DNA polymerase joins deoxynucleotides and RNA polymerase joins ribonucleotides.

3. A 5′-triphosphate and a 3′-OH.

4. The holoenzyme is needed to initiate RNA synthesis at a promoter. The core enzyme is needed for continued synthesis after initiation.

5. (a) An RNA molecule containing one or more contiguous base sequences that are translated into one or more proteins.
(b) A primary transcript is a complementary copy of a DNA strand. It may contain mRNA, tRNA, or rRNA and may be processed before translation can occur.
(c) A coding strand is a segment of a DNA strand that is copied by RNA polymerase. An antisense sense is a DNA strand that is complementary to a coding strand.
(d) A cistron is a DNA segment between and including translation start and stop signals that contains the base sequence corresponding to one polypeptide chain. A polycistronic mRNA molecule contains sequences encoding two or more polypeptide chains.
(e) Leaders, spacers, and the unnamed region following the last stop codon of a mRNA are untranslated regions.

6. A transcription unit is a section of DNA extending from a promoter to an RNA polymerase termination site. It is usually not a gene, but typically includes many genes.

7. Upstream and downstream usually refer to regions in the 5′ and 3′ directions, respectively, from a particular site that is being discussed. Sometimes, the terms are used specifically in the following way: upstream—a region of the DNA before the first base of an mRNA that is transcribed; downstream—in the 3′ direction (of the RNA) from the first base transcribed.

8. (a) It is probably a sequence that existed at very early times and from which others were derived by mutation. Furthermore, it indicates that the biochemical system that uses the sequence existed very long ago.
(b) It is essential for some stage of promotion—in particular, binding of RNA polymerase or initiating polymerization.
(c) The rates of initiation may differ slightly. It is difficult to be certain of this point though because the rate of initiation is mainly determined by the −35 region.

9. Since the triphosphate of the primary transcript is absent, the molecule has been processed. The processing could be extensive or as simple as triphosphate hydrolysis but there is no way of knowing from the information given.

10. (a) A terminal structure in which a methylated guanosine is in a 5′-5′-triphosphate linkage at the 5′ terminus of mRNA.
(b) The 3′-OH end.
(c) All mRNA molecules (except those of several viruses) are capped. Some mRNA molecules lack the poly(A) tail.

11. (a) In eukaryotes all mRNA is monocistronic.
(b) Only the translation start signal nearest to the cap is utilized.

12. (a) Untranslated sequences that interrupt the coding sequence of a transcript and that are removed before translation begins.
(b) Removal of introns.

CHAPTER 10

1. Met Pro Leu Ile Ser Ala Ser.
2. Histidine, glutamine, cysteine, tryptophan, serine, glycine, leucine, proline, isoleucine, threonine, lysine, and methionine.
3. For UAG, the amino acids are Tyr, Leu, Trp, Ser, Lys, Glu, and Gln. For UAA they are Tyr, Lys, Glu, Gln, Leu, and Ser.
4. Arg-2 could be replaced by Gly, Trp, Lys, Thr, or Ser. Arg-3 could be replaced by Ser, Lys, Thr, and Gly. The Arg-1 codon cannot be identified.
5. . . . Val-Cys-Val-Cys-Val-Cys. . . and peptides of various sizes starting with Val or Cys.
6. 3 and 4.
7. Prokaryotic: 70S with 30S and 50S subunits containing 5S, 16S, and 23S RNA. Eukaryotic: 80S with 40S and 60S subunits containing 5S, 5.8S, 18S, and 28S RNA.
8. Steps 1, 2, and 4.
9. (1), (2), (3).
10. Synthesis would have to await the completion of a molecule of mRNA. In the existing system protein synthesis can occur while the mRNA is being copied from the DNA. Thus, protein synthesis can start earlier than would be possible with reverse polarity, and the mRNA is relatively resistant to nuclease attack.
11. Statement 2.
12. Formation of the 70S initiation complex and translocation.

CHAPTER 11

1. A nonsense mutation stops chain growth at the mutational site; a missense mutation causes an amino acid substitution at the mutational site.
2. This mutant could not be isolated because there is no temperature at which it could grow.
3. Many amino acid changes will not yield a mutant phenotype and many mutations will be chain-termination mutants.
4. The ratio of polymerizing activity to exonuclease function will decrease in the antimutator since errors will be removed more often.
5. (1), (3), (7), and (8), as they involve either changes in polarity, charge sign, or chemical properties.
6. A transition is a base-pair change in which the purine-pyrimidine orientation is not altered; a transversion is a base-pair change in which the orientation is changed.
7. No. The frequency of exact replacement should be at least 1000 times lower.
8. There are two possibilities: (1) There are two genes for the original tRNA species, and only one of these has been mutated. (2) Natural chain-termination sequences might usually consist of two or more different termination codons. The nonsense suppressor will suppress only one, and chain termination will still occur. Both possibilities occur.
9. They clearly interact. Since a charge sign-change yields a mutant and a second sign-change in another amino acid yields a revertant, amino acids 28 and 76 are probably held together by an ionic bond.

CHAPTER 12

1. Yes. The simplest plasmid would be a tiny DNA fragment carrying no information other than a copy of the replication origin of the host cell.

2. (a) Rolling circle or looped rolling circle replication.
(b) A nick producing a 3′-OH terminus.
3. Synthesis in the donor provides the single strand that is copied. Synthesis in the recipient converts the transferred strand to double-stranded DNA.
4. $CaCl_2$ transformation using purified plasmid DNA.
5. (a) Lac^-.
(b) $F'(Ts)lac^+$ has integrated into the bacterial chromosome.
(c) Integration has occurred inside a *gal* gene.
6. The transposon itself and the target sequence.
7. Direct repeat: ABCD....ABCD, in which the dots represent nonrepeated bases. Inverted repeat: ABCD....D′C′B′A′, in which X′ is the complement to X.
8. The fact that a replica of the transposon is found at a new position without loss of the original transposon.
9. (a) Integration of the plasmid into the chromosome by homologous recombination; reversion to temperature-insensitivity; transposition of the *amp* gene from the plasmid to the chromosome.
(b) Integration (if homology-dependent).
10. (a) The IS elements are inserted in two different orientations and there is a transcription stop sequence in only one orientation.
(b) Heteroduplexes between the DNA of the two phages would indicate that the transposons are the same or different.
11. The number of copies of the transposon in the new plasmid. These would be two copies if the fusion was mediated by a transposon.
12. A high frequency of adjacent mutations suggests that a deletion including these genes has occurred. The deletion frequency is sufficiently high that one might reasonably suspect that a transposon is adjacent to the genes and that the phenomenon is an example of transposon-mediated deletion.

CHAPTER 13

1. (a) An endonuclease that makes cuts in DNA at one particular base sequence.
(b) To destroy foreign DNA.
(c) Bacteria generally methylate one base in the sequence recognized by their own restriction enzymes and this methylation renders the sequence resistant to the enzyme.
(d) A type II enzyme makes cuts only within the sequence recognized. A type I enzyme makes cuts elsewhere.
(e) Each sequence has rotational (dyad) symmetry.
2. Flush or blunt ends are generated by single-strand breaks that are opposite one another. Cohesive ends are formed by single-strand breaks that are separated by a few nucleotides.
3. They must both recognize the same base sequence.
4. Both enzymes recognize the same base sequence. They might even be the same enzyme.
5. No. The two terminal fragments each have one blunt end.
6. Terminal transferase does not need a template. A poly(dA) strand can be added to one DNA molecule and a poly(dT) strand to another. The two molecules can be joined by hydrogen bonding between the poly(dA) and the poly(dT) tails. This is called homopolymer tail joining.
7. (a) Kanamycin.
(b) Kan-r Amp-r, Kan-r Amp-s.
(c) Kan-r Amp-s.

8.

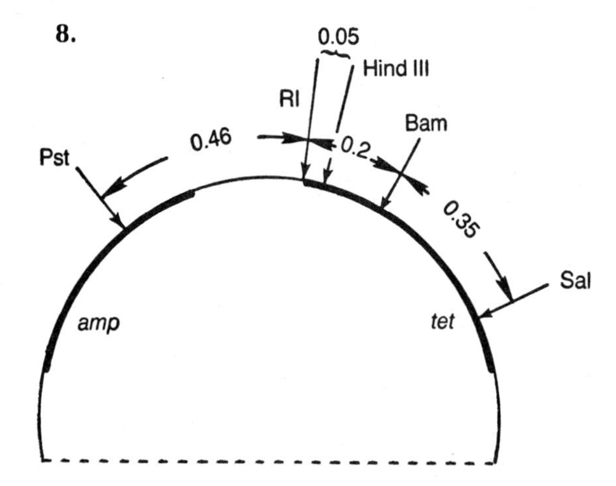

Figure A-13-8

CHAPTER 14

1. Positive—the active regulatory protein enhances transcription above what it would be in its absence (for example cAMP bound to CAP protein). Negative—the active regulatory protein blocks transcription.
2. Repressed—turned off, inactive; induced—turned on, active; constitutive, not regulated; coordinate—regulated together.
3. Regulation is transcriptional and exerted at the initiation and termination (attenuation) steps.
4. (2).
5. Two. One if glucose is present.
6. When the *lac* operon enters the female, no *lac* repressor is present, so transcription of z and y occurs and enzyme is made. However, soon after transfer repressor mRNA and hence repressor are made, which will shut off transcription of z and y and hence enzyme synthesis. If the female is $i^+o^cz^-y^-$, repressor is present, and no enzyme is made.
7. At very low levels all inducible operons are transcribed—for example, a repressor might come off the operator for an instant and RNA polymerase will get on. Hence, each cell will contain a few β-galactosidase and permease molecules. The permease will bring in a few lactose molecules and these will be converted to the true inducer, allo-lactose; then derepression can occur.
8. (a) Constitutive for both.
   (b) Constitutive for β-galactosidase and inducible for permease.
   (c) Constitutive for both.
   (d) Constitutive for β-galactosidase and inducible for permease.
9. There might be a mutation in the gene for adenyl cyclase or for the cyclic AMP receptor protein. Another possibility is a membrane transport mutant, if these sugars were to utilize the same transport system.
10. A transcription termination site whose terminating activity is regulated and which is not far downstream from the promoter.
11. The glucose is irrelevant.
    (a) None.
    (b) One.
    (c) None.
12. (a) G inhibits 5; H inhibits 7; G and H together or E alone probably inhibit 3; J inhibits 8; enzyme 1 could be inhibited by (G, H, and J), (E and J), (G, H, and C), or (C and E); the ( ) enclose the products that act together.
    (b) Step 1, three isozymes; step 3, two isozymes.

CHAPTER 15

1. The bacteria have stopped growing so phage growth is not possible.
   (a) Small plaques will usually result if

the burst size is especially small or if the phage adsorbs very slowly to bacteria. If the phage particle is very small, the plaque will usually be large because smaller particles diffuse further in agar.
(b) If a phage is adsorbed to a bacterium before dilution into agar, new phage will be produced sooner than if there is a time delay before a phage comes in contact with a bacterium in the agar. Since there is only a limited time in agar in which phage can be synthesized (until bacterial growth stops), the size of the plaque depends upon how quickly a phage particle adsorbs to a bacterium.

2. A protein coat to enclose the nucleic acid, a component capable of adsorbing to a bacterium, an element designed to enable the nucleic acid to penetrate the cell wall. The latter could be an injection system or a component that stimulates the bacterium to take in the particle.

3. By destruction of host DNA; by conversion of host RNA polymerase to a form which cannot read host promoters; by inactivation of host RNA polymerase and replacement with phage-coded polymerase.

4. Lysis would probably occur before many (or any) phage were made.

5. In a simultaneous infection the T4 nuclease destroys the T7 DNA. If T4 infection occurs ten minutes after T7 infection, the T4 DNA is not transcribed because T7 has inhibited the *E. coli* RNA polymerase. Therefore, T7, but not T4 phage, are produced.

6. The T4-induced nucleases will destroy all newly synthesized phage DNA molecules since, in cases (a) and (b), glucosylation cannot occur.

7. (1), (3), and (4).

8. T4 makes all of its own proteins and clearly does not need to use *E. coli* DNA as a template since it degrades the host DNA in the normal course of infection. λ apparently needs some host functions.

9. For the dimer, after the first DNA molecule is cut out and packaged, the remaining DNA has no *cos* site and hence cannot be packaged. Thus, only one particle can be formed from a dimeric circle. By the same reasoning, two particles are packaged from a trimeric circle.

10. When a wild-type (cI^+) λ phage adsorbs to one bacterium, the lytic response ensues and many phage are produced that infect other bacteria. In time, the number of phage increases to the point that many bacteria are multiply infected. This favors the lysogenic response. Since the lysogens are immune to subsequent infection by λ, they grow in the plaque and produce a turbid growth of bacteria. A cI^- phage cannot establish repression and, therefore, always enters a lytic cycle; hence, an infected cell can never survive to form a colony. Phage T2 lacks a repressor and would not be expected to produce a turbid plaque.

11. The phage has an amber mutation in one of the genes *cI*, *cII*, or *cIII* and strain B (but not A) has an amber suppressor.

12. The prophage attachment site is between genes *F* and *G*.

13. (a) λ *bio*, *POB′*; λ *gal*, *BOP′*;
(b) λ *gal bio* has BOB′; λ *wild* has *POP′*;

14. (a) The prophage is transferred to a female lacking repressor; the operators are free of repressor and transcription begins.
(b) Once this locus is transferred into the female, the female cells will die. Hence, by using the interrupted mating technique, one can determine at what time the numbers of any recombinant (for example, a locus transferred at an early time) decrease.

15. If enough λ DNA molecules enter a bacterium, the intracellular repressor will be titrated.

CHAPTER 16

1. (a) Eukaryotic genes are not generally organized into polycistronic operons since, without complex splicing events, only one protein molecule could be translated from the primary transcript.
   (b) An operon refers to a set of coordinately regulated genes that are encoded in one or two polycistronic mRNA molecules. If there is a regulatory gene (for example, a repressor), it is usually encoded in a separate monocistronic RNA molecule. A gene family is a collection of genes, each of which yields a distinct mRNA molecule, that encode molecules of similar or related function. The genes may or may not be clustered and they are rarely adjacent, but they are somehow regulated by similar or related signals.
2. There are many instances in which several distinct protein molecules, having like or identical function, are prevalent at various stages of development of an organism. The differences between the proteins are often only a few amino acids. The different forms of the protein are encoded in distinct genes, each of which is active at a particular stage of development. A collection of such genes comprises a developmentally regulated gene family.
3. (a) No. (b) No. (c) No. (d) Between the genes.
4. Limited time—gene amplification; long time—increased lifetime of mRNA with or without gene amplification.
5. An embryonic gene contains a large number of different base sequences that constitute the coding sequences for all antibodies. These sequences are contiguous in the DNA. In the course of development a genetic recombinational event removes large blocks of DNA that include many adjacent sequences. There are many blocks that can be removed, so that many different coding sequences can remain after this recombinational event occurs. One event occurs in a particular cell leaving that cell with a unique coding sequence that enables it to make a particular antibody.
6. (a) Increase the number of *J* genes.
   (b) $150 \times 12 \times 3 \times 5000 = 2.7 \times 10^6$.
7. A cell can respond to a particular hormone only if the cell has a receptor binding site for that hormone on the cell membrane.
8. The hormone is apparently needed to initiate intron excision.
9. (a) The promoters for both transcription units probably have a common sequence acted on by either a positive or negative regulator.
   (b) Both primary transcripts have a common sequence acted on by an element that prevents some stage of processing.
   (c) Both processed mRNA molecules have a common sequence involved in ribosome binding. An effector may remove a protein bound to this sequence, or it may denature a double-stranded region containing the ribosome binding site.
10. No. Note the single-stranded poly(A) tails projecting from the junctions. These are always at the 3′-OH terminus of the RNA yet they are not at the same ends of each loop. Thus, the RNA molecule of the central gene is transcribed from the strand that is complimentary to the coding strand for the outer genes.

Index